P9-CEN-964

Alzheimer Disease

Neuropsychology and Pharmacology

Gérard Emilien
Cécile Durlach
Kenneth Lloyd Minaker
Bengt Winblad
Serge Gauthier
Jean-Marie Maloteaux

T
55
4743
004

Birkhäuser Verlag
Basel · Boston · Berlin

Authors

Dr Gérard Emilien
127 rue Henri Prou
78340 Les Clayes sous Bois
France

Dr Cécile Durlach
23 Rue Raynouard
75016 Paris
France

Dr. Kenneth L. Minaker
Massachusetts General Hospital Geriatric Medicine Unit
Department of Medicine
and the Division of Aging, Harvard Medical School
Boston
USA

Professor Serge Gauthier
Director of the Alzheimer's Disease Research Unit
McGill University
McGill Centre for Studies in Aging
6825 Blvd. LaSalle
Verdun, Quebec, Canada H4H 1R3

Professor Bengt Winblad
Division of Geriatric Medicine
Karolinska Institute
B84 Huddinge University Hospital
SE 141 83 Huddinge
Sweden

Professor Jean-Marie Maloteaux
Université Catholique de Louvain
Laboratory of Pharmacology and Department of Neurology
Cliniques Universitaires Saint Luc
B-1200 Brussels
Belgium

Library of Congress Cataloging-in-Publication Data
Alzheimer disease : neuropsychology and pharmacology / Gérard Emilien (editor) ; with
contributions by Cécile Durlach ... [et al.]
p. cm.
Includes bibliographical references and index.
ISBN 3-7643-2426-0 (alk. paper)
1. Alzheimer's disease. 2. Alzheimer's disease--Chemotherapy. I. Emilien, Gérard, 1952–

RC523.A3678.2003
616.8'31061--dc22 2003062904

Bibliographic information published by Die Deutsche Bibliothek
Die Deutsche Bibliothek lists this publication in the Deutsche Nationalbibliografie;
detailed bibliographic data is available in the Internet at <http://dnb.ddb.de>.

ISBN 3-7643-2426-0 Birkhäuser Verlag, Basel – Boston - Berlin

The publisher and editor can give no guarantee for the information on drug dosage and administration contained in this publication. The respective user must check its accuracy by consulting other sources of reference in each individual case. The use of registered names, trademarks etc. in this publication, even if not identified as such, does not imply that they are exempt from the relevant protective laws and regulations or free for general use.
This work is subject to copyright. All rights are reserved, whether the whole or part of the material is concerned, specifically the rights of translation, reprinting, re-use of illustrations, recitation, broadcasting, reproduction on microfilms or in other ways, and storage in data banks. For any kind of use permission of the copyright owner must be obtained.

© 2004 Birkhäuser Verlag, P.O. Box 133, CH-4010 Basel, Switzerland
Part of Springer Science+Business Media
Printed on acid-free paper produced from chlorine-free pulp. TCF ∞
Cover design: Micha Lotrovsky, CH-4106 Therwil, Switzerland
Printed in Germany
ISBN 3-7643-2426-0

9 8 7 6 5 4 3 2 1 www.birkhauser.ch

Contents

0 53038512

Chapter 3:
Pharmacology of Alzheimer disease

Chapter 4:
Molecular genetics of Alzheimer disease

Glossary

Aβ	Amyloid beta
A2M	α2-macroglobulin gene
AACD	Aging-associated cognitive decline
AAMI	Age-associated memory impairment
ABID	Agitated behavior in dementia
ACh	Acetylcholine
AChE	Acetylcholinesterase
ACMD	Age-consistent memory decline
ACMI	Age-consistent memory impairment
AD	Alzheimer disease
ADAS-cog	Alzheimer's disease assessment scale – cognition subscale
ADCS	Alzheimer disease cooperative study
ADDL	Amyloid β-derived diffusable ligands
ADFACS	Alzheimer disease functional assessment and change scale
ADL	Activities of daily living
ADRDA	Alzheimer disease and related disorders association
ADRQL	Alzheimer disease-related quality of life
AGE	Advanced glycation endproducts
AMPA	α-amino-3-hydroxy-5-methylisoxazole-4-propionate
APA	American Psychiatric Association
APLP	Amyloid precursor-like protein
Apo E	Apolipoprotein E
β-APP	β-amyloid precursor protein
ARCD	Age-related cognitive decline
ARMD	Age-related memory decline
AUC	Area under the curve
BA	Brodmann's area
BACE	β-site APP cleaving enzyme
BAPP	Beta-amyloid precursor protein
BDS	Blessed dementia scale

BEHAVE-AD	Behavioral pathology in Alzheimer diesease rating scale
BIMC	Blessed information-memory-concentration (test)
bp	Base pair
BPSD	Behavioral and psychological symptoms of dementia
BRSD	Behavior rating scale for dementia
BSF	Benign senescent forgetfulness
BuChE	Butyrylcholinesterase
BVRT	Benton visual retention test
CAMDEX	Cambridge examination for mental disorders of the elderly
CaMKII	Ca^{2+}/calmodulin-dependent protein kinase II
CBF	Cerebral blood flow
Cdc	Cell cycle kinase
Cdk	Cyclin-dependent kinase
CDR	Clinical dementia rating
CERAD	Committee of the Consortium to establish a registry for Alzheimer disease
CES	Central executive system
CET	Cognitive estimations test
CGIC	Clinical global impression of change
ChAT	Choline acetyltransferase
CI	Cholinesterase inhibitor
CIBIC	Clinician interview-based impresssion of change
CIBIC-plus	Clinical global impression of change with caregiver input
CMAI	Cohen-Mansfield agitation inventory
CMRgl	Cerebral metabolic rate for glucose
CNS	Central nervous system
CNTF	Ciliary neurotrophic factor
COGDRAS	Cognitive drug research computerized assessment system
COX	Cyclooxygenase
CPMP	(European) committee for proprietary medicinal products
CPP	Common population polymorphism
CPT	Continuous Performance Test
CR	Conditioned response
CREB	Cyclic AMP-dependent response element binding
CS	Conditioned stimulus
CSF	Cerebrospinal fluid
CVLT	California Verbal Learning Test
DAD	Disability assessment for dementia
DAT	Dementia of Alzheimer's type
DBD	Dementia behavior disturbance
DDVP	2,2-dichlorovinyl dimethyl phosphate
DHEA	Dehydroepiandrosterone
DLB	Dementia with Lewy bodies
DQoL	Dementia quality of life
DRAP	Down's region aspartic proteinase

DRN	Dorsal raphe nucleus
DRS	Dementia rating scale
DRST	Delayed recognition span test
DS	Digit span
DSM	Diagnostic and statistical manual (of mental disorders)
DSST	Digit symbol substitution test
EBS	Eating behavior scale
EEG	Electroencephalography
ELISA	Enzyme-linked immunosorbent assay
EOAD	Early onset Alzheimer disease
EPSP	Excitatory postsynaptic potential
EQ-5D	European quality of life instrument
ER	Estrogen receptor
ERAB	Endoplasmic reticulum-associated protein
ERP	Event-related potential
ERT	Estrogen replacement therapy
EURODEM	European community concerted action on epidemiology and prevention of dementia
FAD	Familial Alzheimer disease
FCSRT	Free and cued selective reminding test
FDA	Food and Drug Administration
fMRI	Functional magnetic resonance imaging
FOME	Fuld object memory evaluation
FTD	Frontotemporal dementia
FTDP	Frontotemporal dementia with parkinsonism
GABA	γ-aminobutyric acid
GBS	Gottfries-Brane-Steen
GDS	Geriatric depression scale
GDS	Global deterioration scale
GFAP	Cilial fibrillary acidic protein
GMS	Geriatric mental status examination
GMSS	Geriatric mental state schedule
GOGDRASS	Cognitive drug research computerized assesssment system
HAF	Hippocampus-amygdala formation
HD	Huntington's disease
HERA	Hemisphere encoding-retrieval asymmetry
HNE	4-hydroxy-2-nonenal
HPA	Hypothalamus-pituitary-adrenal (axis)
HSV	Herpes simplex virus
5-HT	5-hydroxytryptamine, serotonin
5-HTT	Serotonin transporter
5-HTTLPR	Serotonin transporter gene-linked polymorphic region
Hu	Human
HUI	Health utilities index
ICD	International classification of disease

i.c.v.	Intracerebroventricular
IDDD	Interview for deterioration in daily living activities in dementia
IL	Interleukin
IMD	Isolated memory decline
IMI	Isolated memory impairment
IML	Isolated memory loss
IST	Isaacs set test
KGDHC	α-ketoglutarate dehydrogenase complex
KS	Korsakoff's syndrome
LCD	Limited cognitive disturbance
LDL	Low density lipoprotein
LLF	Late life forgetfulness
LOAD	Late-onset Alzheimer disease
LRP	Lipoprotein-receptor-related protein
LT	Lymphotoxin
LTD	Long-term depression
LTM	Long-term memory
LTP	Long-term potentiation
MAO	Monoamine oxidase
MCI	Mild cognitive impairment
MD	Minimal dementia
MEG	Magnetoencephalography
mGluR	Metabotropic glutamate receptor
MMSE	Mini-mental state examination
MND	Mild neurocognitive disorder
MOUSEPAD	Manchester and Oxford universities scale
MRI	Magnetic resonance imaging
MTL	Medial temporal lobe
MTR	Magnetization transfer ratio
nAChR	Nicotinic acetylcholine receptor
NC	Normal control
NFT	Neurofibrillary tangle
NGF	Nerve growth factor
NIMH	National Institute of Mental Health
NINCDS	National Institute of Neurological and Communicative Disorders and Stroke
NMDA	N-methyl-D-aspartate
NPI	Neuropsychiatric inventory
NPI-NH	Neuropsychiatric inventory-nursing home version
NSIAID	Non-steroidal anti-inflammatory drug
PCA	Principal component analysis
PCR	Polymerase chain reaction
PDAPP	Platelet-derived amyloid precursor protein
PDGF	Platelet-derived growth factor
PDS	Progressive deterioration scale

PET	Positron emission tomography
PFC	Prefrontal cortex
PHF	Paired helical filament
PKC	Protein kinase C
PS1	Presenilin 1
PS2	Presenilin 2
PSMS	Physical self-maintenance scale
QD	Questionable dementia
QOL	Quality of life
QWB	Quality of well being
RAVLT	Ray auditory verbal learning test
RBM	Rivermead behavioral memory (test)
rCBF	Regional cerebral blood flow
RP	Pre-stimulus readiness
RUD	Resource utilization in dementia
SASG	Self-assessment scale geriatric
SCAG	Sandoz clinical assessment-geriatric
SD	Standard deviation
SIDAM	Structured interview for the diagnosis of dementia of the Alzheimer type, multi-infarct dementia, and dementias of other etiology
SISCO	SIDAM score
SOD	Superoxide dismutase
SP	Senile plaque
SPECT	Single proton emission computed tomography
SRT	Serial reaction time
SRT	Selective reminding test
SSRI	Selective serotonin reuptake inhibitor
STE	Short-term synaptic enhancement
STM	Short-term memory
TIV	Total intracranial volume
TMT	Trail making test
TNF	Tumor necrosis factor
TRH	Thyrotropin-releasing hormone
US	Unconditioned stimulus
VAT	Visual association test
VILIP	Visinin-like protein
vntr	Variable number of tandem repeat polymorphism
VSSP	Visuospatial sketchboard
WAIS	Wechsler adult intelligence scale
WHO	World Health Organization
WMS	Wechsler memory scale
ZCT	Zazzo's cancellation task

Introduction

The most common forms of degenerative dementia are Alzheimer disease (AD) which constitute approximately 60–70% of cases and dementia with Lewy bodies (DLB) that accounts for a further 15–25% of sufferers (Cummings & Benson, 1992; McKeith et al., 1992, 1996). AD is a chronic, degenerative, dementing illness and its onset is typically insidious. It is a diagnosis based on patient history, physical examination, neuropsychological testing, and laboratory studies. Elderly people are the ones commonly afflicted with this disease. However, evidence shows that it can also afflict even individuals as young as 40 years of age. There is yet no definitive diagnostic test for AD. It represents a progressively greater burden on industrialized civilization at the economic, social, and medical levels. A meta-analysis of prevalence studies undertaken in Europe between 1980 and 1990 found similar rates: 0.3% in 60–69 years old, 3.2% in those 70–79 years old and 10.8% of those 80–89 years old (Rocca et al., 1991). In Japan, cerebrovascular diseases affecting about 42% of the dementia population cause the most prevalent type of dementia. AD ranks only second, affecting about 32% of the population (Evans et al., 1989). In the USA, however, statistics show that AD is the leading cause of dementia affecting about four millions of the USA population or 10% of Americans over the age of 65. Due to the absence of revolutionary scientific breakthroughs, the number of persons afflicted with AD in the United States is projected to quadruple in the next 50 years. The prevalence can be expected to rise to 8.64 million by the year 2050 and the illness will affect approximately one in every 45 Americans (Ron Brookmeyer et al., 1998).

Neurological and cognitive dysfunctions

The primary manifestations of brain dysfunction include cognitive impairment, neuropsychiatric disturbances and neurological abnormalities. Cognition refers to higher brain skills including arousal, attention, concentration, learning, memory, concept formation, problem solving, and language skills. Neurological abnormalities include loss of the ability to walk and talk, incontinence, and the emergence of primitive reflexes such as the grasp and suck responses. Neurological impairments typically occur in the late phases of the illness. Secondary effects of AD include progressive inability to perform instrumental activities of daily living (e.g. drive, prepare meals) with eventual loss of the ability to do basic daily tasks such as feeding oneself and toileting. Many patients with AD exhibit motor restlessness, wandering, pacing and carphologia or purposeless finger movements. Typical behavioral symptoms of AD include paranoid and ideational delusions, hallucinations, activity disturbances, aggressiveness, sleep disturbances, affective disturbances, anxieties and phobias. Motor restlessness occurs in 21–60% of AD patients and wandering in 10–61%. Neuropsychiatric alterations are not present in all patients and may vary over the course of the illness. Clinical features are best demonstrated by mental status examination, which usually reveals multiple areas of dysfunction. All patients will have memory impairment, along with language deficits (aphasia), apraxia, agnosia, visuo-spatial dysfunction and/or impaired executive function, including errors of judgment. Each major clinical domain can be measured separately using a specific scale or alternatively can be assessed as part of a multi-dimensional instrument. Many individuals exhibit motor dysfunction such as Parkinsonian signs (Stern et al., 1996).

The amyloidocentric hypotheses

Interest in the molecular basis of AD was aroused in the mid 1980s when the amyloid beta (Aβ) and its gene were defined, and subsequently linked to the causation of AD (Hardy et al. 1998). Amyloid plaques are the most characteristic features of AD. The brains of patients with AD characteristically show a dramatic loss of neurons and synapses in many areas, notably the basal forebrain, amygdala, hippocampus, and cerebral cortex (Eggert et al., 1997). Other pathologic findings associated with AD include granulovacuolar degeneration, Hirano bodies and Aβ deposition in the walls of small cortical blood vessels (Morris, 1995). The amyloidocentric hypotheses of AD propose that Aβ plaque depositions or partially aggregated and/or soluble Aβ trigger a neurotoxic cascade, thereby causing neurodegeneration and AD. The arguments are based on *in vitro* studies suggesting that Aβ is toxic to neurons and on the measurements of increased release of Aβ by cells expressing familial AD mutant genes. The concurrent development of intracellular neurofibrillary lesions in AD also follows a stereotyped pattern. The principal consequence of these lesions is a loss of synaptic function in the affected regions of the brain. However, several lines of evidence suggest that amyloid plaques *per se* might not be the primary mechanism underlying AD neurodegeneration. Soluble oligomeric forms of Aβ, probably complexed with other factors such as apolipoprotein E (Apo E), may be more relevant. As yet, there are no definitive antemortem biological markers for AD; a definitive diagnosis is based on finding neuritic plaques and neurofibrillary tangles (NFT) on postmortem examination of brain tissue (Gearing et al., 1995). Plaques and tangles are not detectable with presently available neuroimaging techniques.

To date, four genes have been confirmed to play an important role in the pathogenesis of AD, and others have recently been implicated (Blacker et al., 1998). A small minority of cases of AD is caused by point mutations in the genes for β-amyloid precursor protein (β-APP), presenilin 1 (PS1) and presenilin 2 (PS2), but most cases occur sporadically, albeit with increased risk for carriers of the apolipoprotein E ε4 (Apo E4) allele (Hutton et al., 1998). The ability to transform the genetic make up of experimental animals such as mice, now provides us with the opportunity to test many of the observations and theories which have developed from *in vitro* experiments. More importantly, the new mouse models of AD allow for the testing of therapeutic concepts.

In recent years, evidence suggests that the cascade of molecular changes leading to the neurodegeneration in AD may begin years before the first clinical symptoms appear.

The concept of mild cognitive impairment

In the early stages of AD, it is difficult to detect the behavioral correlates of the neurodegenerative processes. However, as the neural degeneration spreads, involving an ever-increasing number of neurons, the impact of these losses on behavior and clinical symptoms begin to be seen. Nevertheless, cognitive impairment is more highly correlated with the density of neurofibrillary tangles than with the number of plaques. In view of early therapeutic intervention, the identification of patients at risk of dementia and of patients with early or mild forms of AD is important. This is one main reason why the concept of mild cognitive impairment (MCI) generates a great deal of interest. Many different terms describing very similar patterns of deficits have been used to describe these patients since the early 1960s (Ritchie & Touchon, 2000). According to the 10th revision of the International Classification of Disease (ICD-10), persons with impairment of cognitive functions who do not fulfill the criteria of dementia should be diagnosed as having MCI (F07.8) (WHO, 1992). One often-used criterion for the differentiation of dementia from MCI is the presence or absence of impairment in activities of daily living (Petersen et al., 1999). Although MCI is an ICD diagnosis, its meaning is still a topic of an intense discussion (Christensen et al., 1995; 1997). Competing concepts regard MCI as an inevitable feature of the normal aging process, as a consequence of an underlying (physical) disease, as a separate nosological entity, or as an early-stage dementia (Ritchie & Touchon, 2000; Zauding, 1992). There is at least a great consensus that MCI patients have a significantly increased risk of AD. There is also discussion whether MCI patients have more or less isolated memory impairment or whether they commonly show deficits on tasks of other cognitive functions (Petersen et al., 1999; Ritchie & Touchon, 2000).

The difficult clinical question is in which cases of MCI will there be a rapid cognitive decline to a dementia syndrome. Limited opportunities exist for studying the preclinical stage of AD. The best strategy is to study individuals at risk for dementia with predictably high conversion rates to clinical dementia. Some authors have used a modified form of the Clinical Dementia Rating (CDR) scale to classify their subjects (Daly et al., 2000; Hughes et al., 1982). They followed up the individuals longitudinally to determine which features of the interview best characterized a subject's progression on the CDR scale from a rating of 0.5 (questionable dementia) to a rating of 1 (mild dementia). Several critical questions that will be particularly useful to clinicians in identifying subjects who are at risk of converting to AD have been outlined and among these, the importance of the clinical interview has been emphasized. However, using the CDR scale to characterize subjects from normal function through various stages of dementia, it was pointed out that a CDR of 0.5, while describing very mildly impaired subjects, does not necessarily include or exclude the diagnosis of probable AD (Petersen, 2000). Therefore, a subject with a CDR rating of 0.5 may have either MCI or probable AD. Many subjects with a CDR rating of 0.5 have in fact been diagnosed as having probable AD.

Detecting cognitive decline in persons who, at baseline, had extremely high or relatively low levels of cognitive function is a particular challenge. Recognition of early-stage dementia is emphasized because of the benefits that accrue to patients and families. Early recognition of AD allows time to plan for the future and to treat patients before marked deterioration occurs.

Cognitive and behavioral assessment

A variety of scales have been developed to reliably measure the severity of such symptoms. The use of rating scales helps formalize the evaluation approach, ensures thoroughness, may clarify the presence and absence of mental illness. Clinicians benefit from these tools in being more able to appreciate the presentation, severity, frequency and clinical course of AD and treatment responses. Importantly, appropriate instruments should be used to document efficacy. Hence, these scales help clinicians to make clinical judgments concerning their patient on the simple basis of what they see.

Treatment and prevention

Agents have been designed to increase cholinergic function in the central nervous system to counteract the decrease in cholinergic activity observed in the cerebral cortex and other areas of the brains of patients with AD. One such group of agents, the cholinesterase inhibitors, prevents the enzymatic hydrolysis of acetylcholine by acetylcholinesterase, thus increasing the concentration of acetylcholine. These effects are believed to account for the beneficial effects of cholinesterase inhibitors on cognition and function in AD. Some studies have also shown that cholinesterase inhibitors may have beneficial effects on selected behaviors in patients with AD, such as hallucinations, delusions, agitation, anxiety, depressive features and apathy (Mega et al., 1999; Feldman et al., 2001; Cummings et al., 2000). Cholinergic therapies may, therefore, provide an alternative or complementary approach in the treatment of the neuropsychiatric symptoms of AD, in addition to their positive effects on cognition and patient function.

The concern that treatments for dementia are adequately tested, together with a desire for clear standards, led the Food and Drug Administration (FDA) to propose draft guidelines in 1990 for establishing whether a drug has antidementia efficacy. These guidelines required that clinical trials should be double-blind and placebo-controlled, patients should fulfill established criteria for a diagnosis of AD, that studies should be of sufficient length to appreciate a meaningful effect (Leber, 1990). Drugs must also be demonstrated to be safe in the AD patient population. In addition, studies are necessary to assess drug interactions, measure impact of the drug on other disease symptoms and assess efficacy compared with and in combination with other drugs.

Guidelines for the development of drugs to manage AD are specific for each country's health authority. For example, the US FDA requires two adequate and well-controlled studies to demonstrate the efficacy of a drug (US FDA, 1998). The European Committee for Proprietary Medicinal Products (CPMP) requests two studies and the Japan authorities require one study. The required duration of the study treatment period also varies from 3 to 12 months (CPMP, 1997). The FDA requires demonstration of a statistically and clinically significant benefit on a performance-based measure of cognition (e.g., Alzheimer's Disease Assessment Scale-Cognition Subscale (ADAS-cog) and a measure of global functioning (e.g., Clinician's Global Impression of Change with caregiver input (CIBIC-Plus). The CPMP requires these items in addition to a measure of activities of daily living (ADL) and a definition of responders that incorporates all three domains (cognition, global functioning and ADL). Effective therapeutic strategies for AD ideally should address all three domains of impairment (cognition, psychiatric and behavioral disturbances and activities of daily living) in AD patients. Effective treatment requires monitoring of symptoms, functional impairment, and safety. The use of multiple treatment modalities includes pharmacotherapy, behavioral management, psychotherapies, psychosocial treatments, and support and education for families.

Clinical trials of AD include clinical global measures of change as primary efficacy outcome measures. These multidimensional assessments are based on the premises that a clinically useful treatment must have a clinical effect, not only a cognitive one. The clinical effectiveness of new treatments should be apparent to experienced clinicians (Leber, 1997; Reisberg et al., 1997). In fact, emerging knowledge of the neurodegenerative process has shifted the emphasis of drug discovery efforts towards strategies to modify the progression of disease. Newer AD drugs are targeted toward slowing disease progression (e.g., γ-secretase inhibitors, β-secretase inhibitors, Aβ aggregation inhibitors, anti-Aβ immunization, anti-inflammatory drugs). However, there is currently no accepted definition of what constitutes efficacy in disease progression. Resolution of this scientific-semantic issue remains a central issue in AD drug development. The concept of prevention of disability or delaying the onset of AD has great appeal and it may be technically feasible. The dogmatic viewpoint in this approach is that Aβ is a bad neurotoxic molecule that has to be eliminated from the brain tissue (Schenk et al., 1999). The inability of the toxic concept to prevent the disease became apparent when the anti-amyloid immunization actually exacerbated the condition leading to the trial discontinuation

(Hock et al., 2002). Obviously, further understanding of the physiology of Aβ is required for the innovation of effective and safe AD treatments. It may be that Aβ is a normal and functional component of brain metabolism and synaptic function.

Organization of this book

A book on AD often focuses on either the biological aspects or the cognitive functions involved separately. In discussing both aspects of the disease, the reader will get a rather comprehensive picture of the various pathologies involved in their treatments. This book is organized into three parts. In part I, we discuss the biological correlates of AD. The nature of AD as well its various neurological and neurochemical hypotheses are critically examined. In part II of this work, the neuropsychological aspects of this disease are outlined. In part III, strategies for effective treatment and prevention of AD are discussed.

Acknowledgements
The authors thank all those who have helped in the realization of this work.

Part I
Biological correlates of Alzheimer disease

Chapter 1
Alzheimer disease

1.1 Introduction

In November of 1901, the Bavarian neuropsychiatrist Alois Alzheimer began observing a 51-year-old mentally ill patient whom he referred to as Auguste D. in his reports. He described the behavioral changes of this patient in the years leading to her death five years later. The neuropsychological changes included deteriorating memory, disorientation and occasional hallucinations. He also described the "miliary foci" accumulating extracellularly which are known as senile plaques and the "dense bundles of fibrils" occurring intracellularly which are now known as NFTs. These two pathological lesions that he observed upon postmortem examination remain the key diagnostic features of AD today (Alzheimer, 1907). The disease officially got its name in 1910 when Emil Kraepelin named the disorder after Alzheimer because of his thorough description of the two lesions.

Dementia can be defined as a generalized disturbance in cognition. The degenerative dementias are marked by progressive declines in cognition and functioning. The most prevalent form of dementia is AD, which has been studied more than other, less frequent causes of dementia. AD is characterized early on by neuropathological changes, primarily in the hippocampus and entorhinal cortex. Autopsy findings of clinically demented subjects who do not have predominant AD are distributed primarily among vascular disease, Lewy body disease and frontotemporal gliosis. As a group, these can be classified as the non-AD dementias. In AD, linear monotonic declines can be demonstrated for many clinical processes, including progressive changes in memory, orientation, concentration and calculation ability, language abilities, the ability to draw figures (praxic capacity), loss of various functional capacities. Progressive losses in AD include losses in executive capacities, in the ability to perform basic life activities, and in the ability to perform basic life functions. These progressive changes in AD may be conceptualized as falling into two distinct clinical domains: a cognitive domain and a functional domain. As the disease progresses, cognitive deficits such as remote memory impairment, time/space disorientation and language disturbances become more prominent. Apart from the cognitive and functional domains of symptoms in AD, another category of symptoms can be identified, the emotional changes or behavioral and psychological symptoms of dementia (BPSD) (Finkel, 1996; Finkel et al., 1998; American Psychiatric Association (APA), 1994; Deutsch & Rovner, 1991). The different stages that characterize the disease are summarized in Table 1.1. In the mid-1980s, researchers hypothesized that major categories of BPSD in AD patients would be relatively amenable to both psychological and pharmacological interventions (Reisberg

Table 1.1: Different steps that characterize the stages of AD

Stages	Behavioral changes
1	Decrease in energy and spontaneity, minor memory loss and mood swings, slowed reaction and learning, avoidance of new situations.
2	Slowing of speech and comprehension, loss of train of thought in midsentence, forgetting to pay bills, and getting lost while traveling. Awareness of loss of control may lead to depression, irritability, and restlessness.
3	Short-term memory loss, disorientation to time, place, and possibly person, paraphasic speech.
4	Behavioral disturbances, increasing need for care, incontinence.
5	Loss of ability to chew and swallow, vulnerability to pneumonia and other illnesses. Coma and death.

et al., 1986). The potentially remediable BPSD categories are (a) paranoid and delusional ideation; (b) hallucinatory disturbances; (c) activity disturbances; (d) aggressivity; (e) diurnal rhythm disturbances; (f) affective disturbances and (g) anxieties and phobias.

The specific etiologic origin of AD is unknown. However, genetic factors have been implicated, specifically a defect on chromosome 21. This type of AD is considered a familial form of the disorder. The role of neurotransmitters such as acetylcholine, serotonin and dopamine are also under investigation. Increasing age, environment and metabolic factors play an important role. Head trauma and myocardial infarction have also been shown to be risk factors. Some have postulated that albumin intoxication, disordered immune function and viral infection may be the cause. The unpredictable, relentless progression of AD with resulting perceptual and cognitive deficits renders impaired individuals unable to function independently early in the course of the disease. In later stages, AD causes increasingly greater damage to the higher order association cortices. Most subcortical structures are spared in AD, although neurodegeneration does occur in the basal forebrain and nucleus locus ceruleus. Our knowledge of pathophysiology and natural history of the disease has increased greatly over the past decade, yet the definitive cause remains unclear.

AD appears in two versions, an early onset or familial and late-onset, sporadic, or random form. In the former, onset can range from the fourth decade of life until age 60 to 65 and multiple family members across generations may be afflicted. By contrast, late onset AD (LOAD) which accounts for the overwhelming majority of cases develops after the age of 60 to 65 and other members of the patient's family may or may not become similarly afflicted. AD is characterized by a gradual onset of symptoms and irreversible decline to a near vegetative state. The average course of the disease is approximately a decade, although the duration can vary from 3 to 20 years (Small et al., 1997). Death generally results from illnesses related to progressive debilitation, such as pneumonia (Costa et al., 1996).

1.2 Differential diagnosis

Dementia is often difficult to diagnose correctly in the early stages. Early-stage dementia often goes unrecognized and misdiagnosed in older adults because the deficits are attributed to the consequences of normal aging. As AD is both a clinical and a neuropathological entity, the definitive diagnosis of AD can be made only with a brain biopsy or an autopsy. A diagnostic evaluation for dementia involves a complete history, neuropsychological examination (e.g., the Mini-Mental State Examination (MMSE)), physical examination and selected laboratory studies and neuroimaging (Folstein et al., 1975; Geldmacher & Whitehouse, 1996). In an attempt to quantify the validity of the clinical diagnosis, the "National Institute of Neurological and Communicative Disorders and Stroke" and the "Alzheimer disease and Related Disorders Association" (NINCDS-ADRDA) guidelines provide criteria for assigning the probability of diagnosis as clinically possible, probable or definite on the basis of increasing probability of histopathological confirmation (see Table 1.2) (McKhann et al., 1984). A possible diagnosis is made when symptoms atypical of AD occur but are not attributable to another disease or disorder. The diagnosis of AD is considered to be probable when alternative causes of dementia are excluded. A probable diagnosis requires dementia and a history and pattern of symptoms consistent with those generally seen in AD. Other disorders that mimic AD must be excluded from the history, examination and laboratory tests. Definite diagnosis can only be made with biopsy of brain tissue from a patient who was clinically diagnosed with AD. Microscopic pathology criteria include an age-adjusted minimum number of neuritic plaques and neurofibrillary tangles. Both of these lesions can be seen in brains of older persons who do not have dementia. However, they are found in greater numbers in the neocortex, hippocampus and amygdala in patients with AD. The clinical diagnosis of AD has been confirmed at autopsy in close to 90% of cases (Hogervorst et al., 2000). The NINCDS-ADRDA criteria are widely used as a standard consensus-based classification system for the clinical diagnosis of AD.

Establishing precise operational criteria can help to reliably perform a clinical diagnosis. Hence, several sets of criteria have been proposed for the diagnoses of dementia of the Alzheimer type. Two standardized instruments, the Cambridge Examination for Mental Disorders of the Elderly (CAMDEX) and the Geriatric Mental Status Examination (GMS) were developed to provide diagnostic algorithms that meet criteria of the ICD, 10th Revision (ICD-10) (Roth et al., 1988; WHO, 1992). Another set of criteria is given in DSM-IV (APA, 1994). According to DSM-IV, the diagnosis of dementia requires the development of multiple cognitive deficits manifested by both memory impairment and one or more of the following impairments: aphasia, apraxia, agnosia and disturbance in executive functioning (see Table 1.3). It also requires that the cognitive deficits cause significant impairment in social or occupational functioning and represent a significant decline from a previous level of functioning. The narrow definition of dementia presented in DSM-IV includes only those cases severe enough to interfere with social or occupational functioning. The degree of this interference may depend on the demands placed by the particular culture or occupation of the individual. In many cultures, elderly per-

Table 1.2: Criteria for clinical diagnosis of probable AD. Source: The National Institute of Neurological and Communicative Disorder and Stroke-Alzheimer disease and Related Disorders Association Work Group (McKhann et al., 1984).

Criteria include
- Dementia established by clinical examination and cognitive test (MMSE or Blessed Dementia Scale) and confirmed by neuropsychological tests
- Deficits in ≥ 2 areas of cognition
- Progressive worsening of memory and other cognitive function
- No disturbance of consciousness
- Onset between ages 40 and 90 years
- Absence of systemic disorders or brain disease that could account for progressive cognitive deficits

---

The diagnosis is supported by
- Progressive deterioration of specific cognitive functions such as language (aphasia), motor skills (apraxia), and perception (agnosia)
- Impaired activities of daily living
- Altered behavior
- Family history of similar disorders
- Normal lumbar puncture, normal electroencephalogram or nonspecific changes, progressive cerebral atrophy on computed tomography

---

Features consistent with the diagnosis
- Plateaus in the course of progression
- Associated symptoms including depression, insomnia, incontinence, delusions, illusions, hallucinations,
- Catastrophic outbursts, sexual disorders, or weight loss
- Neurologic signs including increased muscle tone, myoclonus, or gait disorder
- Seizure (in advanced stage)
- Computed tomography normal for age

Features that make the diagnosis uncertain or unlikely
- Sudden, apoplectic onset
- Focal neurologic findings such as hemiparesis, sensory loss, visual field deficits, and incoordination (early in the course)
- Seizures or gait disturbance (at the onset or early in the course)

---

sons live with their families and little is expected from them. Hence, the assessment of impairment in daily living functioning can be difficult.

Detailed neuropsychological testing is helpful in characterizing the pattern of cognitive impairment. However, psychometric methods are useless when patients cannot be aroused or cannot follow simple commands. In the most severe cases of

Table 1.3: Definition of dementia, DSM-IV

Dementia is the development of multiple cognitive deficits manifested by memory impairment and one or more of the following:
- Aphasia - Apraxia - Agnosia - Disturbance in executive function
The cognitive deficits must be sufficiently severe to cause impairment in occupational or social functioning and must represent a decline from a previously higher level of functioning.

dementia, the diagnosis may have to be clinical and not psychometric, that is based on functional status and neurological signs and not on cognitive assessment (Teresi & Evans, 1996). While it is essential for the clinician to evaluate other possible causes of memory loss, a positive diagnosis of probable AD can be made based on a characteristic history from a knowledgeable informant together with a physical and neurological assessment. Hence, key informants play a critical role in identifying roles and the cultural sentiments associated with each role. They may help to describe the behavior of the subject in everyday activities and to report a decline compared to a previous level of function considered as normal. The differential diagnosis for AD includes a broad range of other causes of dementia and nondementing metabolic or psychiatric illnesses.

Despite a considerable degree of accuracy in the diagnosis of AD, the distinction between AD, vascular-ischemic dementia and mixed dementia remains controversial and has been one of the most difficult clinical and pathological challenges. Both the DSM-IV and the NINCDS-ADRDA criteria require exclusion of other causes of dementia such as vascular dementia, DLB, depression, mixed dementia and other DSM-IV disorders. As much as possible, established instruments and diagnostic criteria should be used for the other dementias including measures such as the Hachinski scale (see Table 1.4), the DSM-IV criteria and the criteria for the Consortium on Dementia with Lewy Bodies International Workshop (Rosen et al., 1980; McKeith et al., 1996; Knopman et al., 2001). So far, the most useful of all is the clinical Hachinski Ischemic Score that, although lacking imaging criteria, remains the most sensitive approach to differentiate patients with dementia caused by vascular disease from those with AD. The clinical manifestations of vascular disease or ischemic brain injury depend to a large extent on location. In vascular dementia, memory impairment is more variable, less severe and more likely to respond to cueing than in AD (Looi et al., 1999).

The problem becomes even more complicated when one attempts to differentiate depression from dementia as depression is recognized as the most common psychiatric disorder in later life. Single proton emission computed tomography

Table 1.4: Modified Hachinski Ischemic Score. A score of 4 is suggestive of vascular dementia

	Absent	Present
Abrupt onset	0	2
Stepwise deterioration	0	1
Somatic complaints	0	1
Emotional incontinence	0	1
History or presence of hypertension	0	1
History of strokes	0	2
Focal neurologic signs	0	2
Focal neurologic symptoms	0	2

(SPECT) may be most useful in distinguishing AD from vascular dementia and frontotemporal dementia, but should be used selectively and only as an adjunct to clinical evaluation and computed tomography (Talbot et al., 1998). Positron emission tomography (PET) has the advantage of greater sensitivity and spatial resolution but at a much higher cost. While it is an excellent tool for research purposes, it has limited clinical application. SPECT is simpler and has greater potential in the clinical setting than PET (Waldemar, 1995).

As defined by the NINCDS-ADRDA guidelines, the diagnosis of probable AD requires expensive medical, psychiatric and neurological examinations including laboratory investigations in order to rule out other causes of dementia.

1.2.1 Alzheimer *versus* frontotemporal dementias

Although AD remains by far the most common cause of neurodegenerative dementing illness, there is increasing evidence that other disorders account for perhaps 30% of the dementias. Among them, the frontotemporal dementias (FTD) represent an etiologically diverse group of disorders that share the pathologic feature of primary degeneration (cortical microvacuolation, neuronal loss and gliosis). FTD is the term currently applied to a number of syndromes in which patients present with degeneration of the frontal and/or temporal lobes (previously referred to as Pick's disease) (Hodges & Miller, 2001). In a group of persons with early onset dementia, the FTDs constitute up to 10% of the neuropathological diagnoses (Miller & Cummings, 1999). There are at least two sub-types of FTD (Hodges et al., 1999). The patients at the frontal *versus* temporal ends of the spectrum can be clearly differentiated at a neuropsychological level. Efforts to better characterize the FTD subtypes may result in improved understanding of their etiogenesis and also may allow clinical distinction from AD.

In the frontal variant of FTD, frontal lobe pathology predominantly affects the orbitomedial cortex causing changes in behavior and personality. Abnormal changes are apathy, loss of empathy or emotional warmth, impulsivity, disinhibition, change in dietary preference and stereotyped or ritualistic behaviors (Bozeat et al., 2000; Miller et al., 1995). The preservation of semantic memory in the context of impaired executive function in the frontal variant FTD group was the converse to that found in patients with the temporal variant of FTD (also termed semantic dementia).

1.2.1.1 Semantic dementia

The term semantic dementia typically refers to patients with progressive fluent aphasia, a severe impairment of naming and an inability to understand the meaning of words (Graham, 1999; Graham & Hodges, 1997). These patients have deficits in word retrieval and word meaning (semantics) such that speech, although fluent, is progressively devoid of content words (Hodges et al., 1999). Surface dyslexia and surface dysgraphia are often present while syntax and phonology, visual perceptual and visuo-skills, non-verbal problem-solving abilities and day-to-day memory are often preserved. Other cognitive functions such as episodic memory are relatively well-preserved in early stage semantic dementia. These patients have better preservation of episodic events and semantic facts from very recent life compared with other time periods (Alvarez & Squire, 1994).

Neuroradiological studies have found support for the view that anterior and inferior areas of the temporal lobe are affected early in semantic dementia and that this profile is different from other neurodegenerative conditions that affect long-term memory (LTM) such as AD (Chan et al., 2001; Galton et al., 2001; Harasty et al., 1996). With respect to medial temporal lobe regions, there is increasing evidence that the hippocampi and parahippocampal gyri are involved at later stages in the disease (Chan et al., 2001; Simons et al., 2001). Because different aspects of memory are affected in semantic dementia compared with AD, Galton et al. (2001) proposed that the two disorders differ in volumetric measures of specific temporal areas. For example, because episodic memory loss is greater in AD, they expected greater medial temporal deterioration in AD than in semantic dementia. The results indicated medial temporal lobe atrophy in both groups although hippocampal atrophy was bilateral in individuals with AD but asymmetric (left greater than right) in individuals with semantic dementia. The individuals with semantic dementia also had greater atrophy of the temporal poles. Similar results have been reported by other investigators (Chan et al., 2001).

1.2.2 Alzheimer *versus* dementia with Lewy bodies

In DLB, there is a relative sparing of memory function that is explained by the smaller burden of pathology in the entorhinal cortex compared with what is seen in early classical AD. DLB patients are more impaired than AD patients on various tests of attention. Neuropsychiatric features other than fluctuating cognitive impairment are common in DLB and patients are therefore more likely to be seen in psychiatric clinics or emergency rooms. Clinical presentation is very insidious.

The mean age of onset in 1 post-mortem series was 56 ± 7.6 years (range, 45–70 years) and the estimated duration of the disease was 3 to 17 years (Brun & Gustafson, 1993). Patients with pure Pick's disease live longer than patients with predominant AD.

1.3 Epidemiology

AD is the most prevalent form of the irreversible dementias, which affect an estimated 3 to 4 million patients, accounting for 60% or more cases (Knopman, 1997; Katzman & Rowe, 1992). AD accounts for approximately two-thirds of dementia cases in patients aged 65 years or older (Alzheimer disease and Related Dementias Guidelines Panel, 1996). Estimates of prevalence of dementia vary widely because of differences in research sampling strategies and diagnostic criteria. Representative prevalence rates are: 5–7% at age 75 to 80 years; 15–17% at age 80 to 85 years; approximately 35% at age 85 to 90 years; and 50% at age 90 to 95 years (Evans et al., 1989; Jorm & Jolley, 1998; Graves et al., 1996). One study reported a prevalence of 75% in those aged 95 years and older (Evans et al., 1990). At all ages, AD is reported to cause at least half and usually well over 70% of cases of dementia in the United States.

1.4 Risk factors

Age, family history of dementia and Down syndrome have been consistently established as risk factors. Age represents the most potent risk factor (Breitner & Welsh, 1995). The presence of ApoE ε4 has been shown to increase the certainty of the diagnosis of probable AD. However, its value in the differential diagnosis of dementia is limited because ApoE ε4 may be associated with other dementing illnesses (Slooter & van Duijn, 1997; Small et al., 1997). Of its three forms (ε2, ε3 and ε4), only the ε4 allele increases the likelihood of developing AD. The lifetime risk of AD for an individual without the ε4 allele is approximately 9%; the lifetime risk of AD for an individual carrying at least 1 ε4 allele is 29% (Seshadri et al., 1995). Therefore, while representing a substantial risk of AD, the ε4 genotype is not sufficiently specific or sensitive for the diagnosis of AD to allow its use as a diagnostic test (Relkin et al., 1996). The role of the ApoE ε2 allele remains controversial because some studies suggest that it has a protective effect on late-onset (sporadic) AD while others indicate an increased risk for early-onset AD (Corder et al., 1994; Van Duijn et al., 1995). However, the Apo E alleles are neither necessary nor sufficient for the development of the disease.

Significant risk factors implicated in a variety of studies include head injury, low serum levels of folate and vitamin B12, elevated plasma and total homocysteine levels, family history of AD or dementia, fewer years of formal education, lower income and lower occupational status (Clarke et al., 1998; Guo et al., 2000; Launer et al., 1999). Other risk factors include a history of depression, deprivation during gestation and early childhood brain development, and exposure to known neurotoxins, possibly including organic solvents (Kukull et al., 1995).

1.4.1 Smoking history

The European Community Concerted Action on Epidemiology and Prevention of Dementia (EURODEM) group associated smoking with a higher risk of AD whereas other studies did not observe any significant relation between smoking and the onset of AD (Launer et al., 1999; Hebert et al., 1992). Smokers have been noted to have a decreased risk in cross-sectional studies but an increased risk in a follow-up study (Lee, 1995; Ott et al., 1997). Methodological differences such as duration of follow-up and selection and number of study participants may account for these discrepancies. Given that smoking is a cause of other diseases, any potential benefit to individuals must be demonstrated in studies using alternate routes (e.g., transdermal patches, tablets) of nicotine administration.

1.4.2 Sex

While some prevalence studies have shown a greater number of women affected with AD than men, population-based incidence studies have generally found no difference between the sexes (Rocca et al., 1991; Kokmen et al., 1993). Prevalence studies may be biased by the fact that women tend to live longer than men. Female patients and families that include several affected women may be easier to recruit (Heun et al., 1996).

The EURODEM group also found that female sex and low educational level were associated with increased risks of AD. Conversely, sex and education were not reported as risk factors for incident AD in other studies (Yoshitake et al., 1995; Cobb et al., 1995).

1.4.3 Head trauma

Head trauma came under investigations as a risk factor for AD when amyloid deposits were found in the brains of boxers who were demented before death and in a significant number of patients who died after a single episode of severe head injury. A proposed biological mechanism is that increased expression of Aβ is part of an acute response to neuronal injury, leading to deposition of plaques and the initiation of an AD-type process (Roberts et al., 1994; Graham et al., 1995).

Risk factors such as Apo E4 and a history of head trauma may be synergistic (Payami et al., 1996; Mayeux et al., 1995). Mayeux et al. (1995) examined the association between Apo E genotypes and head trauma. They reported a 10-fold increase in the risk factor of AD associated with the combination of an Apo E4 allele and history of traumatic brain injury.

1.4.4 Cardiovascular risk factors

Epidemiological studies suggest that diseases compromising the vascular system such as hypertension and diabetes mellitus may influence the risk of developing AD. An association between hypertension and AD has been reported in several prospective studies (Meyer et al., 2000; Schmidt et al., 2000; Kivipelto et al., 2001). Chronic hypoperfusion may also induce episodes of hypoxia-ischemia in the brain, triggering free radical-oxidative cascades. Hypertension and hypercholesterolemia independent of the Apo E genotype have been shown to be risk factors for AD (Skoog et al., 1999).

There is growing evidence that cholesterol is of particular importance in regulating α- and β-secretase cleavage (Simons et al., 2001). Levels of total cholesterol and low density lipoprotein (LDL) in serum were reported to correlate with the amount of Aβ in AD brains (Kuo et al., 1998). The level of 24S-hydroxy-cholesterol, a major product of brain cholesterol metabolism, is elevated in plasma of patients with AD (2000). In contrast, patients with advanced AD seem to have low levels of 24S-hydroxycholesterol. The sterol is produced from brain cholesterol through the action of CYP46, a neuronal oxidative enzyme. It is possible that high levels of 24S-hydroxycholesterol reflect marked neurodegeneration or increased activity of CYP46. 24S-hydroxycholesterol has been shown to be neurotoxic and to be implicated in the brain deterioration (Kolsch et al., 2001). A study showed that a high serum total cholesterol level (mean level, >251 mg/dl (>6.5 mmol/l)) at 40 to 59 years of age predicted AD in later life (Notkola et al., 1998). Interestingly, two independent retrospective studies reported a strong decrease in the incidence of AD and dementia in patients treated with 3-hydroxy-3-methylglutaryl-coenzyme A reductase inhibitors (Jick et al., 2000; Wolozin et al., 2000). Statin treatment reduced levels of plasma lathosterol by 49.5%, 24S-hydroxycholesterol by 21.4%, LDL cholesterol by 34.9% and total cholesterol by 25% (Vega et al., 2003). However, despite biological mechanisms and a remarkable convergence of data linking cholesterol levels to the pathophysiology of AD, a recent large, population-based study showed that baseline and long-term average serum total cholesterol levels were not associated with the risk for incident AD (Tan et al., 2003). Several factors may explain the discrepancy in results. Since cholesterol levels decline with age, the age of the subjects at the time of cholesterol level determination can be a significant factor when using single rather than time-averaged cholesterol levels (Wilson et al., 1994). Older subjects are more likely to have low plasma cholesterol levels and also a greater risk for development of AD.

1.5 Psychiatric manifestations

Psychiatric manifestations are common in AD and occur in almost all patients at some point in their illness (Tariot & Blazina, 1993). Agitation becomes increasingly common as the illness advances and is a frequent precipitant of nursing home placement. Depressive symptoms are present in about 50% of patients and approximately 25% exhibit delusions (Mega et al., 1999). In later stages, AD may lead to negative behavioral changes that include mood, anxiety and agitation as well as severe deterioration in learning and memory.

Apathy is a feature of depression but it may also be found in AD without depression. Apathy is apparently early in the clinical course with diminished interest and reduced concern. It has been suggested that apathy and depression are clinically distinct neuropsychiatric syndromes (Levy et al., 1998). In a study, apathy was noted in 37% of patients with AD and it was also observed that apathy in AD was associated with severe impairment in activities of daily living and cognitive functioning, older age and poor awareness of behavioral and cognitive changes (Starkstein et al., 2001).

1.6 Genetic susceptibility factors

Tremendous progress has been made in determining the gene defects underlying early-onset AD (occurring in persons < 60 years of age). These defects include highly penetrant, autosomal dominant mutations in three genes: PS1 on chromosome 14, PS2 on chromosome 1, and the APP on chromosome 21 (Sherrington et al., 1995; Levy-Lehad et al., 1995; Goate et al., 1991). More than 45 known mutations in the gene for PS1 have been observed (Liao et al., 1999). These mutations account for a significant proportion of early-onset autosomal dominant AD. However, overall PS1-related early-onset AD accounts for fewer than 1% of all cases of AD.

Steady but slower progress has been made in deciphering the complex genetic picture of late-onset AD, which is associated with common population polymorphisms (CPPs) that operate primarily as susceptibility factors increasing risk but not guaranteeing onset of AD. Two of these genetic susceptibility factors for AD are the Apo E gene on chromosome 19 and α2-macroglobulin gene (A2M) on chromosome 12 (Saunders et al., 1993; Blacker et al., 1998; 1999). The only known susceptibility gene for LOAD is the gene for Apo E.

The Apo E gene has been widely accepted as risk factor and has been extensively studied. It has three alleles called ε2, ε3 and ε4 and the most common allele is the ε3. The presence of the ε4 allele (chromosome 19) is found to accelerate the age of onset of familial AD (early onset) and to increase the risk of developing the sporadic form of AD (late onset). The ε4 allele is associated in a dose-dependent way with decreased age at disease onset. This finding has been well replicated worldwide, but still only explains 45% to 55% of AD risk (Roses et al., 1995). Those patients with two ε4 alleles are at a higher risk while ε2 alleles have a protective effect. Apo E is involved in the transport of cholesterol and in the normal brain, it efficiently binds to β-amyloid protein (Porer, 1994). This interaction may prevent the toxic aggregation of Aβ. In AD, this property is presumably lost, thus facilitating the accumulation of Aβ and the resulting plaque formation. Apo E also binds to ciliary neurotrophic factor (CNTF) and potentiates its biological activity on hippocampal cells (Gutman et al., 1997). Although the association of Apo E with AD is well recognized, its role in the pathogenesis of AD is not clear (Tsuang et al., 1999).

It has been suggested that PS1 may also be a susceptibility gene for LOAD (Wragg et al., 1996). Homozygosity of allele 1 of a PS1 intron 8 polymorphism (PS1-1) has been associated with doubling of the risk of sporadic LOAD, in some but not all studies. A recent family-based, case-control, genetic-linkage study designed to genotype the PS1 intron 8 polymorphism in predominantly Hispanic families with LOAD to test for association and for linkage between this polymorphism and LOAD could not confirm the relationship between the PS1 intron 8 polymorphism and LOAD (Romas et al., 2000). However, an association between of the ApoE ε4 allele and LOAD (relative risk, 4.05, 95% confidence interval, 1.3–12.5) was observed. The ε4 allele was associated with a 4-fold increase in risk after adjustment for age, education, and sex.

1.7 Herpes simplex virus infections

The etiology of AD is multifactorial and the vast majority of cases cannot be attributed to these genetic factors alone, indicating that environmental factors may modulate the onset and progression of the disease. Although many factors have been proposed (head injury, heavy metal, infectious agents), there has been renewed interest in the neurotropic herpes simplex virus (HSV1) type 1 (Jack, 1996).

The hypothesis that HSV1 might play a role in AD was formerly proposed in the early 1980s (Ball, 1982). It was suggested that the brain regions most severely affected in cases of HSV encephalitis are principally the same as those affected in AD, notably limbic structures and some cases of acute HSV encephalitis have resulted in memory deficits (Barnett et al., 1994; Klapper et al., 1984). Furthermore, HSV1 has the capacity to remain latent in infected neurons and can be spontaneously reactivated in response to various noxious stimuli that could lead to its dissemination into previously uninfected areas. These characteristics of HSV1 are consistent with the chronic, progressive nature of AD.

Using polymerase chain reaction (PCR) methodology along with a number of cautionary controls to guard against cross-contamination, the presence of HSV1 DNA in the temporal and frontal cortex of the majority of tested AD patients and aged-matched controls was observed (Itzhaki et al., 1993; Baringer & Pisani, 1994). These findings indicate that the presence of HSV1 in the CNS is more prevalent than previously thought and that it is brain region specific and age related. The relationship between Apo E genotype and the presence of HSV1 DNA in brain specimens from AD patients and age-matched control patients has been assessed (Itzhaki et al., 1997). It was observed that the frequency of the Apo E4 allele was considerably higher in HSV-positive AD patients. Thus, a diagnosis of AD was more frequent among individuals carrying one or more Apo E4 alleles and whose brains simultaneously harbored HSV1 than those possessing only one or none of these risk factors. Although HSV1 by itself was not found to be a risk factor for AD, these results suggest that infection of the CNS with this neurotropic virus in combination with the Apo E4 allele increases the risk for AD.

If the role of HSV1 is confirmed, a selective subpopulation of AD patients might benefit from antiviral drug therapy. However, the challenge lies in identifying the susceptible subgroup. Although Apo E genotyping is relatively straightforward, detecting viral DNA in the CNS of living patients is clearly more problematic, especially since seropositivity does not necessarily have to correlate with HSV1 infection of the CNS. Since brain biopsy is not feasible for screening a large population base because it presents obvious risks, an alternative approach may be PCR analysis of cerebrospinal fluid (CSF) (Whitley & Lakeman, 1995). However, it remains to be determined whether markers indicative of subacute HSV1 brain infection will likewise be detectable in the CSF. Drug therapies may emerge from research aimed at elucidating the interaction between HSV1 and hosts carrying Apo E4. In this regard, there is evidence that suggests a direct interaction between viral glycoproteins and apolipoproteins (Huemer et al., 1988). Disrupting this interaction may yet present a novel therapeutic target.

1.8 Assessing outcomes

To gain a true overview of the impact of AD on patients, caregivers and society, areas such as cognition, ADL, behavior, caregiver burden and quality of life need to be adequately assessed (Table 1.5). Changes in these domains are currently considered to be of secondary importance by US standards for regulatory approval compared with cognitive and global outcomes. However, European and Canadian regulatory criteria require improvement in ADL as well as clinical and neuropsychiatric effects (Mohr et al., 1995; Committee for Proprietary Medicinal Products, 1997).

1.9 Discussion

In order to be effective in detecting and treating AD, it is important to determine which cognitive changes can be accepted as part of normal aging and to clarify whether brain aging and neurodegenerative dementing illnesses are distinct or continuous processes. Although none of these critical issues have yet been solved, the early detection of mild dementia and precursor states has become a principal target of recent research objectives that suggest the possibility of preventing the pathological process of AD (DeMattos et al., 2001). Patients with AD usually survive 7 to 10 years after onset of symptoms and typically die from bronchitis or pneumonia (Bracco et al., 1994; Beard et al., 1996). Awareness of dementia through use of valid diagnostic criteria can have important benefits. It may help to apply intervention strategies that benefit the patient and family. Appropriate medication can also lessen certain symptoms such as agitation and anxiety, aid sleep and improve participation in activities.

The risk of developing AD is lower in people who have been intellectually active than in those who have not (Friedland et al., 2001). Studies also indicate that higher mental function in one's 20s predicts better cognitive function late in life. Other

Table 1.5: Domains that need to be assessed in a clinical trial for AD. Adapted from Schneider et al. (2001)

Domain	Rating scales	References
Cognitive function	Alzheimer disease Assessment Scale-Cognitive subscale (ADAS-cog)	Rosen et al., 1984
	Mini – Mental State Examination (MMSE)	Tombaug & McIntyre, 1992
Activities of daily living	Blessed-Roth Dementia Scale	Blessed et al., 1968
	Progressive Deterioration Scale (PDS)	Rösler, 1999; Winblad et al., 1999
	Interview for Deterioration in Daily living activities in Dementia (IDDD)	Teunisse et al., 1991
	AD Cooperative Study – Activities of Daily Living (ADCS-ADL)	Galasko et al., 1997
	Disability Assessment for Dementia (DAD)	Gelinas & Auer, 1996
Behavioral symptoms	Neuropsychiatric Inventory (NPI)	Cummings, 1997
	Behavioral Pathology in Alzheimer disease Rating Scale (BEHAVE-AD)	Reisberg et al., 1987
Global clinical assessments	Clinical Dementia Rating scale (CDR)	Hughes et al., 1982; Berg, 1984
	Global Deterioration Scale (GDS)	Reisberg et al., 1982
	Gottfries-Brane-Steen (GBS)	Gottfries et al., 1982
	FDA-Clinician Interview Based Impression of change (FDA-CIBIC)	
	Clinician's Interview-Based Impression of Change plus caregiver input (CIBIC-plus)	Schneider & Olin, 1996
	AD Cooperative Study Clinical Global Impression of Change (ADCS-CGIC)	
Caregiver burden	Caregiver Activities Time Survey	
	Caregiver Activity Survey	
	Caregiver Time Questionnaire	
Quality of life	Dementia Quality of Life (DQoL) instrument	Brod et al., 1999
	Alzheimer disease Related Quality of Life (ADRQL) instrument	
Health economics	Resource Utilization in Dementia (RUD)	

studies have shown that college graduates have a lower risk of eventually developing AD than people with less educational achievements (Del Ser et al., 1999). Mental stimulation or exerting our brains in various ways intellectually may not only improve memory performance but may stave off future cognitive decline. Regular physical activity could be an important component of a preventive strategy against AD (Lindsay et al., 2002).

It is often incorrectly believed that AD is invariably associated with denial of illness (anosognosia) and loss of insight. Many patients with AD have preserved awareness of deficit (Weinstein et al., 1994). AD is uniquely challenging for family caregivers. The illness generally lasts for many years. The symptoms and the resulting demands on the caregiver change as the disease progresses so that the coping strategies that were appropriate at one stage may not have any value at a more severe stage. Caregivers of early stage patients may still be grappling with accepting the diagnosis and understanding its future implications for themselves and for the patients. Later, issues such as driving emerge. Caregivers of middle stage patients frequently are concerned with how to manage agitation and other difficult behavioral symptoms. Caregiver of later stage patients may need information on medical management, nursing home placement and end-of-life decisions. While it is possible to predict the average length of each stage of the illness, there is a great deal of variability from one patient to another. It is not possible to tell an individual caregiver how long each stage will last for his or her own ill relative or whether or when any particular behavioral symptom will occur.

Chapter 2
Neurological dysfunctions in Alzheimer disease

2.1 Introduction

Aging of the brain is accompanied by multiple changes in morphology and functional variations at the cellular and molecular levels of nerve cells. Neuritic plaques, one of the two diagnostic brain lesions observed in Alzheimer's original patient, are microscopic foci of extracellular amyloid deposition. They are associated with axonal and dendritic injury, and are generally found in large numbers in the limbic and association cortices (Dickson, 1997).

The advent of electron microscopy in the 1960s permitted a good description of the ultrastructural changes underlying the two classical lesions linked to AD, senile plaques (SPs) and NFTs. In the early 1980s, Glenner and Wong attempted to purify brain amyloid and eventually identify its main component. They were finally successful in isolating and purifying amyloid from brain vasculature (Glenner & Wong, 1984). They identified two peptides that they named α and β peptides. They continued to study the more abundant b peptide that they designated as amyloid β (Aβ). Using Edman degradation to sequence the peptide, they were able to determine a 24-amino acid residue sequence. Further support of the notion that they had indeed isolated the principal component of the amyloid plaques was confirmed by the observation that antibodies raised against their isolated peptide were able to recognize the SPs in the brain's gray matter and vasculature but failed to recognize the NFTs. It is now known that the Aβ peptides comprise a family of peptides. The surrounding neurites that contribute to any one plaque can emanate from local neurons of diverse neurotransmitter classes. The two most abundant forms of Aβ, termed Aβ40 and Aβ42, are 40 and 42 residues in length. Both forms are capable of assembling amyloid fibrils. Much of the fibrillar Aβ found in the neuritic plaques is the species ending at amino acid 42 (Aβ42), the slightly longer, more hydrophobic form that is particularly prone to aggregation (Jarrett et al., 1993). However, the Aβ species ending at amino acid 40 (Aβ40), which is normally more abundantly produced by cells than Aβ42, is usually colocalized with Aβ42 in the plaque. The cross-sectional diameter of neuritic plaques in microscopic brain sections varies widely from 10 to > 120 μm. The density and degree of compaction of the amyloid fibrils that comprise the extracellular core also show great variation among plaques.

In recent years, it has become apparent that in a variety of neurodegenerative, infectious and developmental disorders, intracellular tau can aggregate and form fibrils, often in the absence of extracellular amyloid deposits. There are six isoforms of tau (352–441 amino acids) in the human brain. They are rich in serine and threonine and can be phosphorylated at more than 20 residues (in AD) by many kinases, including cyclin-dependent kinases (Cdk-5) and cell cycle kinase (cdc-2). When tau is phosphorylated at serine 262 or at serine 214 (in an AD brain), it detaches from the microtubules, causing their breakdown. Tau hyperphosphorylation is believed to occur before aggregation, and hence, it is considered to be the earliest sign of degeneration in AD.

The diagnosis of AD traditionally has been based on histologic findings. This disease is still characterized by the same two classical lesions noted by Alois Alzheimer: the SPs and NFTs.

2.2 Neurochemical studies

2.2.1 Senile plaques and neurofibrillary tangles

SPs contain extracellular deposits of toxic β-amyloid peptides that are produced by aberrant processing of β-APP (Muller-Hill & Beyreuther, 1989). SPs are structurally complex lesions, the temporal development of which is only partially understood. They vary in size, ranging from 10 to over 120 μm in diameter. It is believed that SPs represent amorphous deposits of 39–43 amino acid hydrophobic self-aggregating Aβ peptides (Naslund et al., 2000; Neve et al., 2000; Selkoe, 1998). According to their maturation, SPs are classified as immature or primitive and diffuse plaques, and mature dense-cored and burned-out (compact) plaques. Using end-specific Aβ monoclonal antibodies, it was shown that mature, dense-cored and compact plaques consist of Aβ1–40 (or a mix of Aβ1–40 and Aβ1–42/43) and that immature, diffuse plaques consist mostly of Aβ1-42/43 (Iwatsubo et al., 1994). NFTs, on the other hand, are found intracellularly. The building blocks of the NFTs are paired helical filaments (PHF) that intertwine in neuronal regions typically affected by AD. Immunohistochemical and biochemical analyses of PHFs and NFTs suggest that they are made of a hyperphosphorylated form of tau, a microtubule-associated protein.

Controversy persists as to whether the histopathological hallmarks of AD, the SPs and the NFTs, are disease-specific or also can occur with normal brain aging. For example, SPs and NFTs, were found in 80% of brains from persons aged 65 or more (Gellerstedt, 1933). SPs were present in 78% nondemented patients and NFTs were found in 61% although the NFTs were largely confined to the hippocampal region (Tomlinson et al. (1968). The landmark observations of Tomlinson and collaborators formed the basis of subsequent histologic criteria for AD (Katchaturian, 1985; National Institute on Aging and Reagan Institute Working Group on Diagnostic Criteria for the Neuropatholgical Assessment of Alzheimer disease, 1997). These criteria rest on the assumption that the presence of SPs in the neocortex increases as a function of age and that SPs in AD are distinguished from aging only by their greater age-adjusted density. The fact that neurofibrillary tangles composed of altered, aggregated tau proteins occur in disorders in the absence of Aβ deposition (e.g., subacute sclerosing panencephalitis, Kuff's disease, progressive supranuclear palsy etc.) suggests that tangles can arise secondarily during the course of a variety of etiologically distinct neuronal insults.

The neuropathologic distinction of aging from AD is also rendered complicated by the controversy over the relevance of the two (diffuse and neuritic) SP subtypes. Diffuse SPs are immunoreactive for the amyloid protein associated with AD and morphologically appear as amorphous Aβ deposits with little or no accompanying neuritic pathology. Neuritic SPs are also characterized by deposits of Aβ fibrils but typically the fibrils have coalesced into a central core with dystrophic neuritis arranged around the periphery. Presumably, the Aβ deposits that comprise diffuse SPs progress in fibrillogenesis and induce neuritic degeneration (Davies & Mann, 1993; Morris et al., 1991). Most neuropathologists rely on neuritic plaque densities for the diagnosis of AD and do not include diffuse SPs in the diagnostic criteria (Mirra et al., 1991; National Institute on Aging and Reagan Institute Working Group on Diagnostic Criteria for the Neuropathological Assessment of Alzheimer disease, 1997). It has been suggested that, because SPs initially develop in the neocortex rather than at the site of greatest NFT accumulation such as the limbic cortex, it is likely that SPs and NFTs are produced by different factors (Morris & Price, 2001). NFT formation in the medial temporal structures may be an ubiquitous accompaniment of aging whereas the presence of neocortical SPs indicates AD. Tangle formation, a marker of neuronal degeneration, increases as a function of dementia severity and may be regarded as a determinant of the clinical progression of AD.

2.2.2 The β-amyloid precursor protein

Several groups were successful in their attempts to identify the gene encoding the Aβ peptide (Goldgaber et al., 1987; Tanzi et al., 1987). The gene is approximately 3200 base pairs long and encodes a membrane protein that spans the plasma membrane once. The encoded protein, which has a large extracellular domain and a small intracellular domain, was named the β-amyloid precursor protein (β-APP). The APP is a ubiquitously expressed 100–140-kDa type I integral membrane glycoprotein (Selkoe, 1994). It is a member of a highly conserved gene family, including the amyloid precursor-like proteins (APLP1 and APLP2) in mammals, as well as APP-like proteins (Daigle & Li, 1993; Luo et al., 1990). The APP gene has been assigned to human chromosome 21 whereas APLP1 and APLP2 genes map to human chromosomes 19 and 11, respectively (De Strooper et al., 1993; Lenjkkeri et al., 1998). Numerous studies support the hypothesis that the presence of APP in peripheral cells, i.e., endothelial and blood cells, may contribute to Aβ deposition (Vanley et al., 1981; Li et al., 1995; Rosenberg et al., 1997). A study designed to determine whether a differential level of platelet β-APP isoforms is specifically related to AD and whether it shows a correlation with the progression of clinical symptoms was undertaken (Di Luca et al., 1998). After volunteers were grouped according to diagnosis and severity of dementia, APP isoform levels in platelets were compared. The ratio between the intensity of the 130-kDa and 106-kDa APP isoforms was significantly lower in the AD group (0.31 ± 0.15) compared with both controls (0.84 ± 0.2) and non-AD subjects (0.97 ± 0.4). A statistically significant correlation was observed also when individual MMSE scores and APP isoforms were considered, thus indicating that this value could represent a sensitive measure for the progression of the disease. Therefore, this study suggests that peripheral platelet APP isoforms may be considered as a marker of progression of the disease.

APP can be processed and released from the cell as secreted forms through distinct proteolytic pathways (Checler, 1995). Full-length APP undergoes different cleavages by still unknown proteases termed α-, β- or γ-secretases. The α-secretase cleavage occurs within the Aβ sequence between amino acids 16 and 17 (Esch et al., 1990). This leads to the production of a secreted form of APP termed sAPPα and also to the production of a 10-kDa C-terminal fragment (P10) that may be further cleaved by γ-secretase activity into 3-kDa (P3) and 7-kDa (P7) fragments (Haass & Selkoe, 1993). The β- and γ-secretases are responsible for the production of Aβ40 and Aβ42 (Pike et al., 1995; Citron et al., 1996). Aβ production is a normal process in healthy subjects as Aβ is normally secreted and released from cells. The protease, termed β-site APP cleaving enzyme (BACE1), cleaves APP on the luminal side of the membrane. The γ-secretase cleaves Aβ within the transmembrane domain and it has been shown that PS1 and PS2 mediate γ-secretase activity (De Strooper et al., 1998). There is increasing speculation that they themselves may be actual γ-secretases (Wolfe et al., 1999; Kimberly et al., 2000).

BACE1 is a membrane-bound aspartic protease with all the known functional properties and characteristics of β-secretase. It is a 501-amino acid sequence peptide most closely related to the pepsin aspartic protease family. The expression pattern of BACE1 is highest in pancreas and brain and significantly lower in most other tissues. The enzyme is present in neurons but almost undetectable in glial cells of the brain, as shown by in situ hybridization analysis of BACE1 mRNA (Vassar, 2001). With BACE2, also called Asp-1, memapsin-1 or DRAP (Down's region aspartic proteinase), a second member of the BACE subfamily of membrane-anchored aspartic proteases has been identified, having a high degree of similarity to BACE1. Several groups have now identified the enzymes responsible for b-secretase activity to be BACE1 and BACE2 (Sinha et al., 1999; Lin et al., 2000). Different approaches were followed to obtain the information about the sequence and subsequent characterization of the enzymes.

Apo E appears to play a role in Aβ deposition because a lack of Apo E dramatically reduced Aβ deposition in a transgenic model of AD (Bales et al., 1997; Holtzman et al., 1999). Apo E-deficient mice were crossed with transgenic mice overexpressing a human APP gene. In 6-month-old mice homozygous for the APP mutation transgene and the wild-type mouse Apo E gene, the cerebral cortex and hippocampus had numerous amyloid deposits. In contrast, 6-month-old mice homozygous for the APP transgene gene but lacking the wild-type mouse Apo E gene had no amyloid deposits.

2.2.3 The amyloid cascade hypothesis

There has been a continuing controversy over the significance of the specific pathological changes associated with AD. The amyloid hypothesis contends that the abnormal processing of APP leads to a shift in proportion of β-amyloid 1-40/42 (particularly the relatively more fibrillogenic 1-42 form) produced. This ultimately leads to insoluble protofibril species and plaque formation and subsequently neurodegeneration. A common assumption is that insoluble β-amyloid is "toxic" or promotes the toxicity of other molecules, although a precise mechanism of toxicity has yet to be widely accepted.

The effects of diverse Aβ fragments on learning and memory processes have been investigated in several laboratories with the assumption that Aβ is neurotoxic and may have deleterious behavioral effects. Some of the data are presented in Table 2.1. Thus, acute injections of Aβ have deleterious effects on learning and memory processing. Shortly after the treatment, Aβ does not appear to affect short-term memory (STM) or retrieval processes, but to selectively interfere with memory consolidation, leading to selective deficits in LTM. This effect may be related to a disruptive effect of the peptide on neuronal activity since Aβ has been shown to affect distinct levels of normal neuronal functions. In contrast, at delays of several days post-treatment, Aβ has been shown to affect both STM and LTM, but these effects largely depend on the learning task used and on the protocol of administration.

The amyloid cascade hypothesis, despite its many strengths, has some flaws as fibrillar amyloid is not the only toxic form of Aβ. Evidence also points to a pathogenic role for small toxins that comprise globular Aβ oligomers (Klein et al., 2001). The significance of the correlation between synapse loss and soluble Aβ can also be explained by the "amyloid β-derived diffusable ligands" (ADDL) hypothesis (Lambert et al., 1998). ADDLs (pronounced "addles") are potent Aβ1-42 neurotoxins that are non-fibrillar. Instead of fibrils, their structure comprises relatively small globular oligomers. The presence of high-affinity ADDL binding proteins in hippocampus and frontal cortex but not cerebellum parallels the regional specificity of AD pathology and suggests involvement of a toxin receptor-mediated mechanism (Klein, 2002). The ADDL hypothesis suggests that a major factor in AD is the neurological impact of soluble, globular oligomers of Aβ1-42 ADDLs. By causing specific aberrations in synaptic signaling, ADDLs first cause early-stage memory loss, with broader-scale effects of ADDLs subsequently leading to synapse degeneration and nerve cell death. The properties of ADDLs and their presence in AD-afflicted brain are consistent with their putative role even in the earliest stages of AD. The ADDL hypothesis gives a compelling explanation of why specific brain areas degenerate severely while neighboring areas remain intact, all without any plaques or fibrils in the vicinity. These "ADDLs" may present novel opportunities to develop AD therapeutic drugs.

Table 2.1: Effects of Aβ fragments on learning and memory performance in rodents

Aβ fragments	Injection procedure	Animal	Behavioral effects	Reference
1-40	2 ml of a 10^{-3} M solution hippocampal bilateral injection immediately after acquisition	Rat	Disruption of consolidation processes in an active avoidance paradigm, no effect on short-term spatial working memory	McDonald, 1994
	3, 30 or 300 pmol/day i.c.v. continuous infusions during 2 weeks	Rat	Impaired learning abilities in the water maze and in passive avoidance at the higher dose	Nabeshima & Nitta, 1994
	10 mg in 1 ml unilateral injection in the basal nucleus	Rat	No clear impairment in spontaneous object recognition or passive avoidance response 7 to 60 days following the injection	Giovannelli, 1995
	1, 2 or 3 ml of 10^{-3} M bilateral acute or chronic intrahippocampal injection	Rat	No effect of either acute or chronic injections on short-term memory performance in a delayed conditional procedure	Cleary, 1995
1-28, 12-28, 18-28, 12-20	1 to 6 nmol/mouse post-acquisition i.c.v. or intrahippocampal injections	Mouse	Dose-dependent impairments in an active avoidance paradigm and decreased retention performance in a lever-press operant paradigm	Flood, 1994
12-28	0.1 to 1.0 mg/mouse post-acquisition injection in diverse limbic structures	Mouse	Dose-dependent impairments in an active avoidance paradigm No deficit with a thalamic injection	Flood, 1994
25-35	3 mM bilateral injections in the hippocampi 2 and 1 weeks before acquisition	Rat	Injected mice are able to acquire the task but show longer latencies and swim distances to locate the platform in the water maze paradigm	Chen, 1996

Table 2.1 (continued)-

Aβ fragments	Injection procedure	Animal	Behavioral effects	Reference
25-35 1-28	3 or 9 nmol/mouse i.c.v. Mouse injections	Mouse		Sigurdsson, 1997
	6 days before testing		Impaired spontaneous alternation	
	7 days before training		Affects retention, but not acquisition, in a passive avoidance paradigm	
	1 hour after training		Impaired retention performance in the passive avoidance paradigm	
	4 hours before retention test		No effect on retention performance in the passive avoidance paradigm	
	10–14 days before training		Increased escape latencies in the water maze paradigm	
25-35	5 nmol/mouse, bilateral intra-amygdaloid injections 34–52 days before testing	Rat	No effect in an active avoidance or in the water	Sigurdsson, 1997
Phe($^{24}SO_3H$) -25-35	0.4 nmol/rat, bilateral injection in the nucleus basalis magnocellularis 2–3 before testing	Rat	Impairments in an active avoidance paradigm No effect in the water maze paradigm	Harkany, 1998
1-42	300 pmol/day i.c.v. continuous infusions starting 7–19 days before testing	Rat	Impaired spontaneous alternation after a 8-day treatment; impaired spatial and working memory performance in the water maze after a 2 week treatment; impaired retention performance in a passive avoidance paradigm after 18-day treatment	Yamada, 1999

2.2.4 Selective vulnerability

AD is principally a disease that affects the cerebral cortices including neurons providing connection to and within this brain structure (Vickers et al., 2000; Hof et al., 1999). There appears to be a regional hierarchy with respect to frank nerve cell degeneration with medial temporal lobe/limbic structures such as the amygdala, hippocampus and entorhinal cortex being affected earlier in the disease and typically showing severe damage by later stages of the disease. In other cortical regions, higher-order association neocortices are the principal sites of brain pathology with evolving disease with primary sensory and motor regions typically demonstrating relatively little pathology. This spread of pathology throughout these circuits reflects a general staging of the disease from an early preclinical phase of minimal pathology and isolated neuronal degeneration to later stages involving significant degeneration in other brain areas responsible for higher brain functions. The term "selective vulnerability" was introduced into neuropathology to describe the selective loss of specific populations of neurons, in disorders in which a genetic abnormality or environmental insults affect both vulnerable and resistant neurons (Cooper et al., 1999; Braak et al., 2000). This phenomenon was originally described in studies of the neuropathology of aging and of neurodegenerative diseases (Vogt and Vogt, 1951). The neurochemical mechanisms underlying selective vulnerability are an important and still largely unresolved problem in the neurobiology of disease, although it is now recognized that both "selectively vulnerable" and "selectively resistant" are relative rather than absolute terms.

A recent study examined by immunohistochemistry the cellular distribution of a mitochondrial constituent (the α-ketoglutarate dehydrogenase complex, KGDHC) known to be deficient in AD, in relation to the known selective vulnerability of neurons in areas 21 and 22 of the temporal lobe in this neurodegenerative disorder (Ko et al., 2001). It was found that in normal human brain, cortical layers III and V contain neurons intensely immunoreactive for KGDHC, compared to other cells in these areas. The KGDHC-enriched cells are lost in AD. In layer III, the loss of KGDHC-enriched cells is proportional to total loss of neurons, as determined by immunoreactivity to neuron-specific enolase. In layer V, a higher proportion of the KGDHC-enriched neurons is lost than of other neurons. These data suggest that variations in mitochondrial composition may be one of the factors determining which cells die first when different types of cells are exposed to the same stress.

2.3 Electrophysiology and imaging studies

Methods for the non-invasive measurement of brain activity fall into two main principal categories, depending on whether they measure electrophysiological or hemodynamic variables. These methods can be used to detect transient activity that follows a specific event such as the presentation of a stimulus (event- or item-related activity), as well as the more sustained activity that may accompany engagement in a specific task (state-related activity). Electrophysiological methods detect the time-varying electric (electroencephalography (EEG)) and magnetic (magnetoencephalography (MEG)) fields at the scalp surface generated by synchronously active populations of neurons. Hemodynamic methods measure neural activity indirectly, relying on coupling between changes in neural activity and local changes in variables such as regional cerebral blood flow (rCBF) and blood oxygenation. The methods employ image reconstruction techniques to localize changes in these variables to within a spatial resolution of a few millimeters.

Functional neuroimaging by PET or functional magnetic resonance imaging (fMRI) provides a way to monitor large neuronal populations in awake humans while they engage in cognitive tasks. PET involves measuring regional blood flow using $H_2{}^{15}O$ and allows for repeated measurements on the same individual. fMRI is based on the fact that neural activity changes local oxygen levels in tissue and that oxygenated and deoxygenated hemoglobin have different magnetic properties. It is now possible to image the second-by-second time course of the brain's response to single stimuli or single events with a spatial resolution in the millimeter range.

Electrophysiological studies revealed increases in the theta and delta bands and a decrease in the alpha and beta bands in the baseline EEGs of AD patients (Letemendia & Pampiglione, 1958; Harner, 1975; Stigsby et al., 1981). Serial quantitative EEG studies show a high correlation between the degree of dementia and theta power and mean frequency (Coben et al., 1985). Focal EEG changes with or without generalized slowing suggest either multi-infarct dementia or normal pressure hydrocephalus (Soininen et al., 1982; Brown & Goldensohn, 1973). Neuropsychological and neurophysiological studies that investigated the influences of drugs on AD patients found that the acute administration of some cholinergic drugs which improved memory and attention also exhibited a tendency to shift the EEG into more normal patterns while anticholinergic drugs induced opposite effects (Agnoli et al., 1983; Neufeld et al., 1994).

Atrophy of the hippocampus and amygdala has consistently been shown in AD even in preclinical stages (Golebiowski et al., 1999; Heun et al., 1997). There remains an overlap between AD patients and healthy elderly controls. Many studies have shown age-related reductions of the hippocampus in healthy subjects and AD patients (Golebiowski et al., 1999; Smith et al., 1999; Mu et al., 1999). Reduction of hippocampus volume with age may therefore reduce the ability to use this volume to distinguish AD patients from healthy elderly controls. Recently, a study showed that age transformation of combined hippocampus and amygdala volume may improve the diagnostic accuracy in AD (Hampel et al., 2002). Using an orthogonal rotational transformation of the coordinate system, values of magnetic resonance imaging (MRI)-determined volumes of hippocampus-amygdala formation (HAF) were transformed according to the age of the AD subjects and the sex-matched healthy control subjects. The results showed that the age transformation increased the diagnostic accuracy of HAF volumes in the subjects. It was therefore suggested that age transformation may provide an easily applicable procedure to increase the clinical diagnostic accuracy of hippocampal measurements by considering the effect of aging on hippocampus volume.

Atrophy of the medial temporal lobe in AD has been shown using rating scales, measures of hippocampal width and volumes and measures of entorhinal cortical volume (Scheltens et al., 1992; Killiany et al., 1993; Juottonen et al., 1998). Serial MRI coregistration techniques suggest that the atrophy may be more widespread, involving the cortex diffusely even at an early stage of the disease (Fox et al., 1998). It has been suggested that a larger brain volume provides a greater cerebral reserve against the effects of AD, maintaining cognitive function in the presence of neurodegeneration and thereby delaying the onset of symptoms. This hypothesis was based on the observation that nondemented elderly subjects with histological evidence of AD at autopsy had larger brains and a greater number of large neurons than elderly control subjects without such histological change. However, a study that investigated the relationship between total intracranial volume (TIV) and sporadic and familial AD reported no significant association between TIV and AD (Jenkins et al., 2000). Premorbid brain size did not differ between AD patients and controls and did not delay disease onset.

In functional imaging studies, the most consistently replicated clinical finding associated with the posterior cingulate cortex has been decreased metabolic activity of this region in AD (Nyback et al., 1991; Ishi et al., 1997). Most striking is the observation that the posterior cingulate cortex is the region with the greatest reduction in metabolic activity in patients with very early AD (Minoshima et al., 1997). This region is also hypometabolic in individuals who are at risk for AD (homozygous for the Apo E4 allele and having a positive family history for AD) but currently have no clinical evidence of the disease (Reiman et al., 1996). The translation from fMRI studies of normal memory function to clinically useful assessment tools will require the development of memory tasks that consistently elicit significant activation of specific brain regions in individual subjects. Activations that require group analyses to be clearly detected cannot be used to assess individual patients.

2.3.1 Magnetic resonance imaging

Due to the progressive nature of AD, the question has arisen whether serial MRI, which should be able to document the progressive changes occurring within an individual, would improve the diagnostic accuracy. In a study that documented longitudinal changes in hippocampal volumes using MRI over a period of 3 years in AD patients and normal controls, a statistically nonsignificant trend towards accelerated volume loss in the AD group was observed (Laakso et al., 2000). During the study period, the average shrinkage of the hippocampal volume ranged from –2.2% to –5.8% in control subjects, and from –2.3% to –15.6% in AD patients. However, the observed changes at an individual level were small, and within the accuracy range of the measurements. Serial MRI of the hippocampus did not offer any advantages over a single MRI to support the diagnosis of AD.

A study designed to test the hypothesis that MRI-based measurements of hippocampal volume are related to the risk of future conversion to AD in older patients with a mild MCI showed that hippocampal atrophy at baseline was associated with crossover from MCI to AD (relative risk, 0.69, $p = 0.15$) (Jack et al., 1999). The associations between hippocampal volume and crossover remained significant. It appears that in older patients with MCI, hippocampal atrophy determined by premorbid MRI-based volume measurements is predictive of subsequent conversion to AD.

A recent study designed to investigate whether hippocampal volume is a sensitive and specific indicator of AD neuropathology regardless of the presence or absence of cognitive and memory impairment was performed using postmortem MRI scans obtained for the first 56 participants of the Nun Study who were scanned (Gosche et al., 2002). The results showed that hippocampal volume predicted fulfillment of neuropathologic criteria for AD for all participants. In individuals who remained nondemented, hippocampal volume was a better indicator of AD neuropathology than a delayed memory measure. Among nondemented individuals, Braak stages III and VI were distinguishable from those in stages II or lower ($p = 0.001$). The results suggest that hippocampal volumes may be valuable not only in distinguishing individuals with AD from the nondemented elderly but also in identifying individuals with AD neuropathology who have not yet demonstrated any memory impairment. Thus, volumetric measures of the hippocampus may be useful in identifying nondemented individuals who satisfy neuropathologic criteria for AD as well as pathologic stages of AD that may be present decades before initial clinical expression.

2.3.2 Positron emission tomography

PET measures neuronal activation by estimating local CBF on the basis of detecting the concentration of radioactively labeled $H_2{}^{15}O$ (Fox et al., 1984). Blood flow changes follow the patterns of neural activity by several seconds (Roland, 1993). A number of PET and fMRI studies have measured brain activity during episodic and semantic retrieval using the subtraction method (see Cabezza & Nyberg, 2000). A meta-analysis of four PET studies that used recognition as the episodic retrieval task revealed that increased blood flow in six brain areas could be contributed to episodic retrieval mode, five of which were in the prefrontal cortex including Brodmann's areas (BA) 10, 45/47 and 8/9. While these prefrontal areas were found bilaterally (with the exception of BA 8/9, which was found only on the right), they were clearly lateralized to the right hemisphere. This suggested that the so-called hemispheric encoding-retrieval asymmetry (HERA) pattern of cerebral blood flow, according to which episodic retrieval elicits higher right hemispheric blood flow than encoding, could be mostly contributed to task-related retrieval processes (Tulving et al., 1994; Lepage et al., 2000). Other brain regions have been reported to be activated during episodic retrieval, among which the dorsolateral prefrontal cortex, the medial temporal lobes, the medial parietal and posterior cingulate regions and the parietal cortex are the most frequent (Nyberg et al., 2000; Krause et al., 1999). Semantic retrieval has also been investigated using PET and fMRI. A consistent finding during semantic retrieval has been increased blood flow in ventrolateral (BAs 45/46), ventromedial (BA 11) and mid-dorsal (BAs 9 and 46) areas of the left prefrontal cortex (Kapur et al., 1994; Jennings et al., 1997).

PET studies show that patients with AD have abnormally low measurements of the cerebral metabolic rate for glucose (CMRgl) in parietal, temporal and posterior cingulate cortex. Abnormally low prefrontal and whole brain measurements in more severely affected individuals and a progressive decline in these measurements over time are also noted. Studies suggest that PET has potential to detect these reductions prior to the onset of symptoms in individuals with AD and other degenerative brain disorders (Small et al., 1996).

As a clinical instrument, PET does not solely reveal dysfunctional changes early in the course of the disease but may also provide a deep insight into the functional mechanisms of new potential drug treatment strategies (Nordberg, 1999). The advantage of PET is the capacity not only to measure changes in glucose metabolism, CBF but also to obtain further insight into neuronal communicative processes (transmitter/receptor interactions) in brain and pharmacokinetic events and drug mechanisms. Imaging techniques such as PET and SPECT offer the unique opportunity to study functional effects in the brain induced by drug treatment. So far relatively few drug treatments in demented patients have been evaluated by these techniques.

2.3.3 Magnetization transfer ratio

The technique of magnetization transfer ratio (MTR) was proposed as a novel measure for the early diagnosis of MCI and AD (Kabani et al., 2002). Magnetization transfer is a technique that generates contrast dependent upon the phenomenon of magnetization exchange between semisolid macromolecular protons and water protons (Wolff & Balaban, 1989). Using this procedure, it was shown that the MTR of white matter was significantly lower in the AD group (Kabani et al., 2002). The gray matter volume was significantly lower in the AD group compared to controls (387.29 ± 26.04 cm^3 *versus* 532.93 ± 20.53 cm^3) and MCI (464.64 ± 16.93 cm^3). No significant differences were found in the white matter volume between the three groups. It was therefore suggested that the MTR methodology may be used as a novel approach to improve early diagnosis of AD. The changes in the neuroanatomical substrate of the brain were attributed to changes as a result of disease progression in the MCI and AD groups and not to the age factor. Therefore, these results indicate that MTR measurements in MCI subjects may well be a more sensitive and improved means to measure early changes in the brain tissue indicative of incipient dementia. Determining whether MCI subjects who show abnormal values are destined to deteriorate will require longitudinal follow up.

2.4 Discussion

Both the MCI and AD cases had significantly fewer neurons than non-demented individuals (Kordower et al., 2001). The relationship between amyloid deposits and neurofibrillary lesions remains an important unresolved issue in our understanding of the pathogenesis of AD. There is accumulating evidence that Aβ and its precursor are essential for brain function. Particularly, Aβ is an antioxidant and a molecular sensor of membrane lipid dynamics.

There is a lack of consensus regarding the mechanisms by which Aβ causes neurodegeneration. This highlights one of the key weaknesses of the hypothesis that this molecule causes AD. The concentration of Aβ used in almost all *in vitro* neurotoxic studies is between 1 and 100 μg/ml, which is as much as 10 000 times the concentration of Aβ in human cerebrospinal fluid (CSF). Further problems concern the specificity of the synthetic peptides used; most toxicity studies have used synthetic peptides that might behave differently from the peptides generated *in vivo*. Abundant Aβ deposits can be present in cognitively normal individuals and it is the presence of intracellular neurofibrillary lesions that correlates better with the presence of cognitive changes. More recent studies from the laboratory, however, show that the soluble pool of Aβ may be the major determinant of neurodegeneration (Holsinger et al., 2002).

The correlation of dementia severity with NFTs is consistent with the hypothesis that the Aβ-initiated AD process exacerbates NFT accumulation in vulnerable areas (Bierer et al., 1995). The density and pattern of NFT formation correspond closely to neuronal loss in these regions. Because of neuronal dysfunction, synaptic loss and death produce dementia; the greater the neuronal loss (or the greater the number of NFTs), the greater the severity of dementia.

Functional neuroimaging has provided some new insights in the study of memory dysfunctions. However, interpreting functional neuroimaging of memory is no easier than interpreting lesion studies of amnesics. Although fMRI has steadily improved spatial resolution, its temporal resolution does not approach physiologically relevant levels.

Chapter 3
Pharmacology of Alzheimer disease

3.1 Introduction

AD is characterized by multiple deficits including alterations to the cholinergic system (a major target in the neurotransmitter research because it is central to the cognitive changes observed in AD), the noradrenergic system, the serotonergic system and the dopaminergic system. According to the cholinergic hypothesis, the impairment of cognitive function and the behavioral disturbances that affect patients with AD are mainly due to cortical deficiencies in cholinergic neurotransmission. Deficiencies in the cholinergic system may also be involved in the formation of amyloid plaques and NFTs.

Many laboratories have shown that overstimulation of the activity of excitatory brain cells by introducing the excitatory neurotransmitter glutamate or β-amyloid peptides leads to neuronal destruction (Sigurdsson et al., 2001; Frenkel et al., 2001; Zipfel et al., 2000). Moreover, overstimulation can be initiated by plaque material or other stimuli.

Many studies have shown an altered Ca^{2+} homeostasis in aging (Thibault et al., 1998; Verkhratsky & Toescu, 1998). In aging neurons, Ca^{2+} action potentials are larger and have a longer plateau phase than in young neurons (Moyer & Disterhoft, 1994). This increase in Ca^{2+} influx is at least partially caused by an increase in the functional L-type Ca^{2+} channel density (Chen et al., 2000).

3.2 Cholinergic receptors

Acetylcholine (ACh) is a neurotransmitter that modulates neuronal activity through agonist effects on muscarinic and nicotinic receptors. Cholinergic neurons demonstrate extensive distribution throughout the CNS and the peripheral nervous system. In the CNS they are believed to be associated with cognitive functions.

Nicotinic cholinergic receptors may be especially important in regulating cognitive functions such as attention and in causing the release of more transmitter from cholinergic neurons as well as from numerous other neurons that release dopamine, γ-aminobutyric acid (GABA), glutamate and norepinephrine (Stahl, 2000). Nicotinic receptors can be regulated both by ACh and by allosteric modulators that help ACh.

The loss of the basal forebrain cholinergic system is one of the most significant aspects of neurodegeneration in the brains of AD patients and it is thought to play a central role in producing cognitive impairments (Perry et al., 1978; Whitehouse et al., 1981). There was an extensive loss of cholinergic neurons in the nucleus basalis of Meynert which project widely to neocortex, amygdala, and hippocampus (Whitehouse et al., 1981, 1982). Reduced activity of cortical choline acetyltransferase correlates with the number of SP and with cognitive impairment in patients with AD (Perry et al., 1978). Post-mortem biopsy studies have also shown that as the AD condition worsens, progressive loss of basal forebrain cholinergic neurons is accompanied by incremental loss of nicotinic ACh receptors (nAChRs) in cerebral cortical neurons (Nordberg, 1999; Perry et al., 2000). Two nAChR subtypes are found in abundance in the mammalian CNS. One binds nicotine with high affinity and is composed of α4- and β2-subunits. The other binds α-bungarotoxin and is most probably a homomeric α7-nAChR (Lindstrom, 1997). These receptors are located postsynaptically where they mediate fast synaptic transmission and presynaptically where they modulate synaptic transmission mediated by numerous neurotransmitters, including glutamate, GABA, serotonin, ACh and norepinephrine (Albuquerque et al., 2000). Interestingly, the loss of nAChR is receptor subtype selective. Receptors containing the α4 nAChR subunit (the predominant brain nAChR subtype) are decreased both in the hippocampus and cerebral cortex, whereas cortical levels of receptors containing α7 and/or α3 subunit receptors are maintained (Martin-Ruiz et al., 1999).

Acetylcholinesterase (AChE) is a 76-kDa protein and a member of the α/β-hydrolase family. It is one of the fastest acting enzymes known and its major function is to terminate the action of ACh by hydrolyzing it to choline and acetate. It is found in cholinergic synapses in the CNS and periphery but also in non-cholinergic nerves. It plays an important role in the development of neurons and neurite outgrowths. Butyrylcholinesterase (BuChE) has a structure similar to that of AChE. BuChE is synthesized in the liver and secreted into plasma. It is found with AChE in the gastrointestinal tract and heart and constitutes 1–10% of the total amount of cholinesterase in the adult CNS, where it is present in glial cells. The physiological role of BuChE is unclear, but it may serve as a scavenger in the detoxification of certain chemicals, thus limiting the amount reaching the CNS. It has also been suggested that BuChE may participate in the transformation of the β-amyloid protein from an initially benign form to an eventually malignant form associated with neuritic tissue degeneration and clinical dementia (Guillozet et al., 1997). More work is necessary both at the experimental and the clinical level to elucidate the role of BuChE in normal brains and in brains affected by AD (Giacobini, 2001).

As a result of these observations, substantial pharmacological research focused on attempting to enhance ACh levels in the synaptic cleft, primarily by inhibiting the degradative enzyme. AChE inhibitors enhance neuronal transmission by preventing the hydrolysis of ACh by the enzyme AChE, thus increasing the availability of ACh in muscarinic and nicotinic receptors. These efforts have ultimately led to a few acetylcholinestase inhibitors specifically approved to date for treating AD.

3.2.1 Cholinergic receptors and Aβ deposition

A study that hypothesized that cortical cholinergic deafferentation leads to Aβ deposition showed that cholinergic deafferentation induced in rabbits by a selective immunotoxin leads to Aβ deposition in cerebral blood vessels and perivascular neuropil (Beach et al., 2000). Biochemical measurements confirmed that lesioned animals had 2.5- and 8-fold increases of cortical Aβ40 and Aβ42, respectively. Cholinergic deafferentation may be one factor that can contribute to Aβ deposition. The reason for this is unknown. Longer survival periods may be necessary to allow the development of profuse neuropil deposits.

Very low concentrations of Aβ peptides can directly induce cholinergic hypofunction, particularly in the hippocampal and cortical regions, thus indicating a possible connection between amyloid burden and cholinergic impairment in AD (Pedersen et al., 1996). The potency of Aβ as a cholinergic neuromodulator suggests that it is a very potent inhibitor of ACh release. A greater understanding of the mechanism by which Aβ-related peptides and the cholinergic system interact may lead to significant progress in our comprehension of AD symptomatology.

3.3 Monoaminergic receptors

3.3.1 Serotonin

Recent studies have indicated that the serotoninergic system plays important roles in memory function (Meneses, 2000). Among the serotonin (5-HT) receptor subtypes shown to play a role in various learning and memory models, the 5-HT_{1A} receptors are of particular interest. This receptor is characterized by its high concentrations in the limbic system such as the hippocampus which is known to play an important role in learning and memory (Press et al., 1989; Squire & Zola-Morgan, 1991). It interacts with other neurotransmitter systems such as the glutamatergic and cholinergic systems (Steckler & Sahgal, 1995; Boast et al., 1999).

There is increasing evidence that the functional integrity of the 5-HT system may be compromised in age-related cognitive disorders and that an abnormality exists in the 5-HT system in AD (Palmer & DeKosky, 1993). Histopathological studies have shown the formation of NFTs or SPs in the dorsal raphe nucleus (DRN, a major cell body area for 5-HT) and in the hippocampus as well as modest to severe cell loss in the DRN (Halliday et al., 1992; Brady & Mufson, 1991).

A recent study using PET with WAY-100635 examined the relationship of 5-HT_{1A} receptors with memory function in normal healthy male volunteers (Yasuno et al., 2003). The administration of tandospirone, a specific 5-HT_{1A} agonist, impaired explicit verbal memory in a dose-dependent manner, whereas other cognitive functions showed no significant changes. The change in memory function paralleled those of body temperature and growth hormone that were reported to be induced by the stimulation of postsynaptic 5-HT_{1A} receptors (Blier et al., 1994; Seletti et al., 1995). These data may suggest the possibility that the antagonistic effect of postsynaptic 5-HT_{1A} receptors in the hippocampus leads to improvement of human memory function. 5-HT_{1A} antagonists may be favorable for improved control of memory impairment.

3.3.1.1 Serotonin depletion studies

Acute tryptophan depletion by dietary means reduces central 5-HT levels and 5-HT function (Nishizawa et al., 1997). In young healthy volunteers, acute tryptophan depletion has been shown to cause impairment in learning (Park et al., 1994). Studies that examined the effects of 5-HT depletion induced by acute tryptophan depletion in patients with dementia of the Alzheimer type showed that acute tryptophan depletion significantly impaired cognitive function in AD patients (Porter et al., 2000; Newhouse et al., 2002). The mean difference in score on the Modified MMSE scale after acute tryptophan depletion compared with placebo was 4.6 points or 8% of the placebo score, which is clinically significant. Compromised serotonergic function, in combination with cholinergic deficit, may make an important contribution to cognitive decline in dementia of the Alzheimer type. However, studies of supplementation with L-tryptophan in dementia of the Alzheimer type have failed to demonstrate any consistent improvement in cognitive function (Whitford, 1986). Large studies of treatment with selective serotonin reuptake inhibitors (SSRIs) in AD have failed to demonstrate clear improvement in cognitive function (Gottfries & Nyth, 1991).

3.3.1.2 Serotonin transporter

Because the serotonin transporter (5-HTT) clears the synaptic cleft of the neurotransmitter, thus terminating neurotransmission, it plays a crucial role in serotonergic neurotransmission controlling the concentration of free, active neurotransmitter in the synaptic cleft.

A study that investigated whether 5-HTT sites would be altered in the DRN as well as in the hippocampus in AD compared to controls and using quantitative autoradiography showed that 5-HTT sites were significantly decreased in the DRN in AD (Tejani-Butt et al., 1995). Significant decreases occurred in the lateral wings of the DRN complex. A significant decrease in 5-HTT sites was also observed in the CA2 subfield of the hippocampus and in the entorhinal cortex in AD. Thus, this study suggests that the integrity of 5-HT neurons in the DRN may be compromised in AD and that region-specific alterations in 5-HTT sites may occur in the hippocampal complex. Further studies investigating the mechanisms by which 5-HTT sites are altered in select brain regions will improve our understanding of the link between the 5-HT system and the behavioral cognitive deficits that are associated with AD.

AD-associated reductions in the number of 5-HTTs have been reported in the DRN, hippocampus, entorhinal cortex, and platelets, the latter indicating that this effect is not an epiphenomenon of the disease process. In a population-based association study the frequency of the low-activity allele of the serotonin transporter gene-linked polymorphic region (5-HTTLPR) was increased in patients with LOAD (Li et al., 1997). An excess of the homozygous genotype of the short 5-HTTLPR was found in comparison to controls with a significant odds ratio of 1.7 and a population-attributable risk of 33%. These results indicate that central 5-HT function related to allelic variation in function of 5-HTT expression may be involved in the pathogenetic mechanism of cognitive deficits observed in senile dementias and may be a risk factor for LOAD. Better understanding of the mechanism of action of neurotoxins that target the 5-HTT and the potential impact of this transporter on brain plasticity and neurodegeneration may have significant therapeutic consequences.

Given the fact that the 5-HTT continues to be an important target for drug development, novel therapeutic strategies aiming at 5-HTT gene transcription are currently being explored (Lesch & Mössner, 1998). The 5-HTT gene promoter and its associated regulatory elements may be suitable targets for therapeutic intervention.

3.3.1.3 Serotonergic-cholinergic interactions

There is considerable evidence indicating an extensive loss of both cholinergic and 5-HT neurons in the brain of patients with AD. For example, serotonergic-cholinergic interactions based on 5-HT_{1A} and 5-HT_{1B} receptor-mediated modulation of septal and frontal cholinergic neurons have been suggested to play a central role in learning and memory as well as in the appearance of age-dependent or neurodegenerative cognitive deficits (Kia et al., 1996).

3.3.1.4 Serotonergic receptors and psychotic symptoms

Several studies have shown that psychotic symptoms appear in the course of the disease of a large portion of AD patients. A recent observation has shown that common genetic polymorphisms in the 5-HT_{2A} and 5-HT_{2C} serotonin receptor genes are risk factors for psychotic symptoms in AD (Holmes et al., 1998). According to this study, associations were found between the 5-HT_{2A} 102T/C polymorphism and the presence of visual and auditory hallucinations for genotype frequencies. The distribution of the 5-HT_{2A} receptor gene genotypes and allele frequencies differ significantly between the AD patients with prominent psychotic features compared to those patients without (Nacmias et al., 2001). Homozygosity for the C102 allele in 52% of AD patients with psychotic symptoms as compared to 6.9% of patients without psychosis was observed.

3.3.2 Noradrenaline

3.3.2.1 Noradrenergic-cholinergic interactions

Beside metabolic and myogenic factors, noradrenergic sympathetic and acetylcholinergic parasympathetic innervation of the microvasculature participate in the regulation of the CBF. Noradrenaline causes vasoconstriction whereas ACh induces vasodilatation (Uddmann & Edvinsson, 1989; Suzuki & Hardebo, 1993). The normally existing balance between noradrenergic and acetylcholinergic innervation is dysregulated during aging during which an increase in the sympathetic tone is present. Both synthesis and release of ACh were decreased (Bowen, 1984; Bigl et al., 1987). Otherwise, the noradrenaline concentration increases in cerebral cortex with aging and the stress-induced release of noradrenaline is prolonged (Perego et al., 1993). The imbalance between noradrenergic and acetylcholinergic innervation of microvasculature may contribute to their morphologic abnormalities during aging and in particular in sporadic AD (Mungas et al., 2001).

3.3.3 Dopamine

In non-AD individuals, the dopaminergic and cholinergic pathways are involved in providing balanced motor control and in cognitive and higher functions. Dopaminergic neurons of the ventral tegmental area projecting to the prefrontal cortex are known to modulate working memory (Goldman-Rakic, 1998). The dopaminergic projections to the medial temporal lobe play a role in the encoding and retrieval of sensory information. A microdialysis study in humans established that dopamine is released into the amygdala upon starting a cognitive task and that increase in dopamine release is related to learning performance (Fried et al., 2001).

Dopaminergic dysfunction in AD possibly linked with D_3 dopamine receptor activity is linked to the development of hallucinatory and psychotic symptoms owing to the presence of these receptors in the limbic areas of the brain (Molchan et al., 1991; Schwartz et al., 1993). Some studies have shown that there is increased monoamine oxidase B (MAO-B) in the brains of AD patients that leads to dopaminergic dysfunction and that MAO-B inhibition can mediate cognitive and behavioral improvements in AD patients (Tariot et al., 1987; Sunderland et al., 1992).

3.4 Glutamate and NMDA receptors

Glutamate is the neurotransmitter for most of the excitatory synapses in the mammalian CNS. Three classes of glutamate-gated ion channels (AMPA, kainate and N-methyl-D-aspartate (NMDA) receptors) transduce the postsynaptic signal. Glutamate receptors are divided into two distinct groups, ionotropic and metabotropic receptors (mGluRs) (see Ozawa et al., 1998). Both ionotropic receptors and mGluRs have been implicated in learning and in aging. There are indications that glutamate might be involved in the pathomechanism of neurodegenerative diseases like dementia (Greenamyre & Young, 1989). This issue is still controversial and is largely neglected by supporters of the cholinergic hypothesis. Some findings provide further support for the glutamatergic hypothesis (see Table 3.1).

Table 3.1: Some findings that support the glutamatergic hypothesis

- In postmortem samples from brains of dementia patients, there is a decrease in astroglial glutamate carrier EAA2 in the frontal cortex (Li et al., 1997).
- Some authors observed co-localization of glutamate neurons and pathological alterations (NFT and SP) in postmortem analysis of the brains of dementia patients (Francis et al., 1993).
- Head trauma (associated with glutamatergic dysfunction) has been suggested to be an important risk factor for dementia by some but not other studies (Mendez et al., 1992).
- Acute or chronic NMDA-induced excitotoxicity in neuronal cultures is associated with an augmented immune-labeling of phosphorylated tau proteins at serine 202 (AT8 antibody) as observed in paired helical neuro-filaments (Couratier et al., 1996). NMDA-induced cell death and elevated AT8 tau immunoreactivity is blocked significantly by NMDA receptor antagonists. *In vitro* glutamate toxicity is blocked by antisense of tau mRNA (Pizzi et al., 1995).
- *In vitro* β-amyloid peptide enhances the toxicity of glutamate and augments NMDA receptor-mediated transmission (Wu et al., 1995).
- *In vitro* β-amyloid enhances depolarization-stimulated glutamate release – more in aged animals and inhibits its glial uptake (Harris et al., 1996).
- *In vivo* excitotoxicity increases APP production in glia (Topper et al., 1995).
- Injection of β-amyloid i.c.v. produces long-lasting depression of EPSPs in the hippocampus – an expression of ongoing mild excitotoxicity – that is prevented by the NMDA receptor antagonist (Cullen et al., 1996).

The NMDA receptor is a highly calcium-permeable, ligand-gated ion channel in neurons and a member of the ionotropic glutamate receptor family. NMDA receptors are abundant, ubiquitously distributed throughout the brain, fundamental to excitatory neurotransmission and critical for normal neurologic function. Excessive stimulation of the NMDA receptor leads to excessive intracellular calcium influx, generation of free radicals such as nitric oxide and reactive oxygen species, collapse of the mitochondrial membrane potential, loss of ATP and eventually neuronal apoptosis or necrosis depending on the intensity of the initial insult and the extent of energy recovery. Previous studies have shown an age-related decline in NMDA receptor density as well as several NMDA receptor-mediated plasticities in the hippocampus (Magnusson, 1998; see Segovia et al., 2001). The role of the NMDA receptor in excitotoxicity has driven the search for antagonists as neuroprotective agents. Its role in synaptic plasticity, on the other hand, has inspired research into receptor potentiators to treat cognitive function (Baster et al., 1994; Matsuoka & Aigner, 1997). The key issue is whether drugs that affect NMDA activity can be developed without dose-limiting side effects while retaining some benefit. Work in the NMDA receptor field indicates that developing compounds safe enough to pass the highest hurdles is not easy. Given the many varied opportunities to modify NMDA receptor activity, the potential for finding therapeutically valuable drugs is high. However, the balance between blocking NMDA neurotransmission and potentiating it is a critical one. Will we be able to find effective drugs for improving cognition without increasing the propensity to damage from stroke? Perhaps the recent explosion in genomics, proteomics and in understanding the structure of the excitatory synapse will help further understand glutamate neurotransmission and identify more selective targets for drug development.

3.5 The cytokines

Inflammation may contribute to the neuronal destruction that invariably leads to dementia (Neuroinflammation Working Group, 2000). Aβ induces the release of inflammatory cytokines and these cytokines may in turn enhance Aβ secretion (Mehlhorn et al., 2000). This interplay between Aβ and cytokines could contribute to the chronicity of LOAD. Interleukin (IL)-1β, IL-6 and TNF-α can augment APP expression, increasing the release of Aβ (Blasko et al., 1999). TNF-α and -β (lymphotoxin (LT)-α) are inflammatory cytokines involved in the local immune response occurring in the central nervous system of LOAD patients (Barger et al., 1995).

IL-1, the prototypical acute-phase pro-inflammatory cytokine, mainly of microglial origin in the brain, has emerged as a key player in the orchestration of neurodegenerative events during AD (Griffin et al., 1998). IL-1 has been shown to up-regulate expression and processing of the β-APP in neurons. This phenomenon is likely to contribute to the deposition of Aβ fragment in plaques as well as to the release of secreted β-APP. Brain levels of IL-1 are known to be elevated in AD patients and this cytokine may regulate expression of the APP (Griffin et al., 1989). IL-1 overexpression in the brain correlates with the degree of progression of neuritic plaques, the formation of NFTs and DNA damage in neurons (Griffin et al., 1989; Sheng et al., 1997; 1998a, 1998b). It is stronger in brain regions mostly involved in AD pathology (Sheng et al., 1995). In AD, the expression of cyclooxygenase (COX)-2 is elevated in CNS and higher levels of prostaglandins are found in the CSF (Montine et al., 1999). However, the expression of COX-1 has recently gained attention as a possible source of prostaglandins in AD, which might be important for the different strategies to use anti-inflammatory drugs (Yermakova et al., 1999).

Genetic variation at these genes could contribute to the risk of developing AD or influence the age of onset of the disease. A study genotyped 315 LOAD patients and 400 healthy controls for DNA polymorphisms in the gene encoding TNF-α (–308G/A, –238G/A) and LT-α (Asn26Thr) (Alvarez et al., 2002). Carriers of –308A showed a significant mean age at onset 3 years younger than noncarriers of this allele. These data suggest an effect of the TNF-α-308 polymorphism on the age of onset of LOAD. They also highlight the importance of genetic variation of the proinflammatory components in the origin and progression AD.

3.6 Cyclooxygenase enzymes

Regular use of non-steroidal anti-inflammatory drugs (NSAIDs) is beneficial for the course of AD (McGeer et al., 1996). Most NSAIDs inhibit the activity of COXs (Vane & Botting, 1998). These enzymes are involved in the conversion of arachidonic acid, which is liberated from membrane phospholipids by phospholipase A_2, into prostaglandins and thromboxanes (prostanoids). There are two isoforms of COX, COX-1 and COX-2. Depending on the expression level, neuronal COX-2 may either serve physiological functions through refining excitatory synaptic connections or induce pathological pathways leading to neuronal injury (Nakayama et al., 1998; Nogawa et al., 1997). In AD brain, overall COX-2 expression is increased compared to that in non-demented control brain (Kitamura et al., 1999). Overexpression of COX-2 is associated with inflammation and a variety of proliferative diseases. This observation suggests that the development of anti-inflammatory strategies that may decrease inflammatory neurodegeneration in the brain as a function of the clinical progression of AD may be a treatment approach. Selective inhibitors of COX-2 are currently available.

3.7 Metal-mediated oxy-radical and peroxide formation

There are other factors that are also thought to contribute to the neurodegenerative process of AD; these include neuroinflammation, oxyradicals and oxidative stress. There is an association between increased levels of the oxidative stress-related enzymes, glucose 6-phosphate dehydrogenase, superoxide dismutase (SOD) and heme oxygenase-1 and AD (Behl, 1999). Reactive free radicals and peroxides can be very toxic to cells. Reactive oxygen species have been proposed to cause neuronal injury in several neurological disorders and ischemic brain injury. Vitamin E, a free-radical scavenger, prevents neuronal death induced by Aβ and has been shown to have some beneficial effects in AD.

There is some evidence that the immune system plays a role in the brain inflammatory process. Deposition of Aβ may activate microglial cells (which aggregate along the plaques and tangles), which may be involved in the inflammatory process. In addition, an AD brain presents features typical of chronic neuroinflammation, such as astrocytosis, microglial, oxygen-free radicals, nitrogen species and products of activation of the complement system (The Neuroinflammation Working Group, 2000). The inflammatory reaction is localized mainly in the periplaque area and potentially contributes to neurodegeneration through imbalance of the levels of pro- and anti-inflammatory mediators. Region-specific imbalance of neurotrophin, which supports the survival, differentiation and maintenance of neurons, is likely to be another factor contributing to the degeneration of specific neurons in the hippocampal and cortical area (Leturman et al., 2000). Consistent with this hypothesis is the observation that inflammatory cytokines are linked to the clinical progression of AD, and anti-inflammatory drugs can enhance cognitive performance.

3.7.1 Neuroinflammation

A complex relationship seems to exist between Aβ deposition and the Aβ-induced inflammatory response. Because Aβ accumulation appears to trigger inflammation, factors that either promote or inhibit Aβ aggregation would be expected to similarly affect inflammation. If once triggered, the inflammatory response could result in removal of Aβ aggregates, which could prevent the harmful effects of chronic Aβ accumulation. Alternatively, if the inflammatory response does not clear the aggregated Aβ, a chronic inflammatory condition is likely to result. This chronic inflammation will probably damage the CNS. In developing pharmacological strategies directed at Aβ, it is important to know whether Aβ is neurotoxic, whether it is pro- or anti-inflammatory and whether it is up-regulated during inflammation. As Aβ interacts with many other molecules, unambiguous answers to these questions are difficult to obtain.

3.7.2 Oxidative stress

There is increasing evidence that free radical-induced oxidative damage plays a role in the pathogenesis of AD. Free radicals are reactive oxygen compounds that may attack and damage lipids, proteins and DNA. The brain is especially sensitive to oxidative damage because of its high content of readily oxidized fatty acids, high use of oxygen and low levels of antioxidants.

Evidence suggests that brain tissue in patients with AD is exposed to oxidative stress during the course of the disease. Oxidative stress is characterized by an imbalance in radical production and antioxidative defense. The latter is considered to have a major role in the process of age-related neurodegeneration. For example, increased levels of DNA strand breaks have been found in AD (Stadelmann et al., 1998). They were initially considered to be a part of apoptosis but it is now accepted that oxidative damage is responsible for DNA strand breaks and this is consistent with the increased free carbonyls in the nuclei of neurons and glia in AD. The induction of heme oxygenase-1, an antioxidant enzyme involved in the conversion of heme to bilirubin, is increased in AD brains. This augmentation is tightly correlated with NFTs that contain adducts of malondialdehyde and 4-hydroxy-2-nonenal (HNE), the most highly reactive lipid peroxidation product (Smith et al., 1996).

In addition to HNE, advanced glycation endproducts (AGEs), a sugar-derived non-enzymatic posttranslational modification of long-lived proteins, is also a potent neurotoxin. AGE accumulation has been shown in SPs in different cortical areas and some glial cells of AD brain. Long-lived protein deposits such as β-amyloid plaques with bound transition metals create an ideal chemical environment for the formation of AGEs (Loske et al., 2000).

A further reason for oxidative stress is caused by imbalance among the radical detoxifying enzymes in AD. The ratio of SOD to catalase decreases, leading to a buildup of hydrogen peroxide, which is particularly dangerous in the presence of free iron due to hydroxyl radical formation (Gsell et al., 1995).

3.8 Aβ and endoplasmic reticulum-associated binding protein

Recent evidence suggests that Aβ binds an intracellular polypeptide known as endoplasmic reticulum-associated binding protein (ERAB) which is overexpressed in neurons affected in AD. The ERAB sequence indicates that it might be a hydroxysteroid dehydrogenase or an acetoacetyl-CoA reductase. ERAB normally functions in cellular metabolism and biosynthesis, and these functions may be perturbed by Aβ. By interacting with intracellular Aβ, ERAB may therefore contribute to the neuronal dysfunction associated with AD. ERAB may contribute to the pathogenesis of AD by being an intracellular target for Aβ, mediating cellular stress and, ultimately, apoptosis due to increased amounts of Aβ.

3.9 A neurochemical marker for preclinical Alzheimer disease?

The clinical neurochemical laboratory findings might help to identify a neurodegenerative disease process and to define the preclinical state of the disease. By the time of diagnosis, probably a large percentage of the neurons at risk may be already degenerated (Hörtnagl & Hellweg, 1997). Accordingly, specific hippocampal volume reductions have been reported for individuals with MCI who were regarded to be at high risk for future AD, separating the MCI individuals from both AD and normal elderly (Convit et al., 1997). Against this background, biochemical findings that are related to the assumed progressive pathophysiological processes should be tested if they may be even faster and more sensitive in depicting clinically subthreshold dementing disorders than neuropsychological testing.

3.9.1 CSF tau

The microtubule-associated protein tau is a major protein subunit of NFTs, and it has been shown that tau protein in AD brains is abnormally phosphorylated (Cacabelos, 1992; Cacabelos et al., 1993). Abnormal phosphorylation of tau protein in the affected neurons leads to the breakdown of the microtubule system and consequently to neuronal dysfunction and degeneration. Neuronal and synaptic degeneration in the surroundings of the neuritic plaque is accompanied by neuritic sprouting and synaptic remodeling (Cacabelos et al., 1993; Samuel et al., 1994). The appearance of SPs containing dystrophic neurites, tau protein and abundant condensed Aβ deposits lead to the ultimate loss of neurons and synapses that correlate with the intellectual decline of AD patients (Samuel et al., 1994).

Although elevated levels of CSF tau can also be detected in patients with other acute and chronic neurological diseases, including dementing disorders which resemble AD clinically, CSF tau assays have been reported to be useful as a predictor of progression to AD in individuals with memory impairments but who do not meet clinical criteria for dementia (Trojanowski et al., 1996; Mecocci et al., 1998; Arai et al., 1997). Because the CSF tau level did not correlate with age, duration, or severity of AD, CSF tau measurement might be useful in revealing the disease in its preclinical phase, more than in confirming overt AD (Mecocci et al., 1998; Arai et al., 1997; Riemenschneider et al., 1996). A recent trial that studied CSF-tau and CSF-Aβ42 in 16 patients with MCI who at follow-up investigations 6-27 months later had progressed to AD with dementia showed that in comparison to aged-matched healthy individuals, at baseline 14/16 (88%) of MCI patients had high CSF-tau and/or low CSF-Aβ42 levels (Andreasen et al., 1999). AD is associated with a significant increase in CSF tau levels and a significant decrease in CSF Aβ1-42 levels (Sunderland et al., 2003). These differences are found in patients with mild AD as well as in patients with moderate to severe AD. While these CSF measures may have a potential clinical utility as biomarkers of disease, future studies should address the usefulness of these CSF measures for predictive, diagnostic or treatment purposes for AD.

3.9.2 CSF cytokines

Whereas some conflicting results have been reported for CSF levels of cytokines in AD, CSF microglial antibodies are usually found in early stages of AD, suggesting that microglial antibodies could be of value in detecting neurodegenerative processes before the onset of dementia (McRae et al., 1993). Further studies including follow-up studies with neuropathological examinations are still needed to confirm whether these assays are in fact useful, i.e., specific and sensitive as biomarkers for early, subthreshold dementing disorders.

It also appears that the strongest biochemical predictor to indicate patients who were more likely to progress to AD was their Apo E status (Petersen et al., 1997).

3.9.3 Choline acetyl transferase activity

A study examined the presynaptic cholinergic marker choline acetyltransferase (ChAT) in the brains of individuals with no cognitive impairment, with MCI and individuals with mild to moderate AD (DeKosky et al., 2002). In patients with MCI and mild to moderate AD, ChAT levels generally were not different from those found in nondemented aging subjects. The surprising exception was elevated ChAT activity (suggesting up-regulation of cholinergic systems) in the frontal cortex and hippocampus of individuals with MCI. This study suggests that cholinergic systems may be up-regulated in MCI individuals. The loss of this apparent compensatory response may mark the conversion of MCI to diagnosable AD and hence could be a therapeutic target. In addition to involvement of the cholinergic system, MCI individuals share other features of AD (memory impairment, increased frequency of the Apo E4 allele, medial temporal atrophy and cerebral metabolic abnormalities) (Morris & Price, 2001). Amnestic MCI individuals progress to more overt AD at an accelerated and predictable rate and the neuropathology of amnestic MCI overwhelmingly is that of AD (Petersen et al., 1999; Morris et al., 2001; Kordower et al., 2001).

3.9.4 Isoprostanes

Isoprostanes, sensitive and specific markers of *in vivo* lipid peroxidation, were observed to be increased in CSF, blood and urine of patients with a clinical diagnosis of AD (Pratico, 1999). These levels were highly correlated with other biomarkers of AD pathology and with the severity of the disease (Pratico et al., 2000). A recent study that investigated levels of this biomarker in subjects with MCI reported significantly higher 8, 12-iso-i$PF_{2\alpha}$-VI levels in cerebrospinal fluid, plasma and urine of subjects with MCI compared with cognitively normal elderly subjects (Pratico et al., 2002). These data suggest that individuals with MCI have increased brain oxidative damage before the onset of symptomatic dementia. Therefore, measurement of this isoprostane may identify a subgroup of patients with MCI with increased lipid peroxidation who are at increased risk to progress to symptomatic AD.

3.9.5 Calcium

Calcium ions act as messenger molecules in many tissues and organs regulating a variety of important cellular events, i.e. cell metabolism, gene expression, programmed cell death, signal transduction, neurotransmission and others. Several lines of evidence point to disturbances of calcium signaling in the aging human brain and in neurodegenerative diseases (Heizmann & Braun, 1992; Peterson, 1992). In AD brains, the intracellular immunostaining for neuronal calcium sensor proteins, visinin-like proteins (VILIP)-1 and -3, was reduced in comparison to controls (Braunewell et al., 2001). It is conceivable that VILIP-1 is involved in the pathologic changes either by affecting trk expression or being itself affected by a changed trk expression. Further studies are required to elucidate the signal cascades from calcium to neurotrophins in AD.

3.10 Serum anti-amyloid peptide antibodies

The appearance of amyloid deposits is thought to depend on impaired clearance catabolism of Aβ peptides (Selkoe, 2001). It has been thought that anti Aβ antibodies act directly on cerebral plaques (Bard et al., 2000). However, it now appears that antibody-mediated clearance of Aβ peptides in the periphery can also reduce cerebral amyloid deposits (DeMattos et al., 2001).

A possible role of autoimmunity in the pathogenesis of AD has also been explored (Singh, 1997). There is evidence that cerebral amyloid deposits are linked to an immunodeficient response to peptides (Aβ) derived from APP (Du et al., 2001). The level of anti-Aβ peptide antibodies in patients with AD is decreased in CSF, although the level of total anti-Aβ peptide antibodies were reported by another group to be comparable in plasma from AD patients and age-matched control subjects (Hyman et al., 2001). It appears that AD patients have a specific defect that leads to a lower serum level of IgG anti-Aβ42 antibodies and not the age-associated generalized decline in immune responses or increase in polyclonal, autoreactive IgM antibodies (Selkoe, 2001).

3.11 Discussion

In addition to the cholinergic deficit, loss of several other neurotransmitter systems has been reported in AD, including norepinephrine, dopamine and serotonin. This therefore provides other opportunities for therapeutic intervention, either alone or in combination with compounds acting on the cholinergic system. The numerous subtypes of 5-HT receptor that have been described allow a variety of ways in which serotonergic compounds might affect cognition (Hoyer et al., 1994).

Increased sAPP release from different cell types has been shown to occur after stimulation of m_1 and m_3 muscarinic ACh receptors, serotonergic 5-HT_{2a} and $5HT_{2c}$ or mGluR (Farber et al., 1995; Nitsch et al., 1992; 1996; 1997). In addition, stimulation of epidermal growth factor receptor or neuropeptide receptors, such as vasopressin and bradykinin receptors, also increases the release of sAPP into the media of cell cultures (Nitsch et al., 1998). Most of these cell surface receptors are coupled to G proteins and act through the phospholipase C/protein kinase C (PKC) pathway, thus suggesting that sAPP release from the cell depends on PKC activation.

A high titer of anti-Aβ42 antibodies may protect humans from AD. The titer of anti-Aβ42 antibodies in serum from individuals with and without late onset AD was assessed using an ELISA (Weksler et al., 2002). The titer of Ig (IgM, IgG and IgA) and IgG anti-Aβ42 peptide antibodies was significantly higher in serum from elderly controls than in AD patients. Furthermore, IgG but not Ig anti-Aβ42 antibodies distinguished sera from AD patients and elderly controls that did not have the Apo E4 allele. The lower titer of serum anti-Aβ42 peptide antibodies in AD patients may reflect the reported specific impairment of helper T cell activity for B cells that produce anti-Aβ42 peptide antibodies in APP-transgenic mice.

Chapter 4
Molecular genetics of Alzheimer disease

4.1 Introduction

AD is a multifactorial disease implicating interactions between environmental and genetic factors. However, there are various genetic defects that may predispose an individual to AD. AD cases can be divided into familial and sporadic cases as well as early onset (EOAD: < 65 years) and late onset (LOAD: > 65 years) (see Table 4.1). Familial AD (FAD) cases are rare, comprising only 5–10% of all cases. Mutations in several genes have been described as causes of AD: the PS1 gene in chromosome 14, the PS2 gene in chromosome 1, and the APP in chromosome 21 (Sherrington et al., 1995; Levy-Lahad et al., 1995; Goate et al., 1991). All known mutations that predispose a person to EOAD either increase the overall level of Aβ production or increase the ratio of Aβ42 to Aβ40 (Suzuki et al., 1994; Borchelt et al., 1996).

To date, the only recognized risk factor for the most common forms of AD-defined as sporadic or complex (i.e., without obvious Mendelian inheritance) is the Apo E gene on chromosome 19 (Strittmatter et al., 1993). This gene exists as three major alleles in the general population: ε2, ε3 and ε4 resulting from amino substitutions (Arg and Cys) at positions 112 and 158 of the protein. It is now well recognized that carrying an ε4 allele increases the risk for AD in an allele dose-dependent manner and is associated with an earlier age of onset of the disease (Corder et al., 1993). Conversely, bearing an ε2 allele confers protection against the disease (Corder et al., 1994; Chartier-Harlin et al., 1994). Thus, Apo E genotype appears to be an important biologic marker for AD susceptibility accounting for between 45 and 60% of AD variability (Farrer et al., 1997). However, possession of the ε4 allele is not sufficient to develop the disease because as many as 50% of individuals who have two copies of ε4 and survive to age 80 years are not cognitively impaired. It has been suggested that other environmental determinants such as serious head injury, smoking and cholesterol level may modify the Apo E-related risk (Mayeux et al., 1995; Van Duijn et al., 1995; Jarvik et al., 1995). While severe head trauma and lower educational achievement seem to increase the risk for AD, other factors may be protective. Unproved but possible protective factors include use of the non-steroidal anti-inflammatory drugs, postmenopausal estrogen in women, anti-oxidant vitamins and cholesterol-lowering statin drugs, low fat diets and aerobic conditioning (Small, 2002).

FAD refers to families that have more than one case of AD and usually implies multiple affected persons in more than one generation. Early-onset FAD refers to those families in which the age of onset is consistently before the age of 60–65 years and often before age 55 (Levy-Lahad and Bird 1996). EOAD is mostly an inherited dominant disorder. Although causes of sporadic AD, the most common form, are not yet understood, three genes have been found to play an important role in familial autosomal dominant early onset AD which is a relatively infrequent but devastating form of the disorder (Selkoe, 2001). On each of these chromosomes is a mutated gene, in other words, an autosomal-dominant gene that will likely produce an abnormal and potentially harmful protein and thereby cause AD in the patient. β-amyloid peptide, the primary component of the SPs unique to the brains of AD patients, is produced from a larger protein, β-amyloid precursor protein (BAPP). A mutated gene on chromosome 21, found in a small number of families with AD, is responsible for an abnormal form of BAPP (Motter et al. 1995). There is a high incidence of AD in families that carry PS2 (Roses, 1995). Mutations in genes of BAPP and PS1 and PS2 have been shown to segregate in familial autosomal dominant EOAD cases (Yasuda et al., 1999; Dermaut et al., 1999). Mutations in BAPP (chromosome 21) and PS2 (chromosome 1) are believed to account for less than 1% of all cases, whereas mutations in PS1 (chromosome 14) may account for more than 40% of early onset cases of FAD (Campion et al., 1999).

Thus the various risk factors associated with AD that have been described are the presence of variants of Apo E on chromosome 19, α-antichymotrypsin, the LDL receptor, butyrylcholinesterase K, A2M, and a subunit of the α-ketoglutarate dehydrogenase complex (Okuizumi et al., 1995; Lehmann et al., 1997; Blaker et al., 1998).

4.2 Early onset Alzheimer disease

The PS1, PS2 and APP genes have been found to cause familial EOAD whereas the Apo E gene and other risk factors are involved in sporadic or familial AD with different ages of onset. PS1 gene mutations are involved in 18–50% of the autosomal dominant EOAD cases (Sherrington et al., 1995). A novel missense mutation (Leu166Arg) at an atypical site associated with EOAD (range, 32-44 years) has been identified in a Spanish family (Ezquerra et al., 2000). Exon 6, corresponding to TMIII and part of hydrophilic loop II, was not considered a cluster of mutations, where initially only amino acid position 163 had been found to be mutated (Sherrington et al., 1995). However, two additional novel mutations have been found in this exon: Ser169Pro and Leu171Pro (Ezquerra et al., 1999; Ramirez-Duenas et al., 1998). Thus, together with the reported Leu166Arg mutation, at least four different mutated positions have been described in exon 6.

4.2.1 PS1 gene

Testing for PS1 mutations is available clinically and detects 20-70% of cases with early onset FAD (Campion et al 1999). More than 40 mutations that result in early onset FAD have been described in more than 50 families (AD Collaborative Group

1995; Campion et al 1995; Cruts et al 1998; Poorkaj et al 1998). All mutations except for one are missense mutations. The single exception is a mutation destroying a splice site in which exon 9 is lost but the reading frame is unaltered and the protein is predicted to be 29 amino acids shorter. At least nine mutations occur in a cytosolic domain between transmembrane domains 6 and 7 and the rest of the mutations are within the other hydrophobic domains or immediately at the hydrophilic/hydrophobic junctions, especially of transmembrane domain 2. The relative frequency of mutations in the cytosolic domain that is encoded by the alternatively spliced exon 8 suggests that this region of the protein is functionally important.

4.2.2 PS2 gene

Mutations in PS2 have only been found in a few families, most of which are of Volga German ancestry living in the United States (Bird et al 1988), Italian kindred (Rogaev et al 1995), and a Spanish patient (Beyer et al 1998). Testing for PS2 gene mutations is available on a research basis only. Thus far, two mutations in the PS2 gene resulting in generally early-onset FAD, but with a relatively broad range of onset age, have been described. A single mutation (Asn141Ile) has been found in several Volga German FAD pedigrees, confirming the founder effect in this population (Levy-Lahad et al 1995).

4.2.3 APP gene

The APP is a type I transmembrane glycoprotein containing a large extracytoplasmic region, a transmembrane domain and a small cytoplasmic tail. There are several isoforms composed of 672 to 714 amino acids derived by alternative splicing of a single gene on chromosome 21. Aβ and associated pathogenic peptides are released from APP through the proteolytic actions of α-, β-, and γ-secretases.

Mutations in the APP gene account for 5–20% of cases of early-onset FAD (Lendon et al., 1997). The first FAD mutation to be identified was APP-692, causing a variant form of AD with prominent congophilic angiopathy. Then the more common ("London") codon V717I mutation was found. Cells expressing APP with substitutions of the Val residue at codon 717 secrete a higher proportion of Aβ42 relative to cells that express wild-type APP. Additional FAD-linked APP mutations have been reported. APP harboring a double mutation at codons 670 and 671 that results in a Lys-Met to Asn-Leu substitution proximal to the b-secretase site (APP (Swe), the "Swedish" mutation) is subject to processing in a manner that results in elevated secretion of both Aβ40 and Aβ42 peptides. Most of the FAD mutations influence APP processing in a manner that elevates production of the more amyloidogenic Aβ42 peptides. However, it is unknown whether the mutations have any effect on the normal function of β-APP. The deposition of Aβ42 in the neural parenchyma occurs early, and this species predominates in mature plaques. Significantly, Aβ42 appears to be more toxic to neurons but the mechanisms of Aβ fibril-

Table 4.1: Genes involved in AD

Type of AD	Chromosomal localization	Mutant gene product or genetic susceptibility	Primary pathway leading to degeneration	Clinical features
Early onset (<65 years old, ~1–3% of all AD	21q21.2	APP	Increased total Aβ production: ↑ ratio of Aβ42: Aβ40.	Aggressive, rapidly advancing disease (duration less than 7 years), often with abnormal movements.
	14q24.31q41	PS1, PS2	PS mutations may directly affect γ-secretase activity, resulting in ↑ Aβ42:40 ratio.	Chromosome 1 (PS2): linked to the Volga German pedigrees.
Late onset (Most common form)	19q13.2	Apo E	The ε2,3, 4 alleles may interact differentially with Aβ, and thereby affect toxicity or clearance from the brain.	More indolent progression of disease (duration greater than 10 years)
	12p 12	α2M LRP	May interact with extracellular clearance of Aβ through LRP	

logenesis/toxicity are not well understood. Mutations at codons 692 and 693, corresponding to the α-secretase site, lead to decreased cleavage by this enzyme and evidently more flux down the β- and γ-secretase pathway, resulting in a higher production of Aβ. Diverse mutations at the b-secretase site lead to increased cleavage by this enzyme and excess Aβ production. Mutations at and around the γ-secretase site lead to increased production of the Aβ 42 (Lichtenthaler et al., 1999).

Testing for these APP mutations is available on a research basis only. The most common APP mutation is a substitution of isoleucine for valine at codon 717 (Goate et al 1991). Substitutions of phenylalanine and glycine may also occur at this codon. Each of these mutations results in early onset FAD. A double mutation at codons 670 and 671 in exon 16 also results in early onset FAD (so-called Swedish mutation; Axelman et al 1994). Combined cerebral hemorrhage and presenile dementia have been caused by a mutation in codon 692 with substitution of glycine for alanine. A substitution of glutamine for glutamic acid at codon 693 results in cerebral hemorrhagic amyloidosis of the Dutch type, a disease in which dementia and brain amyloid plaques are uncommon. Transgenic mice containing some of these mutations have been produced.

4.2.4 Interleukin genes

CPPs in the IL-1 genes (IL-1A and IL-1B) are reported to be associated with increased risk not only for AD but for EOAD in particular (Grimaldi et al., 2000; Nicoll et al., 2000). IL-1 is a pro-inflammatory cytokine that in AD brain is markedly overexpressed in microglia. Some forms of IL-1 have been reported to carry functional consequences. For example, homozygosity for allele 2 of IL-1B (at position +3953) has been associated with a four-fold increase in production of the cytokine (Pociot et al., 1992). Homozygosity for the IL-1A*2 (T)-allele at position –889 in the promoter region of the gene conferred a three-fold increased risk for AD (Nicoll et al., 2000). Together, these findings strongly suggest that inheritance of these IL-1 polymorphisms confers increased risk for AD. The IL-1A and IL-1B polymorphisms associated with AD may directly operate as the pathogenic variants, perhaps as by increasing the amount of cytokine-mediated inflammation in the brain. A recent study reported that IL-1A 2,2 genotype is noted in 12.9% of 232 neuropathologically confirmed AD patients and 6.6% of 167 controls (Nicoll et al., 2000). Homozygosity for both allele 2 of IL-1A and allele 2 of IL-1B conferred even greater risk (odds ratio, 10.8). IL-1 genotypes may confer risk for AD through IL-1 overexpression and IL-1-driven neurodegenerative cascades. These findings also suggest that risk for both LOAD and EOAD may be associated with susceptibility genes that affect the inflammatory cascade in the disease progress. Consequently, anti-inflammatory therapies and especially ones aimed at modulating IL-1-mediated (and other inflammatory cytokine-mediated) inflammation may prove to be useful in treating and preventing AD.

Another inflammatory cytokine, IL-6, has been reported to affect the age of onset of AD (Papassotiropoulos et al., 1999). Several studies have reported that the C allele of IL-6 variable number of tandem repeat polymorphism (IL-6vntr) delayed initial onset of AD and also decreased its risk *per se* (Shibata et al., 2002). Another polymorphism, G/C allele of IL-6 gene promoter region (IL-6prom), is also a candidate because it has an influence on the regulation of plasma IL-6 concentration. Recent findings suggest that the IL-6prom G allele might be a risk factor for sporadic AD in Japanese (Shibata et al., 2002).

4.3 Late-onset Alzheimer disease

The origins, genetic and other, of LOAD are much more poorly known, probably because this form of AD is heterogeneous and hence does not follow simple Mendelian inheritance patterns. Not only are multiple gene mutations involved, but they probably act interdependently and in interaction with environmental factors as well as such factors as age, sex, other susceptibility genes and head trauma.

A number of risk factors have been identified that are associated with late-onset disease. The most significant one is Apo E4 that occurs in 35–40% of AD (Saunders et al., Strittmatter et al., 1993). Apo C1 A also occurs in 35–40% of patients. Apo E4 and Apo C1 A are in linkage disequilibrium and are found on chromosome 19 (Poduslo et al., 1995; 1998). There is broad consensus among scientists that, at present, the predictive value of the Apo E genotyping for asymptomatic individuals is extremely limited.

4.3.1 Apo E gene

The Apo E gene at 19q13.2 has three common alleles and is involved in lipid transport. It is a polymorphic lipoprotein involved in the transmembrane transport of cholesterol. Apo E is thought to play an important role in neuronal growth and in the central nervous system response to injury, particularly in the hippocampal region (Poirier et al., 1994). These alleles, Apo E2, Apo E3 and Apo E4, influence lipid levels. Apo E is also involved in the mobilization and redistribution of cholesterol and phospholipids during the membrane remodeling associated with synaptic plasticity (Poirier et al., 1991; 1993). Apo E4 carriers show reduced Apo E levels in the hippocampus and cortex as compared to normal controls and Apo E2/3 AD patients (Bertrand et al., 1995). Thus, it is possible that lowered Apo E levels in the brain and in the CSF of AD patients will lead to reduced transportation or homeostasis of lipids and other membrane components which consequentially will constitute the ground for impaired synaptic plasticity (Poirier et al., 1995).

The effect of Apo E on susceptibility has been confirmed in multiple racial groups and in numerous studies. Subsequent analyses showed that the E4 allele acts in a dose-related manner to increase risk and decrease age of onset both in late-onset familial and in sporadic AD, and in early-onset sporadic AD, so that those with two E4 alleles are at greatest risk. Apo E4 promotes the early appearance of Aβ and NFTs in the elderly. Apo E also forms a tight complex with Aβ in the extracellular space (Russo et al., 1998). The pathophysiological changes associated with inheritance of Apo E4 in AD have not been fully elucidated but have been suggested to be attributable to decreased activity in the cholinergic system (Allen et al., 1997; Poirier et al., 1995). However, some other studies have found no relation between Apo E4 and the cholinergic system (Anderson & Higgins, 1997; Corey-Bloom et al., 2000).

4.3.2 Alpha-2 macroglobulin

The A2M gene has two polymorphisms that, in some studies, have been associated with AD. A 5-base-pair (bp) deletion at the 5' splice site of exon 18 was found to be associated with the disease in the families from National Institute of Mental Health (NIMH) Genetics Initiative AD sample (Blacker et al., 1998). The analysis was based on affected probands *versus.* unaffected sibs. The second polymorphism (a variant Val-Ile in exon 24) was increased in the NIMH samples combined with samples from the Massachusetts Memory Disorder Unit and from a European consortium (Poller et al., 1992; Liao et al., 1998). These data suggested that polymorphisms in the A2M gene may contribute to the risk for AD. However, a recent study does not support the A2M polymorphisms as a risk factor for AD (Poduslo et al., 2002). Using association studies and sib transmission/disequilibrium tests, no association of the two polymorphisms with AD was noted.

4.4 Tau immunoreactivities

The gene for tau is located at 17q21-22. Abnormally phosphorylated tau aggregates to form PHF and straight filaments, within NFTs and in dystrophic neurites. In contrast to AD, frontotemporal dementia with parkinsonism, chromosome 17 type (FTDP-17), a recently defined disease entity, is clinically characterized by disturbed executive function, personality changes, bradykinesia and rigidity (Hutton et al., 1998). Several mutations in the tau gene have now been linked to FTDP-17. The effect of these mutations is to lower the affinity of tau for tubulin, and to allow the 4-repeat isoform to accumulate. In AD, however, only hyperphosphorylated forms of tau are found, and presumably are a consequence of the toxic effect of Aβ.

4.5 Genetic testing for Alzheimer disease susceptibility

Discovery of the Apo E gene as a major risk for AD has opened doors for many new lines of research. It has led to epidemiological studies showing that the age of onset can vary by as much as 20 years, depending on which form of the gene a person carries. Finding the relationship between Apo E and AD provides a crucial biologic marker for epidemiological studies. Another important challenge is to use this genetic information to devise new and more effective interventions.

Genetic testing for two of the genes, Apo E and PS1, is commercially available. People who have a certain allele of Apo E are more likely to develop AD than are those who have other alleles of Apo E. However, it is important to note that there are people who carry the "susceptibility" allele who will never develop AD and conversely, individuals without the allele who will develop AD. Genetic testing for Apo E is currently limited to diagnostic testing for people who have symptoms of dementia. Carrying a mutation in PS1, on the other hand, confers near certainty of developing AD. However, mutations in PS1 are found in only a small fraction of all AD cases (Lendon et al., 1997). Thus, while the predictive value is high, the usefulness of this test for most people is limited. A vast majority of the people who develop AD would test negative for PS1 mutations. Diagnostic and predictive testing for PS1 is currently available (Tobin et al., 1999). Probably, additional kinds of predictive testing will become available in the future.

Although the relative risk of AD is increased with the presence of one or two Apo E4 alleles, it is not possible to predict whether or when disease will develop in a cognitively intact person (Roses, 1995). It is quite possible for someone with the Apo E3/4 genotype to live more than 100 years without the disease and for a significant proportion of people with the Apo E2/3 genotype to have onset before 90 years of age.

4.5.1 Ethnic and racial differences

Apo E4/4 homozygotes, constituting approximately 2% of the general population, have the greatest risk of AD. However, there are differences in the prevalence of each genotype in various ethnic and racial groups (Farrer et al., 1997). In Japan, most of the Apo E4-bearing AD patients have the Apo E3/4 genotype. Consequently, there are fewer Japanese patients with disease onset before the age of 70 years. Another example comes from Finland where the Apo E4 allele frequency in the general population is quite high, namely 22%. Age-matched controls for the AD cases in the population had an allele frequency of only 16.5%. The high incidence of fatal myocardial infarction in Finland contributed to the attrition of many Apo E4-bearers before the age of risk for AD.

The association of Apo E4 and AD in African-Americans and Hispanics remains controversial and may be absent in a sample of Nigerian AD cases and controls. Both the frequency of the E4 allele and the strength of the association between Apo E and AD vary in people from different ethnic and cultural backgrounds (Farrer et al., 1997; Tang et al., 1998). Age possibly also influences the association (Blacker et al., 1997; Combarros et al., 1998). These differences have significant implications for the clinical application of Apo E genotyping and requires further investigation if genetic test results are to be interpreted accurately in heterogeneous populations.

4.6 Discussion

Since the above mutations account for less than 7% of AD cases and most AD cases are of late onset and sporadic, additional genes are likely to play roles in AD (Daw et al., 2000). In this respect, it is interesting to note that genetic polymorphism in the promoter region of the phenylethanolamine N-methyltransferase gene is associated with early but not late onset AD (Mann et al., 2001). The role of PS1 protein in AD remains unclear, but it is proposed that it acts through a gain of function or through dominant loss of function. Detection of Leu166Arg mutation in exon 6, together with the mutations found in PS1 protein positions 163, 169 and 171, suggests that this site of the PS1 protein should be reassessed as an additional cluster of pathogenic mutations. Subsequent to the discovery of linkage of some EOAD families to chromosome 21, further molecular biological research in AD identified a locus on chromosome 14 as being responsible in about 70% of early-onset families. However, the discovery that neither the chromosome 21 or 14 loci were linked to AD in the Volga German families suggested the presence of another locus. This linkage led to the identification of chromosome 14 gene (PS1) and the Volga German locus on chromosome 1 (PS2) (Sherrington et al., 1995).

Mutations in PS1 are the most common cause of FAD; mutations in the PS2 gene are a rare observation. More than 50 missense mutations have been identified in PS1, and these mutations account for up to 25–30% of early cases of FAD. All mutations represent a toxic gain-of-function. Many of these mutations occur within transmembrane domains or immediately adjacent to the predicted cytoplasmic loop domain. The most provocative insights pertaining to the mechanisms by which mutant PS1 causes AD emerged initially from studies of the conditioned medium from fibroblasts or the plasma from affected members of pedigrees with PS1/PS2-linked mutations. Surprisingly, FAD-linked PS1/PS2 variants influence processing at the γ-secretase site and may cause AD by increasing the extracellular concentration of highly amyloidogenic Aβ42 species, thus fostering Aβ deposition in the brain (Sheuner et al., 1996). The age of onset of AD in families harboring PS1 mutations is quite young, ranging from 30–60 years, whereas the age of onset in cases with PS2 mutations is somewhat older (55–70 years). Thus mutations in APP, PS1 and PS2 cause AD and may, therefore, illuminate a pathway for the development of therapeutic targets. A study designed to determine the prevalence of EOAD and of autosomal dominant forms of EOAD in the city of Rouen in France showed that PS1 and APP mutations account for 71% of autosomal dominant forms of EOAD families (Campion et al., 1999). Nonpenetrance at age < 61 years is probably infrequent for PS1 or APP mutations.

Even if cures for AD are not now available, professional counseling for persons at risk and their families exists and often is beneficial to them psychologically and otherwise. Genetic test results that inform individuals of their enhanced risk for EOAD may increase the probability that they will seek out such counseling. Finally, some individuals may use genetic information regarding their own future propensity to be afflicted with EOAD to make reproductive decisions. This is both because they fear that their own severe disability relatively early in life would jeopardize the welfare of offspring they would be raising and to spare potential offspring of a life spent in the shadow of probably developing EOAD themselves. Individuals in this category may choose to protect against pregnancy by using contraception or even to abort fetuses that come into being.

A recent study that used a covariate-based linkage method to reanalyze genome scans from affected sibships collected by the Alzheimer disease Genetics Initiative of the National Institute of Mental Health reported that the families with AD that have the oldest current age and no Apo E alleles exhibit linkage to APP (Olson et al., 2002). An increase in the LOD score from 0.4 to 5.54 after current age as a covariate in the linkage model was noted. These data suggest the existence of a genetically identifiable subtype of LOAD limited to the most elderly and characterized by joint linkage to 21p and 20p. The development and rate of progression of AD in such families requires the presence of high-risk alleles at both genes that likely interact biologically to increase the risk. However, such data are exploratory and need confirmation. Probably there are still undiscovered genes responsible for a predisposition to AD as there are several families whose AD genotype does not link to any of the five known AD-associated genes (PS1, PS2, APP, Apo E4, A2M). More research is required on the inheritance of AD in Western and non-Western populations. For example, some studies indicate that the Apo E4 genotype is not associated with AD among the Yoruba or East Africans (Osuntokun et al., 1995; Ogeng"o et al., 1996). It would be of interest to determine if the underrepresentation of the Apo E4 allele or the lack of its association with AD in a given population is correlated with overrepresentation of the Apo E2 allele in the same population. Comparative research should also focus on the distribution of other polymorphisms that have been claimed to serve as genetic risk factors for AD and the putative peripheral biomarkers for this disease.

The presence or absence of an E4 allele do not provide diagnostic certainty and the proper interpretation of either result in heterogeneous population requires further investigation. Whether Apo E genotyping provides sufficient information to change patient management decisions has not been determined. These tests present foreseeable, significant psychosocial consequences for family members that must be weighed against any psychosocial benefits. The diagnostic use of Apo E genotyping outside research settings may be premature until testing is shown to be of significant practical value.

Some families have a relatively low risk of AD throughout the life span. If this is so, it presents opportunities to identify genes involved with a reduction of risk of AD. The only well-established genetic protective factor is Apo E2 but there may be other as yet unidentified protective alleles against AD that are present at other loci. The investigation of families of aged (e.g. > 90 years old) nondemented probands may help to narrow these alternatives and aid in identifying individuals within families who are more likely to carry genetic protective factors against AD. Future genetic studies (e.g., an unaffected sibling pair approach in which both siblings are required to be free of AD) aimed at identifying such genes might benefit from targeting such probands and their families.

Chapter 5
Promises of animal models of Alzheimer disease

5.1 Introduction

The hypothesis that a cholinergic deficit might be responsible for memory impairment was based on studies showing learning and memory deficits in animals and humans whose cholinergic system was blocked by atropine or scopolamine (Drachman, 1977). Confirmation was sought by animal investigations in which the forebrain cholinergic nuclei were destroyed, initially by electrolytic lesions, then by excitotoxins and finally by the selective immunotoxins 192 IgG-saporin (Wenk et al., 1994). Intracerebroventricular (i.c.v.) injection of 192 Ig-G-saporin brings about a widespread degeneration of cholinergic neurons including the nucleus basalis of Meynert and the septum (Lin et al., 1999). The reversion of the learning and memory deficits induced by nucleus basalis of Meynert lesions has become a classical preclinical test for the screening of drugs aimed to correct the cholinergic hypofunction in AD (see Pepeu, 2000).

Aging animals have been and are the most common for investigating drugs potentially active on AD. Old mice and rats are used most frequently since they are easy to obtain and relatively cheap. Old monkeys have been used and remain the last preclinical step before clinical trials. However, there are two main limitations of the aging animal model. First, aging animals do not develop the neuropathological picture of AD. Rare plaques have been described in the monkey, but none of the histopathological alterations typical of AD occur in aging rodents. Second, in aging animals, it is relatively easy to obtain an improvement of the cholinergic hypofunction and the cognitive deficits with drugs whose efficacy is difficult to demonstrate in clinical trials. This is the case of phosphatidylserine which administered to aging rats restores well the cholinergic hypofunction and the cognitive deficits. Unfortunately, the clinical trials have been less successful even if some therapeutic activity has been suggested (see Pepeu et al., 1996).

Despite intensive efforts in both academia and industry, a fully authentic transgenic mouse model of AD has not yet been created. Nevertheless, overexpression of Aβ in various mouse lines has produced a variety of phenotypes that, as a first approximation, are remarkably similar to the human AD condition (Table 5.1). These first generation models will undoubtedly be crafted into successive generations in which all features of AD will be present. Even at this stage, the first generation models are providing an assay system in which selected details of pathogenesis and therapeutic intervention can be evaluated. Although these transgenic models are relatively recent, the models have proven to be valuable and have led to the development of an experimental drug to decrease amyloid burden in the brain (Helmuth, 2000; Marwick, 2000).

5.2 Transgenic models of mutant human (Hu) APP with Aβ deposition

To generate animal models of Aβ amyloidogenesis and the associated lesions of AD, many groups have created transgenic mice that overexpress wild-type APP, FAD-linked APP variants or C-terminal fragments of APP. A number of transgenic lines with these constructs have been published (see Table 5.2). These efforts have resulted in three lines that recapitulate some of the key neuropathological features of human AD (Games et al., 1995; Hsiao et al., 1996; Sturchler-Pierrat et al., 1997).

5.2.1 The Games mice

In the "Games mice", the platelet-derived growth factor (PDGF) β-promoter was used to drive expression of a human APP minigene that encodes the FAD-linked APP (V717F) mutation in an outbred strain (Games et al., 1995). The construct contained portions of APP introns 6-8, which presumptively enhanced alternative splicing of exons 7 and 8. Levels of HuAPP mRNA and protein significantly exceeded levels of endogenous APP. At approximately 6–9 months of age, transgenic animals began to exhibit deposits of human Aβ in the hippocampus, corpus callosum and cerebral cortex. As the animals aged (≥ 9 months), the density of the plaques increased until the Aβ staining pattern approached that of AD. The majority of plaques were intimately surrounded by glial fibrillary acidic protein (GFAP)-positive reactive astrocytes, and also compressed and distorted the surrounding neuropil as seen in the AD brain. There was also some dystrophic neuritic components and loss of synaptic density with regional specificity resembling that of AD. Unfortunately, no behavioral and cognitive assessments have yet been published.

5.2.2 The Hsiao mice

In a second line of "Hsiao mice", the hamster prion protein promoter was used to overexpress human APP with Lys-Met to Asn-Leu (Swedish) mutations (HuAPP695swe) (Hsiao et al., 1996). The brains of one of these lines (Tg2576) showed elevated levels of Aβ40 (five-fold increase) and Aβ42 (14-fold increase). There were some dystrophic neurites around moderate numbers of Aβ deposits as plaques and around vessels in amygdala, hippocampus, and cortex. The Tg2576 mice showed impairments at a young age on several memory tests including the Morris water maze, a spatial reference memory task, and the Y-maze alternation task. At 3 months of age, these mice showed normal learning and memory in spatial reference and alternation tasks, but by 9–10 months, they were impaired.

5.2.3 The Novartis mice

In the third model the “Novartis mice”, a combination of human APP mutations (Swedish KM670/671 NL and the V717I) is driven by the Thy-1 promoter (Sturchler-Pierrat et al., 1997). Abundant amyloid plaques with very convincing neuritic changes have been described, but in the absence of typical tau-positive NFT.

Each of the above three principal lines of transgenic models has been restricted in distribution to the wider research community because of commercial considerations. Hopefully this will change in the near future as better models emerge. Nevertheless, the present three lines should be sufficient to convince all but the most resolute sceptic that the APP/Aβ pathway is at the center of AD. Some features of the disease, however, most noticeably the formation of tau containing PHF, activation of the complement pathway and robust clinical phenotypes have yet to be achieved. It also appears that the impairment of cognitive functions observed in some transgenic mice is not necessarily due to the generation of Aβ but might result from either overexpression of wild-type APP or the accumulation of Aβ precursors or both.

While the β- and γ-secretases are essential for the production of Aβ, a third secretase (α-secretase) is known and its activity precludes the formation of Aβ due to cleavage within the Aβ sequence (Anderson et al., 1991; Hendriks et al., 1995). To assess the specificity of α-secretase in AD, transgenic mouse strains expressing human APP/F1 and APP/Du from the neuron-specific mouse-thy-1 gene promoter were generated (Moechars et al., 1996). A recent study that assessed the contribution of the APP gene of the Flemish (APP/A692G) and Dutch (APP/E693Q) model to the pathogenesis of AD indicated that these transgenic mice showed the same early behavioral disturbances and defects as the APP/London (APP V7171), APP/Swedish (K670N, M671L), and other APP transgenic mice (Kumar-Singh et al., 2000). Pathological changes included intense glial reaction, extensive microspongiosis in the white matter, and apoptotic neurons in select areas of the brain, while amyloid deposits were absent, even in mice over 18 months of age. This contrasts with extensive amyloid deposition in APP/London transgenic mice and less pronounced amyloid deposition in APP/Swedish transgenic mice generated in an identical manner. It showed, however, that the behavioral deficiencies and the pathological changes in brain resulting from an impaired neuronal function are caused directly by APP or its proteolytic derivative. These accelerate the aging process and amyloid deposits *per se* are not necessary for this phenotype.

Several laboratories have bred transgenic mice that produce Aβ and develop plaques and neuron damage in their brains (see Van Leuven, 2000). Although these animals do not develop the widespread neuron death and severe dementia seen in the human disease, they are used as models for the study of AD. Production of anti-Aβ antibodies by immunization with the fibrillar Aβ of the mouse model of AD led to inhibition of the formation of amyloid plaques and the associated dystrophic neu-

Table 5.1: Cardinal pathologic features of AD and the transgenic mouse models.

	Human disease	Transgenic mouse model
1. Extracellular Aβ amyloid plaque/ perivascular deposition	+	+
2. Intracellular P-tau as PHF/NFT formation	+	–
3. Intracellular P-tau as neurites/neuritic dystrophy	+	±
4. Circumscribed synoptic loss and neurodegeneration	+	±
5. Reactive gliosis	+	+
6. Widespread oxidative stress response	+	+

ritis in the mouse brain, suggesting the feasibility of vaccination against AD (Schenk et al., 1999). Appropriate animal models were used to test the effects of anti-Aβ antibodies on both brain damage and the cognitive losses caused by AD. Indeed, immunization with Aβ peptide improves learning and memory, and diminishes brain damage in animal models (Janus et al., 2000; Morgan et al., 2000). The results support a previously observed reduction in the formation of amyloid deposits but they go further to show that immunization also gave mice some protection from the spatial learning deficits that normally accompany plaque formation. Both groups of investigators suggest that either a small or a selective reduction of Aβ deposition may be sufficient to protect against dementia. These findings indicate that Aβ overexpression or Aβ plaques are associated with disturbed cognitive function and importantly suggest that some but not all forms of learning and memory are suitable behavioral assays of the progressive cognitive deficits associated with AD-type pathologies.

5.2.4 The CT100 mice

Mice engineered with the C-terminal 100 residues of APP (CT100) are designed for the study of γ-secretase and its inhibitors. Several lines are now available, with considerable variation in phenotypes (Oster-Granite et al., 1996; Li et al., 1999). None exhibit yet the extent of extracellular Aβ deposition seen with the full-length APP constructs, and indeed there may be an accentuation of intracellular Aβ accumulation (Li et al., 1999), suggesting that the CT100 construct is aberrantly targeted in the intracellular pathway.

5.2.5 Mutant PS mice

The preliminary analysis of mice expressing PS constructs shows that mutant PS1 selectively increases brain Aβ42, and suggests that the PS mutations probably cause

AD through a gain of deleterious function that increases the amount of Aβ42 in the brain (Duff et al., 1996). However, these mice do not show AD pathologic lesions. This is intriguing, and suggests that overexpression of the rodent Aβ sequence alone is insufficient for Aβ amyloid aggregation and plaque/perivascular deposition. This interpretation is consistent with the findings from two groups that double transgenics (mutated PS × Hu (APP)) result in an acceleration of cerebral Aβ deposition. In the first study, transgenic mice were generated that express either wild-type or mutant PS1 (A246E) and these were crossed with mice that express a murine APP transgene with a "humanized" Aβ domain (Mo/Hu-APPswe) (Borchelt et al., 1997). In the second study, the doubly transgenic progeny were derived from a cross between Tg2576 line and a mutant PS1M146L transgenic line (Holcomb et al., 1998). Both studies demonstrated that mice coexpressing mutant PS1 with mutant APP develop Aβ deposits much earlier than age-matched controls. Thus, crossing mice transgenic for human APP with mice expressing a PS1 missense mutation leads to a substantially accelerated AD-like phenotype in the offspring, with Aβ42 plaques (first diffuse and then mature) occurring as early as 3-4 months of age (Holcomb et al., 1998).

5.2.6 Hu APP overexpression on the Apo E null background

The effects of Apo E on amyloid deposition have been tested by mating Apo $E^{-/-}$ mice with APP-V717F transgenic mice (Bales et al., 1997). At six months of age, APP-V717F × Apo $E^{+/+}$ mice showed robust amyloid deposition, whereas APP-V717F × Apo E exhibited only sparse, diffuse Aβ deposits. These studies suggest that Apo E may either directly or indirectly may influence the aggregation or influence the clearance of Aβ peptides. A full description of the kinetics of Aβ biogenesis, aggregation and degradation in this model is awaited with interest.

5.2.7 Hu APP overexpression combined with oxidative stress

In an attempt to investigate the role of oxidative stress responses in the pathogenesis of AD, it is relevant that transgenic animals show the same type of oxidative stress responses that are found in AD and that these directly correlate with the presence of Aβ deposits (Smith et al., 1998). Several groups are now engaged in programs in which the Hu APP transgenic lines are being modulated through the SOD-1 and glutathione peroxidase pathways.

5.2.8 APP knockout mice

Mice with functionally inactivated alleles of APP were generated by deleting its promoter and first exon (Zheng et al., 1995). The mutant animals weighed 15-20% less than age-matched wild-type controls. The mutant mice also showed an impaired

Table 5.2: Transgenic mouse models of AD

Construct	Neuropathology	Behavioral deficits	References
A. Mutant human APP with robust amyloid plaques			
1. APP (V717F) (PDGF promoter) The "Games mouse"	Aβ deposits, neuritic plaques, synaptic loss, astrocytosis and microgliosis	Not described	Huang et al., 1997
2. APP (K670N, M67IL) (PrP Promoter) The "Hsiao mouse"	A five-fold increase in Aβ40 and a 14-fold increase in Aβ (1-42/4. Elevated amounts of Aβ in cortical and limbic structures. Moderate amyloid plaque and perivascular amyloid deposition in cerebral cortex and hippocampus	Learning and memory performance significantly impaired in the Y-maze and the Morris water maze.	Behl, 1999
3. Combinations of APP (K670N, M671L) with (V717I) Thy-1 promoter The "Novartis mouse"	Very convincing neuritic plaques	Not described	Yan et al., 1997
B. γ-secretase test mice			
4. APP – CT100	Intra- and extracellular Aβ immunoreactivity, microglial activation, gliosis, neuronal loss	Variable	Games et al., 1995 Hsiao et al., 1996
C. Mutant presenilin –APP			
5. PS1 (M146L & M146V) (PDGF promoter)	Overexpression of mutant PS1 selectively increases Aβ42 in the brain and in peripheral cells.	Not described	Sturchler-Pierrat et al., 1997
6. APP (Swe)+ Presenilin (PS1-A246E)	A selective 41% brain increase in Aβ42. Aβ plaques develop earlier than aged-matched animals that express APP (Swe) alone.	Not described	Oster-Granite et al., 1996; Li et al., 1999

Table 5.2 (continued)

Construct	**Neuropathology**	**Behavioral deficits**	**References**
D. APP, PS and Apo E knockouts			
7. APP knockout (Deletion of APP gene promoter and its first exon)	Extensive astrogliosis in the hippocampus CA1 region and in the molecular layer, reactive gliosis throughout the cortical layers. Onset of reactive gliosis may be age dependent	Decreased locomotor activity and reduced grip strength	Holcomb et al., 1998
8. PS1 knockout	Embryonic lethal	Not applicable	Bales et al., 1997 Holcomb et al., 1998
9. Mutant APP on Apo E knockout background	Reduction of amyloid burden compared to Apo E wild type background.	Not described	Duff et al., 1996

neurological and muscular function (reactive gliosis, decreased locomotor activity and forelimb grip strength). However, since neuronal cell damage or loss in the brains of APP-deficient mice were not observed, the mechanisms responsible for the reactive gliosis in these mice could not be determined. It was postulated that the absence of substantial phenotypes in APP knockout mice may be related to functional redundancy provided by the homologous amyloid precursor-like protein molecules (APLP1 and APLP2) that are expressed at high levels and have developmental and cellular distributions similar to APP. This has now been confirmed through the analysis of double knockouts which showed that some combinations (e.g. APP × APLP2) are not viable.

5.2.9 PS1 knockout mice

To examine the *in vivo* role of PS1 in mammalian development, mice with a targeted disruption of the PS1 gene were generated (Shen et al., 1997). Homozygous mutant mice failed to survive beyond the first 10 mins after birth. The most striking phenotype observed in $PS1^{-/-}$ embryos was a severe abnormality in the development of the axial skeleton and ribs. Fibroblast cultures established from these embryos revealed a remarkable phenomenon: these cells accumulate the C-terminal fragments of APP, suggesting that the PS molecules play a direct or indirect role in the activity of γ-secretase (De Strooper et al., 1998). It may yet transpire that PS forms a direct complex linking γ-secretase directly to APP, and thereby provide an elegant solution to the effect of PS mutations on Aβ 40/42 processing.

5.3 Immunization with Aβ

Active immunization can prevent or reverse both amyloid deposition and the associated memory impairment in transgenic mouse models of AD (Janus et al., 2000; Morgan et al., 2000). Passive immunization, consisting of prolonged treatment with monoclonal or polyclonal anti-Aβ antibodies can prevent the development of amyloid deposits in platelet-derived amyloid precursor protein (PDAPP) mice (Bard et al., 2000; DeMattos et al., 2001). Researchers have used APP transgenic mice as animal models to investigate the effect of immunization with Aβ42 at early ages (Schenk et al., 1999). After 13 months, the immunized mice did not display any neuropathological abnormalities. However, the non-immunized mice showed typical amyloid plaques. This vaccination was successful even when it was given to mice beyond the age at which the pathology in this model is fully developed. A reduction in the dense core plaque burden as the plaques began to decrease in size and quantity was observed. These data suggest that the Aβ vaccination was successful not only at preventing the development of amyloid plaques but was able to promote the disaggregation of previously formed plaques. Other laboratories have confirmed the effect of vaccination on the clinical symptoms of AD by using the mouse models and behavioral paradigms (Morgan et al., 2000; Janus et al., 2000).

They subjected the mice to Aβ vaccination at various time points and studied their performance in a Morris water maze. The results showed that immunization with Aβ42 improved performance in the water maze whether the vaccination was given at early ages or even at late ages when the pathology had already started. The precise mechanism of action of reduced AD-like pathology and improved cognition in transgenic models of AD is still under investigation.

Recently, Dodart et al. (2002) showed that administration of the monoclonal antibody m266 to PDAPP mice can rapidly reverse memory deficits in both an object recognition task and a holeboard learning and memory task but without altering brain Aβ burden. This study suggests that passive immunization with this anti-Aβ monoclonal antibody can very rapidly reverse memory impairment in certain learning and memory tasks in the PDAPP mouse model of AD, owing perhaps to enhanced peripheral clearance and sequestration of a soluble brain Aβ species.

5.4. Discussion

Although these mice have provided some valuable information on human AD and can be used as an assay system, none of them are considered to be a fully equivalent model of human AD. This is supported by the fact that the chemical structure and morphology of transgenic mouse plaques are not equivalent to those of human AD. If such differences are found to be widespread between animal systems and human AD, then the development of therapeutic drugs using animal models may be hampered (Kuo et al., 2001).

Further understanding of the critical age-dependent factors that confer vulnerability to Aβ neurotoxicity may provide better insight into the neurodegenerative mechanisms involved in AD, with potentially significant therapeutic innovations. It is widely believed that mutations and polymorphisms in other genes, as yet unknown, will modulate susceptibility to AD and will therefore be found to be additional trait-dependent markers. Neuroprotective growth factors such as the neurotrophins, basic fibroblast growth factor and insulin-like growth factors may play such a role.

Whether therapeutic agents that affect the concentration, deposition, aggregation, degradation, clearance, or toxicity of Aβ will influence the clinical and pathological features of AD is still unclear. Nevertheless, it seems likely that approaches that reduce the concentration of Aβ or the rate of amyloid aggregation and deposition in proximity to synapses and neuronal cell bodies will be beneficial for patients with AD. Even with existing transgenic mouse models, a start for screening and testing therapeutic efficacy of lead compounds has commenced. The results of initial endeavors will be known shortly.

Part II
Neuropsychology of Alzheimer disease

Chapter 6
What is memory?

6.1 Introduction

Memory is a theoretical construct that may explain current behavior by reference to events that have happened or will happen in the future. Memory may be conceived as a complex, dynamic, recategorising and interactive process based on actual sensory-motor experiences and manifests itself in the behavior of an organism. Repetition in learning is a prerequisite for the formation of accurate and long-lasting memory. Practice is most effective when widely distributed over time, rather than when closely spaced or massed. But even after efficient learning, most memories dissipate with time unless frequently used (Spear, 1978; Spear & Riccio, 1994). Memory always has a subjective and an objective side. The subjective side is observed by the individual's history (developmental perspective). The objective side is noted by the neural patterns generated by the sensory-motor interactions with the environment. It has been well established by many memory researchers that memory is not a unitary function. This notion is supported by evidence that different brain lesions in patients produce highly specific changes in some aspects of memory functions while sparing other ones. Research has attempted to delineate patterns of impaired and spared memory functions in order to assess whether there is a particular class of processing system that is not functioning while other systems function normally.

Memory may be broadly divided into explicit (conscious) and implicit (unconscious) components (Squire, 1987). In the explicit memory tasks, subjects memorized (encoded) and recalled (retrieved) words and pictures (Camp et al., 1996). Retrieval may be further subdivided according to the nature of the cue used to provoke it. For example, in one task, subjects simply see a symbol indicating that they should recall the item that they had learned. In the other (recognition), they are presented with an item they have seen several days earlier. They are required to indicate whether or not they recognized it from the pre-operative study phase. Researchers agree that implicit memory is more preserved in dementia patients than explicit memory (Vanhalle et al., 1998). Implicit memory has been defined as an unconscious form of retention that may be measured with tasks that do not require conscious recollections (Schacter, 1992).

Since the development of new methods in the neurosciences (MEG, PET, functional magnetic resonance imaging (fMRI)), interdisciplinary memory research has opened fascinating possibilities to study cognitive processes in the living brain. A large number of studies using neuroimaging have discovered different memory systems in the brain (Schacter, 1989; Tulving, 1985; Pfeifer & Scheier, 1999; Meares, 2000). Although the brain weighs less than three pounds, it is subdivided into many highly specialized zones, each responsible for a particular function – word recognition, speech, vision, sensation, locomotion, coordination and so forth. New research using the technique of fMRI suggests that at least two areas of the brain play key roles in memory. The hippocampus appears critical for long-term memory (LTM) while the frontal lobes contribute more to short-term memory (STM). A good memory requires many things including a healthy heart and blood vessels, the right amount of oxygen and glucose and the proper balance of brain chemicals. Medications that influence neurotransmitters such as dopamine, serotonin and acetylcholine may influence memory function.

The distinction between recollection and familiarity underlies several contemporary theories of human memory (Jacoby, 1991; Jacoby & Dallas, 1981). It has played a critical role in characterizing memory-impaired populations such as amnesics and it has proven useful in accounting for results from recent neuroimaging studies of memory (Gabrieli et al., 1997). Central to all of these dual-process theories is the claim that recognition memory judgments can be based on two distinct memory retrieval processes: familiarity and recollection. The familiarity process reflects the assessment of the memory strength or experimental familiarity of a test item. Recollection reflects a search of memory whereby qualitative information about the study event is retrieved. Jacoby (1991) has distinguished between recollection and familiarity in terms of intentional control. If an individual can recollect information about a previous event, then he or she should be able to accurately discriminate between items from different episodes or sources. In contrast, assessments of familiarity should support recognition judgments (i.e., old items are more familiar than new items), but should not be useful in discriminating between equally familiar items from different sources. On the basis of this distinction, Jacoby developed the process dissociation procedure to measure recollection and familiarity. Recollection is measured as the ability to retrieve a specified aspect of the study event (e.g., where or when an item was presented) and to use this as a basis for intentionally controlled responding. In contrast, Tulving (1985) argued that the fundamental difference between the components underlying recognition memory is in the nature of the conscious experience associated with each component. Recollection is associated with autonoetic consciousness (i.e., self-knowing or remembering), in which the episodic aspects of the study event are consciously experienced. In contrast, familiarity is associated with noetic consciousness (i.e., knowing) whereby the individual knows that the item was studied but does not re-experience any specific information about the study event. To measure these different types of recognition, Tulving developed the remember-know procedure, in which individuals are required to introspect about the basis of their recognition judgments and to report whether they recognize items on the basis of recollection or familiarity.

Some studies have suggested some differences in cognitive abilities between women and men (Macoby & Jacklin, 1974; Ionescu, 2002). Macoby & Jacklin

(1974) advocated that women are superior to men in verbal ability while men surpass women in spatial and mathematical ability. In a recent study conducted in Sweden, women outperformed men on verbal tests while men outperformed women on visuospatial tasks (Herlitz et al., 1997). This study employed 1000 participants aged 35–80 years. In addition, the authors noted that women consistently outperformed men on a battery of episodic memory tasks.

6.2 Short-term memory, working memory and long-term memory

Just as memory involves different regions of the brain, forgetting involves the loss of different types of memory. LTM retains information learned in the past while STM stores information from the present. LTM is designed to be highly durable but STM can be temporary, such as forgetting a phone number as soon as you have written it down. If one did not forget information in STM, ones mind would soon be cluttered with all sorts of useless information. More than simply normal, that type of forgetting is adaptive and helpful. Forgetting items in the LTM bank is different. LTM can be episodic (e.g., remembering when you last rode a bicycle), semantic (e.g., remembering facts and principles such as knowing what a bicycle is) or procedural (e.g., remembering how to ride a bicycle). It is perfectly normal to forget episodic memories but semantic and procedural memories should be much more deeply entrenched.

6.2.1 Short-term memory

STM involves a temporary synaptic modification, presumably consisting of transient alterations in the concentration, binding or uptake of various neurotransmitter substances. Short-term synaptic enhancement (STE) refers to a model of such STM processes and suggests a way that neurochemical activity might briefly maintain memories by modifying the synapse for brief periods of time (Fisher et al., 1997). The mechanisms mediating STE appear to occur primarily in the presynaptic neuron, perhaps involving an increased number of synaptic vesicles releasing neurotransmitter substance into the cleft in response to an action potential.

6.2.2 Working memory

The term working memory is generally used in cognitive psychology to refer to a limited capacity system that allows the temporary storage and treatment of information necessary for such complex tasks as comprehension, learning and reasoning. The concept of working memory proposed by Baddeley & Hitch (1974) provided such a framework for conceptualizing the role of temporary information storage in the performance of a wide range of complex cognitive tasks. It has proposed a three-compartment model of working memory. This approach represented a development of earlier models of STM such as those of Broadbent (1958) and Atkinson & Shiffrin (1968); however, it abandoned the concept of a unitary store in favor of a multicomponent system. Moreover, it emphasized the function of such a system in complex cognition rather than memory *per se*. The model comprised an attentional control system, the "central executive" supported by two subsidiary slave systems, the "phonological loop" and the "visuospatial sketchpad". The loop is assumed to hold verbal and acoustic information using a temporary store and an articulatory rehearsal system which imaging studies suggested are principally active with Broadmann areas 40 and 44, respectively. Working memory is a limited-capacity system (able to hold only about seven recognizable items) capable of storing and manipulating information for only short periods of time (about 20–30 sec) without rehearsal (Baddeley and Hitch, 1974).

Behavioral as well as functional neuroimaging data in healthy humans and neuropsychological evidence in brain-damaged patients indicate separate working memories for the brief retention of verbal and visual data (Smith et al., 1996; Vallar & Shallice, 1990). In the visual working memory domain, a further fractionation between short-term retention of shape and spatial location of objects is suggested by the identification in the monkey of two neural pathways, the ventral (occipito-temporal) pathway and the dorsal (occipito-parietal) pathway, specialized in the transmission of the former and the latter aspect of visual information, respectively (Stark et al., 1996). Behavioral studies in healthy humans demonstrate that the execution of arm movements in space during the delay interval of visual-spatial STM tasks specifically interferes with memory of the target spatial location whereas the exposure to irrelevant stimuli during the same interval selectively disrupts memory of the visual shape of objects (Baddeley & Lieberman, 1980; Logie & Marchetti, 1991). Finally, functional neuroimaging investigations confirm the anatomical segregation of neural circuits underlying visual-object and visual-spatial working memory (Smith et al., 1995).

Caplan and Waters (1999) discussed the view that language processing makes use of specialized memory resources. They argued that the working memory resources used for sentence processing are separate from those used for consciously controlled processes such as remembering a list of words, and cited evidence from neurological patients, individual differences in the normal range, language comprehension in elderly adults and the comprehension of language while simultaneously maintaining a memory load. These observations are consistent with the view of Lewis (1999) that memory for syntactic relations is independent of the memory used for list of words. Ericsson and Kintsch (1995) argued that the ability of working memory to support skilled performance does not stem from the efficiency that it derives from representing only a small number of items but rather from how its high degree of organization supports efficient retrieval of the appropriate information.

6.2.2.1 The central executive system

The central executive system (CES) is critically responsible for the planning, organization and other strategic aspects of memory that facilitate both the encoding and the retrieval of information in LTM. It processes the information on hand by using and allocating (limited) attentional and cognitive resources. The central executive is also assumed to be fractionable and frontal areas appear to be strongly implicated (Gathercole, 1999). For example, the CES is active when someone is driving while talking to a passenger in the car. This latter function of the CES, measured with a dual task paradigm, has been shown to be impaired in patients with minimal and, to a greater extent, mild AD (Baddeley et al., 1991; 1997; Greene et al., 1995). The CES is also active when someone has to monitor information when performing a mental operation, as can be measured with the digit span (DS) subtest of the Wechsler adult intelligence scale (WAIS) (Wechsler, 1981). In AD, the progressive impairment of performance on the DS subtest is generally attributed to a deterioration of the CES (Flicker et al., 1993; Linn et al., 1995).

6.2.2.2 The phonological loop

The phonological loop (labeled as such by Baddeley (1992) but formerly known as the articulatory loop) contributes to the retention of verbal information. Its function is to maintain phonologically encoded material in primary memory through short-term storage and to translate, when applicable, visually presented material into verbal material (Morris & Baddeley, 1988). It is assumed to comprise a temporary phonological store in which auditory memory traces decay over a period of a few seconds unless revived by articulatory rehearsal. The loop is assumed to have developed on the basis of processes initially evolved for speech perception (the phonological store) and production (the articulatory rehearsal component). It is particularly suited to the retention of sequential information and its function is reflected most clearly in the memory span task whereby a sequence of items such as digits must be repeated back immediately in the order of presentation.

The function of the articulatory loop is spared in patients with early stages of AD, as measured by repetition of digits, and by the presence of phonological similarity, length and recency effects on tests of verbal learning (Morris & Baddeley, 1988). However, recent findings suggest that the recency effect is disrupted in moderate AD (Riekkinen et al., 1998).

6.2.2.3 The visuospatial sketchpad

The visuospatial sketchpad (VSSP) is principally of use to remember the spatial positions of objects and its action lasts a few seconds (Baddeley, 1991). The VSSP is assumed to hold visuospatial information, to be fractionable into separate visual, spatial and possibly kinesthetic components and to be principally represented within the right hemisphere (areas 6, 19, 40 and 47). It is particularly specialized for tasks involving generation and manipulation of mental images, under the control of the CES.

The functioning of the VSSP as assessed by the Corsi's blocks test is altered in patients with, even mild AD (Morris & Baddeley, 1988).

6.2.2.4 The episodic buffer

There are, however, a number of phenomena that are not readily captured by the original model. Therefore a fourth new component of the model, the episodic buffer, has been proposed (Baddeley, 2000). The episodic buffer is assumed to be a limited-capacity temporary storage system that is capable of integrating information from a variety of sources. It is assumed to be controlled by the central executive which is capable of retrieving information from the store in the form of conscious awareness, of reflecting on that information and, where necessary, to manipulate it. The buffer is episodic in the sense that it holds episodes of the information that is being manipulated across space and time. In this respect, it resembles Tulving's concept of episodic memory. Thus, this new model of working memory comprises a limited capacity system that provides a temporary storage of information held in a multimodal code, which is capable of binding information from the subsidiary systems and from LTM into a unitary episodic representation. Conscious awareness is assumed to be the principal mode of retrieval from the buffer. This revised model differs from the old principally in focusing attention on the processes of integrating information rather than on the isolation of the subsystems. Therefore, this new model of working memory provides a better basis for talking the more complex aspects of executive control in working memory.

6.2.3 Long-term memory

LTM appears to alter the structure of the synapse permanently through experience-dependent gene expression. The dominant models of synaptic plasticity for LTM are referred to as long-term potentiation (LTP) and long-term depression (LTD) (Bliss & Lomo, 1973). Although not universally acknowledged, a good deal of evidence supports them as viable models of the cellular basis of learning. LTP refers to a persistent increase in synaptic efficacy resulting from a number of neural mechanisms. LTD involves persistent, experience-dependent decreases in synaptic efficacy. LTP may involve several processes, including an increase in the number of synapses between axons and dendrites, changes in neurotransmitter release and modification of the structure of dentritic spines (Kauer et al., 1988).

6.3 Declarative memory

Declarative memory supports the ability to acquire new facts and events and depends on the integrity of the hippocampus and anatomically related structures in the medial temporal lobe and diencephalon. Declarative memory encompasses the acquisition, retention and retrieval of knowledge that can be consciously and intentionally recollected (Cohen & Squire, 1980). Such knowledge includes memory for events (episodic memory) or facts (semantic memory) (Tulving, 1983). In contrast, nondeclarative or procedural kinds of memory encompass the acquisition, retention, and retrieval of knowledge expressed through experience-induced changes in performance. These kinds of memory are measured by indirect or implicit tests where no reference is made to that experience. Skill learning, repetition priming, and conditioning are classes of implicit tests that often reveal procedural memory processes dissociable from declarative memory.

6.3.1 Episodic memory

Episodic memory is the kind of memory that allows one to remember past happenings from one's life. It represents memory for personally experienced events stored in spatiotemporal and emotional contexts (Tiberghien, 1991). Episodic memory has to do with one's "autonoetic" awareness of one's experiences in the continuity of subjectively apprehended time that extends backward into the past in the form of remembering. It also allows viewing the future in the form of imagining or planning for it. This definition emphasizes the conjunction of three ideas: self, autonoetic awareness and subjectively sensed time. Identifying similarities and differences between episodic memory and both semantic memory and priming will require careful componential analysis of episodic memory (Mayes & Roberts, 2001).

Episodic retrieval is thought to involve an interaction between a "retrieval cue" (self-generated or provided by the environment) and a memory trace, leading to the reconstruction of some or all aspects of the episode represented by the trace. This interaction and its sequelae were termed "ecphory" by Semon (1921). Whether an episodic retrieval attempt is successful or not is influenced by numerous factors, not least of which is the way the event was initially "encoded" into memory. Also important are the cues available, and the processes engaged during the retrieval attempt. The importance of retrieval cues and the nature of their processing is emphasized in the principle of "transfer appropriate processing" according to which memory performance is a function of the degree to which cognitive operations engaged are recapitulated at retrieval (Morris et al., 1977). Tulving (1983) has also proposed that a further prerequisite for successful episodic retrieval is that the rememberer is in the appropriate cognitive state, which he termed "retrieval mode". According to this proposal, only when the rememberer is in retrieval mode will a stimulus event be treated as an episodic retrieval cue. Retrieval mode is also necessary for retrieval to be accompanied by the experience of "reliving the past" or "autonoetic" remembering (Wheeler et al., 1997). These two conceptions of retrieval mode share the notion that mode is manifest as a "tonically" maintained cognitive state.

Functioning of episodic memory is principally assessed in the laboratory with methods such as learning and recall of lists of words or paragraphs, presented either orally or visually, learning and recall of abstract drawings and facial recognition tests. For instance, reading a list of words to a patient and then asking the patient to recall the words a minute later is an episodic memory test, since the patient must retrieve a specific list of words presented in a specific temporal and spatial context. Another method is to test the ability to remember autobiographical events that occurred prior to a lesion/disease (Sanders & Warrington, 1975). Patients with AD show encoding and retrieval deficits in free and cued recall situations, together with recognition deficits very early in the course of the disease (Deweer et al., 1994; Goldman et al., 1994; Herlitz et al., 1997).

6.3.2 Semantic memory

Semantic memory is defined as memory for general knowledge that is context free and includes memory for vocabulary, as well as facts about the world and oneself. It refers to one's fund of general knowledge that is not dependent upon contextual cues for retrieval (Tulving, 1983). It is the system that processes, stores and retrieves information about the meaning of words, concepts and facts. Asking a patient the name of the first president of the United States or the capital of France all involve retrieval from semantic memory. In each case, the specific information can be retrieved without recalling the particular episode in which that information was acquired. Semantic memory has been proposed to exist as a representation of knowledge based on an organized network of interrelated categories, concepts and attributes. Semantic or category fluency tasks (e.g., animals, vegetables, clothes), the vocabulary subtest of the WAIS and the Pyramids and Palm tress test are good examples of semantic memory tasks (Wechsler, 1981; Howard & Patterson, 1992).

A model of semantic memory may posit two major components: long-term knowledge of the features contributing to word meaning and the processes that operate on this knowledge. Some feature knowledge is likely to be concerned with the perceptual appearance of an object or an action such as the shape, color or motion of a word's exemplar (Jackendorff, 1990; Miller & Johnson-Laird, 1976). Knowledge that is neither perceptual nor functional is likely to be retained in semantic memory for word meaning as well, such as an associative network of prepositional knowledge that is related particularly to abstract words that have few perceptual features.

The neural organization of semantic memory is not very clear. Some researchers argue for a non-fractionated distributed semantic system in which input modality produces a regionally specific effect but the particular category of semantic knowledge does not (Tyler & Moss, 2001). Others argue for regional specialization according to knowledge category. For example, tools and animals are represented in different areas of temporal cortex (Martin & Chao, 2001). According to this approach, the neural representation of a feature contributing to semantic memory is thought to be in the processing stream most relevant for the feature. The neural representation of visual-perceptual feature knowledge, for example, may be associated with activation of the fusiform gyrus in ventral temporal-occipital visual association cortex (Martin et al., 1995; 2000). This hypothesis holds that the representation of a specific category of knowledge in semantic memory is tightly linked to the kind of feature contributing importantly to the category.

On the basis of neuroanatomical connectivity patterns, the neural substrates for these integrative processes are hypothesized to include heteromodal association cortices in the dorsolateral temporal-inferior parietal (BA 22 and 39) brain regions. Heteromodal cortical areas such as these have reciprocal projections with modality-specific association regions where feature knowledge may be represented and with paralimbic region such as medial temporal areas important for the long-term consolidation of knowledge in semantic memory. Heteromodal association cortex is implicated in semantic processing, on the basis of structural imaging studies of patients with insult to this region (Hillis et al., 2001). Functional neuroimaging studies of lexical semantic memory frequently show lateral frontal (BA 9, 44 & 46) and posterolateral temporal (BA 21, 22 & 39) activation for multiple semantic categories (Whatmough et al., 2002).

Some data indicate that semantic memory is disrupted very early in the course of the disease in both minimal and mild stages of AD (Dalla Barba & Goldblum, 1996; Howard & Patterson, 1992; Persson & Skoog, 1992). However, there is heterogeneity in the results with some patients showing severe impairment on all semantic memory tests while others performed variably on specific tests (Hodges & Patterson, 1995).

6.3.3 Autobiographical memory

The term autobiographical memory is used to focus attention upon an individual's record of his own personal experiences as opposed to his performance on standard laboratory tests such as free recall. Autobiographical memories are recollections of personally meaningful events that are used to construct one's life history. As such, the memories formed are a reflection of a person's self-concept as well as their relationship with significant others. Autobiographical memory may be best described as involving the remembrances of events that have been personally experienced (Neisser, 1986).

Personally experienced events are structured in autobiographical memory by hierarchically ordered types of memories for events that vary in their scope and specificity and this structure is organized along temporal and thematic pathways that guide the retrieval process. Although there have been slightly different views regarding the various types of memories that people recover when remembering their personal past, in general there are three main types (Barsalou, 1988; Conway, 1996). The most general type of memory, which comprises the top and middle of the hierarchy, is for extended events. Memories for extended events reveal the temporal nature of autobiographical memory, as such events are extended in time for periods as long as many years to as short as just a few days. Comprising the middle of the hierarchy are memories for summarized events that emphasize the thematic aspects of autobiographical memory. In remembering summarized events, individuals are considering the common themes that underlie events of the same kind. Finally, the most detailed type of memory at the bottom of the hierarchy is for specific events. Memories of specific events include the perceptual and episodic information that provides a sense of reliving a particular episode as it originally occurred. Therefore, autobiographical memory structure may be characterized as a hierarchical network that includes extended, summarized and specific events (Belli, 1998). It permits retrieval of past events through multiple pathways that work top-down in the hierarchy, sequentially within life themes that unify extended events and in parallel across life themes that involve contemporaneous and sequential events.

Autobiographical memories are traditionally indexed by providing participants with cue words and requesting recall of a specific personal memory in response to them (Rubin, 1986). Specificity of memories is defined as the participant's being able to "give a date, day of the week or time of the day when the episode occurred" (Williams & Scott, 1988). With respect to remembering when events happened, people tend to report events as having occurred more recently than in actuality, a phenomenon known as forward telescoping (Bachman & O'Malley, 1981; Thompson et al., 1988; Rubin & Baddeley, 1989). Finally, accuracy in reporting how often events happened has been found to be a complex interplay of such factors as the frequency, similarity and regularity of events, the length of the retention interval and the use of closed or open-ended questions (Brown, 1995).

The Autobiographical Incidents Schedule, the questionnaire of Flicker and collaborators and the autobiographical fluency task can assess autobiographical memory performance (Kopelman et al., 1989; Flicker et al., 1987; Dritschel et al., 1992). Patient groups with minimal and mild AD were found to be impaired on autobiographical memory. Patients in the minimal and mild stages show a temporal gradient, with stronger deficits for recent events (late adult) than for events of childhood and early adult life (Greene et al., 1995; Sagar et al., 1988).

6.3.4 Prospective memory

A common everyday memory task is to remember to perform an intended action at some appropriate point in the future. This type of memory has been termed prospective memory. It is an important aspect of cognition that is necessary for individuals to function in their everyday life. Prospective memory tasks require planning and keeping the cue-action association activated during ongoing performance of a background task, functions that both presumably involve the frontal lobes.

A characteristic of laboratory paradigms for examining prospective memory is that participants are instructed to perform an action whenever a target event occurs. For example, in the context of some ongoing activity (e.g., rating the pleasantness of a list of words), participants might be instructed to press a special key on a computer keyboard each time a particular target word appears on the screen (Einstein et al., 1997). The Rivermead behavioral memory (RBM) test contains subtests measuring prospective memory such as remembering an appointment, remembering a belonging and remembering to deliver a message. Prospective memory has been shown to be impaired early in the course of AD (Huppert & Beardsall, 1994; McKitrick et al., 1992).

6.4 Procedural memory

Procedural memory (implicit memory) is quite different from both episodic and semantic memory, because it pertains to an unconscious form of remembering that is expressed only through the performance of the specific operations comprising a particular task. Procedural memory is responsible for the acquisition and retention of motor, perceptual and cognitive skills. Playing pool, solving arithmetic problems are examples of procedurally learned skills. Procedural learning is also an essential contributor to the aspect of personality often referred to as character – the remarkable consistency displayed by people in their behavior over time (Grigsby & Stevens, 2000). Procedural memory is independent of the medial temporal lobe and diencephalic structures important for declarative memory.

6.4.1 Skill learning

In skill-learning tasks, subjects perform a challenging task on repeated trials in one or more sessions. The indirect or implicit measure of learning is the improvement in speed or accuracy achieved by a subject across trials and sessions. Preservation of sensorimotor, perceptual, and cognitive skill learning in amnesia indicates that such learning for some skills is not dependent upon declarative memory. Some of the neural systems underlying such skill learning have been identified in neuropsychological and neuroimaging studies.

Intact sensorimotor skill learning in amnesia is well documented for three tasks: mirror tracing, rotary pursuit, and serial reaction time (SRT). Rotary-pursuit skill learning is intact in amnesia and in AD (Heindel et al. 1989). SRT learning is intact in amnesia and intact in some but not all AD patients (Ferraro et al., 1993). Variability in AD performance may reflect severity and perhaps specific impairment in spatial working memory. Patients in the mild and moderate stages of AD showed preserved performance on skill learning measured with pursuit-rotor and mirror-reading tasks (Sahakian et al., 1988; Deweer et al., 1994; Grafton et al., 1992; Deweer et al., 1993).

6.4.2 Repetition priming

Priming is a form of memory for specific factual and episodic information that depends on automatic retrieval processes and does not need to involve feeling of familiarity for what is remembered. A widely held view is that priming depends on storage changes in the brain regions that represent the primed information and that these changes ensure that, when some or all of the information is next re-encoded, the representation is reactivated more fluently (i.e. more rapidly and strongly) (Mayes, 2001).

One important distinction is between perceptual priming, which reflects prior processing of stimulus form, and conceptual priming, which reflects prior processing of stimulus meaning. Perceptual priming occurs in visual, auditory, and tactual modalities. It is maximal when study-phase and test-stimuli are perceptually identified, and reduced when there is a study-test change in modality or symbolic notation. Conceptual priming is maximal when study-phase processing enhances semantic analysis of stimulus meaning, and reduced when study-phase processing diminishes semantic analysis. Priming procedures include tasks of perceptual priming which essentially rely on the analysis of the physical properties of stimuli and require that their format remains identical between study and test (e.g. perceptual identification of words or pictures) and tasks of conceptual priming in which retrieval cues provide information conceptually related to the target information, without identical format presentation of stimuli between study and test being necessary (e.g., paired-associates procedures) (Schacter et al., 1993).

Several lines of evidence indicate that priming is mediated by neocortical areas, with perceptual priming being mediated by modality-specific cortical regions and conceptual priming by a modal language areas. One source of evidence is the performance of AD patients who exhibit severely reduced conceptual priming but intact perceptual priming on visual tasks (Monti et al., 1996; Fleischman et al., 1995). This pattern of impaired conceptual and intact perceptual priming may be interpreted in terms of the characteristic neocortical neuropathology in AD.

In explicit memory tests such as recognition and free recall, subjects are instructed to actively remember previous events, whereas in implicit memory tests, subjects are not instructed to explicitly use memory. Tasks of repetition priming usually fall into two categories: word identification/incomplete-picture and word-stem completion tasks. For example, in a word stem completion test, subjects are first exposed to a series of words (e.g., SALMON) and are then shown word stems such as SAL___ and are instructed to complete the stem with the first word that comes to mind. Subjects are more likely to complete a stem with a word if it was previously studied (i.e., exhibit priming effects) even if they do not consciously recollect having studied that word. Moreover, amnesic subjects who perform more poorly than normal subjects on tests of explicit memory perform normally on tests of implicit memory, indicating that implicit and explicit forms of memory rely on partially distinct neuroanatomical substrates (Schacter et al., 1993). These two forms of memory also exhibit functional properties. For example, changes in the perceptual format of words between study and test, such as from uppercase letters to lowercase letters or from auditory to visual presentation, greatly reduce the priming effects observed in implicit tests such as stem completion. In contrast, similar changes in perceptual format often do not affect performance on explicit tests like recognition and recall.

Priming for line drawings of real and nonreal objects was examined in an object decision task for AD patients and normal elderly control participants (Fleischman et al., 1998). It was observed that the classification of real and nonreal objects in AD patients, albeit mildly impaired compared to the control group, was reasonably accurate. The explicit memory for real and nonreal objects was substantially impaired in the AD patients but the repetition priming for real objects was intact. Repetition priming for nonreal objects was robust when measured by latency, but opposite to the control group. The AD patients were slower in classifying repeated nonreal objects compared to novel nonreal objects, whereas control participants were faster. Although both groups were less accurate in identifying repeated normal objects, the decrease in accuracy was significantly greater for the AD group. Therefore, the results of this study support the claim that AD patients with MCI show normal perceptual priming.

Although there is some agreement about the spared performance of patients with mild and moderate AD when compared with controls on data-driven (perceptual) process tasks using word-identification or incomplete-picture paradigms, controversy on results of conceptually-driven process tasks using word-stem completion still persists (Russo et al., 1994; Fleischman et al., 1996; Gabrielli et al., 1994).

An important question concerns the capacity of showing priming for materials without pre-existing representations in memory in normal and pathological ageing. A study which assessed volunteers (20 patients with mild AD, 20 elderly controls and 20 young control subjects) with a paradigm of priming for new verbal associations reported that contrary to young subjects, neither AD patients nor normal elderly subjects demonstrated priming effects for new associations (Ergis et al. 1998). These results indicate that the absence of priming for new verbal associations is more related to an effect of aging than to a specific effect of AD.

6.5 Controlled and automatic memory process

A defining characteristic of dementia of Alzheimer's type is a profound impairment on any direct or explicit tests of memory (Butters, 1984). Thus, when demented patients are instructed to recall to consciousness previously studied information, their performance is almost invariably abnormal. Nevertheless, it is apparent that on some tasks patients with dementia of Alzheimer's type (DAT) do show a degree of preserved capacity for learning. In particular, on tests measuring facilitation of performance by a previous learning episode, where the test is indirect and does not require conscious recall, AD patients may show normal or near normal performance (Monti et al., 1994). However, the dissociation between impaired direct memory and preserved indirect memory, which has been consistently demonstrated in persons with circumscribed amnesia, is not as clear-cut in AD. For example, on the widely used word-term completion test of implicit memory function, there are reports of both abnormal and normal performance (Deweer et al., 1994; Keane et al., 1991).

Jacoby (1991) described a method for separating conscious from unconscious (or automatic) processes in memory that allows the estimation of controlled memory performance without contamination by unconscious or automatic influences, which he termed the process dissociation procedure. One advantage of this method is that it allows examination of differences in the processes impaired and nonimpaired individuals use to recall or recognize previously studied information. Using the process dissociation procedure of Jacoby, a recent study examined the contribution of controlled (conscious) and automatic (unconscious) memory processes to the performance of a stem-completion recall task by AD patients and a matched group of healthy elderly individuals (Knight, 1998). The results of this study provided an understanding of the manner in which AD patients perform stem completion tasks used as direct tests of cued recall. Recollection was noted to be severely impaired in the AD patients. Further, the estimates of the automatic processing were also observed to be reduced, although there was considerable overlap in the performance of the two groups on this parameter. It was concluded that the residual capacity of AD patients to recall previously learned information was supporter to a great extent by their automatic memory processes. Therefore, this study indicates that at the practical clinical level, results from process dissociation tasks may demonstrate that when controlled processing is impaired by old age, amnesia or even dementia, automatic processes may continue to function at a normal or near normal level. This observation may be significant in planning effective learning contexts in the development of rehabilitation strategies. As an assessment tool in the practice of clinical neuropsychology, the process dissociation procedure shows considerable promise. The application of Jacoby's model permits the assessment of an estimate of controlled memory uncontaminated by automatic processes, thereby increasing the precision of measurement of memory dysfunction in amnesic or demented patients.

6.6 Neuroanatomy and physiology of memory

Combined behavioral and molecular genetic studies suggest that despite their different logic and neuroanatomy, declarative and nondeclarative forms of memory share some common cellular and molecular features. In both systems, memory storage depends on a short-term process lasting minutes and a long-term process lasting days or longer. STM involves covalent modifications of pre-existing proteins, leading to the strengthening of pre-existing synaptic connections. LTM involves altered gene expression, protein synthesis and the growth of new synaptic connections. In addition, a number of key signaling molecules involved in converting transient short-term plasticity to persistent LTM appear to be shared by both declarative and nondeclarative memory. A striking feature of neural plasticity is that LTM involves structural and functional change (Greenough & Bailey, 1988).

Human memory includes multiple systems with different anatomic substrates. At the neuropsychological level, memory begins with the registration of information in cortical and subcortical processing structures. According to the consensus view of the neural mechanisms supporting autobiographic or episodic memory, the ability to recollect past events consciously is mediated by multiple systems distributed throughout the cortex (Tulving, 1983; Tranel & Damasio, 1995; Damasio, 1994; Rubin, 1998). Encoding new experiences for later recollection is mediated by areas in the medial temporal lobes, especially the hippocampus and surrounding areas and the diencephalons that includes the mammillary bodies and dorsomedial thalamus. The long-term storage of information for conscious recall is mediated by these encoding centers (Damasio, 1994). The extrastriate occipital regions are likely to mediate perceptual priming while basal ganglia seem to be critical for mediating procedural memory.

Theories of episodic memory need to specify the encoding, storage and retrieval processes that underlie this form of memory and indicate the brain regions that mediate these processes and how they do so. Primarily those parts of the posterior neocortex that process perceptual and semantic information probably mediate representation and retrieval of the spatiotemporally linked series of scenes that constitute an episode. However, establishing where storage is located is very difficult and disagreement remains about the role of the posterior neocortex in episodic memory storage.

Synaptic plasticity, including LTP and LTD, are thought to be important for learning and memory (LeDoux, 2000; Bliss & Collingridge, 1993). Several reports have suggested the existence of endogenous molecular suppressors that negatively control the efficacy of neuronal transmission and memory formation (Abel et al., 1998). One proposed category of such memory suppressors is constituted by protein phosphatases. These molecules, together with protein kinases, regulate many cellular processes by the reversible phosphorylation/dephosphorylation of specific substrates (Sweatt, 2001). For instance, the Ca^{2+}/calmodulin-dependent protein phosphatase calcineurin (PP2B) was reported to block learning, memory storage and memory retrieval (Mansuy et al., 1998; Malleret et al., 2001). Protein phosphatase 1 (PP1) determines the efficacy of learning and memory by limiting acquisition and favoring memory decline (Genoux et al., 2002). When PP1 is genetically inhibited during learning, short intervals between training episodes are sufficient for optimal performance. The enhanced learning correlates with increased phosphorylation of cyclic AMP-dependent response element binding (CREB) protein, of Ca^{2+}/calmodulin-dependent protein kinase II (CaMKII) and of the GluR1 subunit of the AMPA receptor. It also correlates with CREB-dependent gene expression. Inhibition of PP1 prolongs memory when induced after learning, suggesting that PP1 also promotes forgetting (Genoux et al., 2002). This finding emphasizes the physiological importance of PP1 as a suppressor of learning and memory and as a potential mediator of cognitive decline during ageing.

CREB is a major transcription factor that is centrally involved in the formation of LTM in both invertebrates and vertebrates (Josselyn et al., 2001; Silva et al., 1998). CREB is activated by phosphorylation of the serine 133 residues, potentially within a short period of time. Two major pathways, the cAMP signaling and calcium-calmodulin protein kinase pathway, mediate CREB activation. Inhibition of CREB impairs behavioral performance in various memory tests across species whereas overexpression of CREB facilitates long-term fear memory (Yin et al., 1995; Josselyn et al., 2001).

The aging brain loses weight, volume and neurons. Morphological study evidence of a general decrease in gross brain volume reaching significance in the seventh decade of life has been well documented (Haug & Eggers, 1991). The greatest regional differences in the degree of volume reduction, estimated to be 10–17%, are found in the prefrontal cortex. This is compared to volume decreases in the temporal, parietal and occipital cortices, all estimated at 1% (Guttman et al., 1998; Raz et al., 1997). With advancing age, an increased number of pathological structures in the brain, such as SPs and tangles accompany the reduction in cortical volume throughout the human life span (West, 1996). Functional changes in the aging brain also occur such as declines in rCBF, particularly in the prefrontal cortex and decreased prefrontal activity in older *versus* younger adults during verbal encoding tasks (Cabeza et al., 1997; Tulving et al., 1994). Bilateral prefrontal activation in older adults as compared to hemisphere-specific prefrontal activation in younger adults during tasks involving verbal recognition memory and spatial memory tasks is also noted (Reuter-Lorenz et al., 2000).

6.6.1 Medial-temporal and diencephalic systems

The medial temporal lobe (MTL), which includes the hippocampal formation (the dentate gyrus, fields CA1, CA2 and CA3 and the subiculum), entorhinal cortex, perirhinal cortex, parahippocampal cortex and amygdala, is postulated to guide memory consolidation processes that yield stable representations of memory in neocortex (McGaugh, 2000; Squire & Alvarez, 1995). Most researchers believe that the hippocampus and MTLs are involved in storing episodic information and that long-term potentiation may indicate the mechanism of storage (Morris & Frey, 1997). Lesions to medial-temporal and diencephalic brain regions yield amnesia, a selective deficit in declarative memory with sparing of short-term memory, remote memories, and motor, perceptual and cognitive capacities. In patients with large MTL lesions and severe amnesia, the capacity for new semantic learning seems quite limited (patient HM, patient SS and patient EP; Marslen-Wilson & Teuber, 1975). When the damage is more restricted and the amnesia correspondingly less severe, considerable semantic learning is possible although the degree of learning seldom approaches what can be achieved by normal individuals (Van der Linden et al., 2001).

The MTL has also been related to processing that leads to contextual memory (see Mayes, 1988). Damage to this area also disrupts attentional orienting towards novel information (Honey et al., 1998). Many functional neuroimaging studies have shown MTL activations when subjects carry out memory encoding tasks (Gabrieli et al., 1997; Stern et al., 1996). There is some evidence that these MTL activations are greater when associations rather than items are encoded (Henke et al., 1997; 1999). There is also some evidence that these MTL activations are greater when novel rather than familiar information is encoded (Dolan & Fletcher, 1997; Tulving et al., 1996). However, although this region is often activated when episodic information is encoded, it is unclear whether the activation reflects the processes involved in representing context and other kinds of episodic information or the early consolidation processes that put this information into LTM. Even if the activation does reflect representational processes, this does not prove that either the hippocampus or the medial temporal lobe cortex plays a critical role in making episodic representations.

Studies in the monkey suggested that large MTL lesions do impair auditory recognition memory (Fritz et al., 1999). Tests of auditory recognition memory in patients with MTL lesions showed that auditory recognition, like recognition memory in other sensory modalities, is dependent on the MTL.

The first lesions in most cases of AD may occur in the medial temporal lobe and this may account for amnesia being the most common initial problems in AD (Hyman et al., 1984). Unlike patients with pure amnesia, however, AD patients have a dementia defined by the compromise of at least one additional, nonmnemonic function. Further AD patients also have early damage to cholinergic neurons in the basal forebrain and lesions in that area cause declarative memory impairments. Therefore, it is difficult to ascribe the amnesia in AD exclusively to medial-temporal injuries.

6.6.2 The amygdala

It is now clear that the basolateral amygdala is involved in negative and positive effect as well as spatial and motor learning and memory. Several structures surrounding the basolateral amygdala, including the central, medial and cortical nuclei are traditionally included in the "amygdaloid complex". These surrounding structures, together with the basolateral amygdala, have come to be called the amygdala. The amygdala has been implicated in emotion and emotional memory in humans and experimental animals (LeDoux, 1995; LaBar et al., 1995; Morris et al., 1996). Emotional arousal and arousal generally is known to strengthen episodic memory (McGaugh, 2000). There is evidence that such arousal slows forgetting in the first hour or so following learning that is consistent with the arousal enhancing the protein synthesis-dependent consolidation process (Kleinsmith & Kaplan, 1963). A number of recent studies have investigated the role of the human amygdala in the processing of emotions in facial emotions (see Schmolck & Squire, 2001). Many of these patients tended to confuse sadness with disgust and anger and fear with surprise and anger.

The emergence of neuroimaging technologies such as PET and fMRI allows the study of the normal amygdala in humans. Overall, one can see that amygdala activation appears to be reliably produced by presentation of biologically-relevant sensory stimuli. For example, fMRI signal intensity is greater when subjects view graphic photographs of negative material (e.g., mutilated human bodies) compared to when they view neutral pictures (Irwin et al., 1996). Human amygdala fMRI signal intensities have been shown to be increased during Pavlovian fear conditioning in response to stimuli that predict an aversive event (Büchel et al., 1999; LaBar et al., 1998).

Anatomical and imaging studies have shown that the amygdala is a major locus of pathology associated with AD (Cuénod et al., 1993; Maunoury et al., 1996; Hyman et al., 1990; Van Hoesen et al., 1999). However, rather than affecting the entire structure, the neuropathological changes are found mainly in the basolateral group, which is extensively connected to the hippocampal area and the temporal lobe. There is relative sparing of the phylogenetically older corticomedial group that maintains connections with phylogenetically older CNS regions such as the olfactory bulbs and the hypothalamus. This may explain the relative preservation of emotional memory found in AD patients. A study designed to investigate whether the emotional content of a text can influence memory in patients affected by AD and whether this effect is related to attentional processes as measured by event-related potentials (ERP) was performed (Boller et al., 2002). The results show relatively preserved emotional processing in patients with AD. They further show that the emotional content of a context can influence memory performance. No evidence was found that this effect is mediated by attention as measured by ERP.

6.6.3 The hippocampus

The hippocampus is a part of a MTL system necessary for the formation of a stable declarative memory in humans and spatial memory in rodents (Scoville & Milner, 1957; Squire & Zola-Morgan, 1991; Eichenbaum, 2000). This seahorse-resembling structure has one of the highest concentrations of receptors for corticosteroids (glucocorticoid hormones whose levels elevate in response to stress) in the mammalian brain and participates in the glucocorticoid-mediated negative feedback of the hypothalamus-pituitary-adrenal (HPA) axis (McEwen, 1982). Recent memory for specific information (declarative or explicit memory) depends on the temporal lobes, with the language-dominant temporal lobe being important for memory of verbal material and the non-dominant temporal lobe for visuospatial materials (Milner, 1972). Within the temporal lobe, this memory function is usually related to structures such as the hippocampus. However, the lateral temporal neocortex is also part of the neural substrate for declarative memory, based on effects of lesions and neuroimaging (Milner, 1967; Ojemann & Dodrill, 1987; Kirchoff et al., 2000). Using fMRI, hippocampal activity during both traditional and associative recognition memory tasks was reported (Stark & Squire, 2001). It seems likely that recall memory, recognition memory, episodic memory, semantic memory and indeed all of declarative memory relies on and benefits from the processing afforded by the hippocampal region.

The human hippocampus has long been associated with episodic memory while the hippocampus in rodents has been associated with spatial navigation. O'Keefe and Nadel (1978) proposed that a possible link between topographical and episodic memory is the existence of an allocentric (world-centered) cognitive map stored in the hippocampus. They suggested that a spatial system in rats might have developed into an episodic memory system in humans with the addition of verbal and temporal inputs. This hypothesis predicts that hippocampal damage in humans should impair both topographical and episodic memory. According to Moscovitch (1992), the hippocampus and related limbic area in the medial temporal lobe and diencephalon represent another specialized processing structure. This structure automatically encodes previously consciously apprehended information and in response to an appropriate cue automatically delivers information back to consciousness. This output is perceived as a memory, but a memory that is not placed in its proper spatiotemporal context. This output gives rise to the feeling of familiarity.

Over time, however, a consolidation process may take place, allowing reactivation to occur without medial temporal or diencephalic support (Squire, 1987; McClelland et al., 1995; Schmajuk & DiCarlo, 1992; Nadel & Moscovitch, 1997). It has been argued that the role of the hippocampus in memory is time-limited. During a period of consolidation, other brain regions such as the neocortex are said to acquire the ability to support memory retention and retrieval on their own. An alternative view, based on recent evidence using more sensitive scoring techniques to assess remote memory loss following hippocampal complex lesions in humans was suggested (Nadel & Moscovitch, 1997; Moscovitch & Nadel, 1998; Nadel et al., 2000). To account both for extensive retrograde amnesia and for temporal gradient observed in some studies, the multiple trace theory posits that a new hippocampally mediated trace is created when old memories are retrieved. Old memories are represented by more or stronger traces than are new ones, making them more resistant to partial lesions of the medial temporal lobe.

It has been hypothesized that hippocampal damage produces a deficit only in recall memory whereas perirhinal pathology implicates recognition memory as well (Aggleton & Shaw, 1996). The possibility that completely spared recognition memory in the hippocampal cases simply reflects relatively mild memory impairment cannot be excluded. Several human, monkey and rat studies that employed object-recognition tasks indicate that the hippocampus plays an important role in recognition memory. For example, amnestic patients with relatively limited brain damage that includes the hippocampus exhibit impaired recognition memory with long, but not short delays (McKee & Squire, 1993). Another study found an equivalent impairment in spatial and nonspatial object recognition memory in human patients with hippocampal damage (Cave & Squire, 1991). Hippocampal atrophy in AD has been associated with impairment of declarative memory functions that are characteristic symptoms of AD (Squire & Zola-Morgan, 1991). Cholinergic deficiency contributes to the memory disturbance of AD (Coyle et al., 1983). Loss of cells in the nucleus basalis of Meynert in the basal forebrain in AD produces a decrease in choline acetyltransferase and a subsequent deficiency in acetylcholine (Whitehouse et al., 1981; Kasa et al., 1997).

6.6.4 The frontal lobes

Whereas the results of neuroimaging investigations are largely convergent with the clinical findings in amnesic patients that suggest a pivotal role of MTL structures, in particular the hippocampal formation, the involvement of the frontal lobe in the episodic memory is more controversial. Lesions of the frontal lobes are not usually associated with clinically evident amnesia. However, a consistent activation of the prefrontal cortex has been found not only during working memory tasks but also during long-term episodic learning (Haxby et al., 2000; Cabeza & Nyberg, 2000; Fletcher & Henson, 2001). The prefrontal cortex receives a dense innervation from the brainstem aminergic nuclei, including the serotonergic dorsal and median raphe nuclei of the midbrain (Azmitia & Segal, 1978). The prefrontal cortex contains a very large density of 5HT1A and 5HT2A receptors located on pyramidal neurons.

The frontal lobes are commonly associated with executive functioning. For instance, the frontal lobes are assumed to play a critical role in mental activities such as formulating plans, initiating actions, monitoring ongoing behavior and evaluating outcomes (Glisky, 1996). Maintaining and manipulating information in working memory are frequently listed as key functions subserved by the frontal lobes. Neurophysiological and neuroanatomical evidence suggests that changes in old age may occur earlier and progress more rapidly in frontal areas than in most other parts of the brain (West, 1996). According to the frontal-lobe hypothesis of cognitive aging, these distinct patterns of brain aging should be reflected in correspondingly distinct time courses of changes in mental abilities, with mental functions that involve the frontal lobes being particularly susceptible to the effects of normal aging (Craik, 1986; West, 1996). The evidence on age effects in prospective memory performance appears to be reasonably consistent with predictions derived from the frontal-lobe hypothesis of cognitive aging.

Retrieval of information involves the frontal lobes but like the role of the diencephalons and MTLs in encoding, the frontal lobes are not thought to contribute the retrieval strategies needed to access it (Moscovitch, 1995; Wheeler et al., 1997). For some time after encoding, retrieval or reactivation also requires the involvement of the hippocampal region. Imaging studies of episodic memory, mostly for verbal stimuli, suggest a hemispheric encoding-retrieval asymmetry. The left prefrontal cortex is crucial in encoding and the right prefrontal cortex in retrieval. The HERA model developed from these observations has been the focus of a number of imaging studies that have tried to characterize other factors affecting both the hemispheric asymmetry and the functional neuroanatomical subdivisions of frontal activation. Prevalent right-sided or bilateral activations have been observed during the encoding of non-verbal items such as unfamiliar faces or complex scenes (McDermott et al., 1999; Kirchhoff et al., 2000). However, the left prefrontal cortex is also activated in response to non-verbal stimuli such as unfamiliar faces or complex figures (Fletcher et al., 1997). Prevalent right prefrontal activation has been associated with successful retrieval, retrieval effort or monitoring of the retrieval information (Shallice et al., 1994). Left prefrontal activation has also been observed in studies dealing with recognition and source memory (Schacter et al., 1997; Rugg et al., 1999). Thus, both the material and the type of memory process may affect the lateralization of frontal activation during memory tasks.

The prefrontal cortex has been afforded the highest functions of the brain, including the capacity for insight, abstraction, self-awareness and complex problem solving. Many of the changes seen in aging, including general slowing, decreased drive, forgetfulness and deteriorated cognitive functioning, may be frontal lobe based. Memory alterations in at least some older adults might be considered a frontal lobe system disorder. Most current research demonstrates that damage to the frontal lobes does not produce classic amnesia; rather it results in frontal cognitive impairments that influence the successful functioning of these memory systems by allowing information to be remembered in its appropriate context. Luria (1973) proposed that memorizing (putting information into storage) is not affected by frontal lobe damage; rather it is affected by essential, associated, frontal lobe abilities of motivation, programming, regulation, attention and verification.

6.7 Amnesia

Temporally graded amnesia refers to a phenomenon of premorbid memory loss whereby information acquired recently is more impaired than information acquired more remotely. One account of this phenomenon suggests that memory is stored in the same neocortical structures that were involved in processing the relevant information during learning. Initially, the hippocampus serves to bind these cortical regions and to allow memory to be reactivated for retrieval. Over time and through a process of reorganization, the connections among the cortical regions are progressively strengthened until the cortical memory can be reactivated and retrieved independently of the hippocampus (Squire and Alvarez, 1995).

An alternative suggestion is that memories that are initially hippocampus dependent remain dependent on the hippocampus. Following this hypothesis, older memories have a more redundant and spatially distributed representation within the hippocampus than recent memories. Temporally graded retrograde amnesia occurs because a partial lesion of the hippocampus is more likely to spare a remote memory than a memory acquired recently (Nadcl & Moscovitch, 1997). This hypothesis predicts that temporally graded retrograde amnesia will not be observed when hippocampal lesions are complete.

Psychological forms of retrograde amnesia can be relatively "pure" in that there is no known evidence of cerebral pathology or they can occur in the context of either minor or major brain disorder. However, even in the "pure" cases, which are the ones often referred to as "psychogenic amnesia", there is commonly a past history of a transient organic memory loss (Markowitsch, 1996). Psychogenic amnesia can be situation-specific, e.g. loss of memory for an offence by either the perpetrator or the victim: in these cases, there is a brief gap in memory for the episode presumably as a result of compromised anterograde memory encoding (Kopelman, 1987). Alternatively, psychogenic amnesia can be "global" encompassing the whole of a person's past, as occurs in a soc-called "fugue" episode, also known as "functional retrograde amnesia".

6.8 Mental imaging

Most mental imaging appears to arise when LTM information is activated but both STM and LTM can be involved in this process (Farah, 1984; Kosslyn et al., 1995). For all age groups, words rated high in imagery produced older memories and faster reaction times (Rubin & Schulkind, 1997).

Word memory has been reported to activate event-related brain potentials (Ferlazzo et al., 1993). With respect to localization of imagery, it does not appear to be exclusively a left or right hemisphere phenomenon although some investigators considered a posterior lesion in the left cerebral hemisphere a cause for inability to evoke images from memory (Farah, 1984). Others reported that left-sided lesions impaired verbal memory, with right-sided lesions impairing nonverbal memory (Milner, 1971). Gender differences in imaging tasks were usually not observed (Kolb & Whishaw, 1990). Some investigators consider memory retrieval to be under the control of the prefrontal cortex (Hasegawa et al., 1998).

Patterns of cerebral activation during mental rehearsal of a motor act were reported as similar to those produced by its actual execution (Sirigu et al., 1996). Localization of these imagery or memory-related responses has been in the same brain areas where the actual sensory stimuli were perceived with enough frequency that they have been considered to involve similar representational and neural mechanisms (Phelps & Mazziotta, 1985; Behrmann et al., 1994). In general, less cerebral glucose utilization was measured in frontal cortical regions using PET scanning when a cognitive task involved passive rather than active stimulus perception (Mazziotta & Phelps, 1984; Phelps & Mazziotta, 1985; Mazziotta et al., 1982).

6.9 Memory alterations in aging

Memory changes occur as a part of normal aging, as well as part of dementing processes. Many older adults experience subtle yet disturbing symptoms of forgetfulness while continuing to lead active and intellectually demanding lives. Attempting to understand and interpret age-related memory decline, Craik and Lockhart (1972) proposed the level-of-processing hypothesis. According to this hypothesis, a series or hierarchy of processing stages in memory exists, where greater levels or "depth" implies a greater degree of semantic or cognitive analysis. Memory is seen as a product of processes of information extraction and elaboration carried out for the purpose of perception and comprehension of environmental information and events. Depth of processing determines the nature and durability of the information in memory. Along these lines, it can be hypothesized that normal, older adults engage in less extensive and less efficient initial processing of the material to be learned, resulting in degraded memory traces that are more difficult to retrieve. Some evidence suggests that cognitive speed and memory performance decline with age but that crystallized abilities remain largely intact in those who survive for long-term follow-up (Christensen, 2001). Variability in test scores for memory and speed increases with age. Poor health, fewer years of education, lower activity, the presence of the Apo E4 allele and blood pressure appears to predict faster cognitive decline (Christensen, 2001).

Decrements with age are observed in measures of episodic memory (verbal and nonverbal), visuospatial ability, confrontation naming and especially speeded psychomotor performance (Salthouse, 1992; 1994). In light of reduced processing resources that older adults are thought to have, age-related deficits in remembering to perform an action could be due to older adults having fewer resources available for the prospective memory task and not to inherent deficit in retrieving prospective

(Craik, 1986; Craik & Jennings, 1992). However, the deficits usually are small in magnitude and appear not to result in impaired function. Age-associated cognitive changes do not appreciably affect the ability to carry out usual activities of daily living (Rubin et al., 1993).

Research in which the effects of age on working memory were assessed has shown consistent age differences in working memory performance (Craik & Jennings, 1992; Salthouse, 1994). Children, adolescents and older adults all have smaller memory spans than do young adults. Relative to young adults, both children and older adults show more pronounced age deficits in spatial memory than in verbal memory (Hale et al., 1997). Although the mechanisms underlying the relationship between processing speed and working memory are not fully understood, it has been suggested that faster processors may activate and rehearse information quicker than slower processors (Dempster, 1981; Salthouse, 1994).

A potentially serious confound for aging studies relates to the possibility that study samples are note entirely normal but instead are contaminated with subjects who have unrecognized early-stage dementia. Because the incidence of AD increases dramatically with age, some supposedly "normal" subjects older than 75 years almost certainly include some with cognitive impairment of AD. This group of very mildly affected persons exaggerates the cognitive decline that is attributed to age. Underestimation of normal cognitive function in elderly subjects owing to the inclusion of poor performances of those with preclinical dementia limits the recognition of mild impairment because it falls in the normal range (Silwinski ct al., 1996). Subjects who are studied longitudinally and who later are known to have dementia may be excluded from these assessment in order to yield uncontaminated data.

The findings from several studies are consistent with the hypothesis that substantial cognitive decline does not occur in the majority of elderly individuals who escape developing a dementing disorder (Ganguli et al., 1996; Haan et al., 1999; Howieson et al., 1997). However, although subjective memory loss alone is unlikely to be a useful clinical predictor of dementia, memory complaints may predict the development of dementia and clinicians should monitor these elderly subjects more closely (St John & Montgomery, 2002). Further research is needed into further refining clinical predictors of those who will develop dementia.

6.10 Discussion

Whereas declarative memory is tied to a particular brain system, nondeclarative memory refers to a collection of learned abilities with different brain substrates. For example, many kinds of motor learning depend on the cerebellum, emotional learning and the modulation of memory strength by emotion depend on the amygdala and habit learning depends on the basal ganglia. These forms of nondeclarative memory that provide for myriad unconscious ways of responding to the world are evolutionarily ancient. By virtue of the unconscious status of these forms of memory, they create some of the mystery of human experience. From them arise the dispositions, habits, attitudes and preferences that are inaccessible to conscious recollection, yet are shaped by past events, influence our behavior and our mental life.

Cognitive decline related solely to aging remains a controversial topic. A clear conclusion from studies of neuropsychologic function in the elderly is that aging-related declines are not inevitable and that when they do occur, it is age-related diseases that are often responsible (Howieson et al., 1993; Rubin et al., 1998). Patients with memory complaints need a careful cognitive assessment. It is not appropriate to assume that memory complaints, at any age, are caused by senescence.

The working memory model has been used to explain STM deficit both in normal ageing and in Alzheimer patients and to determine which component or components might be affected in either case (Anderson & Craik, 2000; Craik, 1977, 1994; Vallar & Papagno, 1995). In the first case, the elderly have a poorer performance when, for example, they are asked to carry out two tasks (dual task procedure) indicating that the central executive is performing less adequately in the elderly than in the younger group. It has been suggested that low performance rates attained by AD patients in the Brown-Peterson task is due to a dysfunction of the central executive which would otherwise allow the subject to review the items while doing the distracting task (Morris, 1986). Manipulating the complexity of the task, it was observed that the more difficult it was, the lower was the performance of AD patients in serial recall of the items. Some studies have also examined the phonological loop and central executive functioning in AD and have noted a deficit in both components.

Results of studies performed with PET suggest left hemisphere preponderance for the retrieval of semantic information and a right hemispheric preponderance for the retrieval of episodic information (Fletcher et al., 1995; Tulving et al., 1994; 1996). The HERA hypothesis was based on findings that the left frontal region was particularly involved in the encoding of episodic memories whereas the right frontal region together with the precuneus was of particular importance in episodic retrieval. Lateralization of memory-related activations may not depend on whether encoding or retrieval are primarily being engaged but more on whether verbal or hard-to-verbalize materials are being processed and whether this processing relates primarily to encoding or retrieval may not matter (Kelley et al., 1998). Nolde et al. (1998) have also argued that the extent of left PFC activations during retrieval probably increases as the executive demands of retrieval increase. For example, "Yes/No" recognition increases left-sided activation relative to forced choice recognition and the same applies when free recall is compared with cued recall or when more complex kinds of information such as the temporal location or source of items have to be retrieved.

The ability to remember previously learned words relies on a distributed neural network that includes modules in the prefrontal cortex and MTL (Buckner & Wheeler, 2001). Neural neuroimaging studies of word retrieval in episodic memory tasks have documented several neural substrates underlying simple recognition, cued recall or free recall of previously studies words. One mechanism common to all three retrieval modes is the activation of anterior (Broadmann's area (BA10) and posterior (BA 44/6) dorsolateral prefrontal cortex modules, each of which contributes to distinct aspects of the recollective experience (Buckner & Wheeler, 2001). Activation of MTL structures is correlated with retrieval success and hippocampal recruitment is linked to the detailed recollection of an episode in contrast to a general sense of familiarity (Nyberg et al., 1996; Eldridge et al., 2000). A recent PET study suggests that word repetition and semantic encoding increase recall accuracy during subsequent word retrieval *via* distinct hippocampal mechanisms and that ventral tegmentum activation is relevant for word retrieval after semantic encoding (Heckers et al., 2002). These data confirm the importance of hippocampal recruitment during word retrieval and provide novel evidence for a role of brainstem neurons in word retrieval after semantic encoding.

Chapter 7
Mild cognitive impairment

7.1 Introduction

Identifying older people at high risk for AD is important for both the patients and their carers as it may allow therapeutic intervention in the very early stages of the disease which in turn may delay or even prevent the onset of the disease process. In the worst case it allows for the planning of patient care. Such identification is currently hampered by the use of inadequate and inconsistent criteria for MCI. Nine such criteria have been developed and a number of other terms have been used to describe cognitive dysfunction in older individuals. Table 7.1 describes these terms.

The term "preclinical" has been used to describe a period of mild cognitive decline prior to dementia diagnosis (Elias et al., 2000; Small et al., 2000). The concept of preclinical AD has been derived from epidemiological studies that showed that more than 10 years can pass between the onset of MCI and the diagnosis of dementia. Although preclinical AD cases have substantial SPs and NFTs, the disease process has not yet produced measurable neuronal loss and these individuals remain nondemented without cognitive impairment or decline (Morris & Price, 2001). There is converging evidence that preclinical AD is characterized by a common behavioral phenotype, with cognitive decline in several domains, predominantly in episodic memory (Almkvist et al., 1998). The decline appears to start many years before the clinical onset of AD. Moreover, the progression of the impairment appears to be continuous. The main opportunities for studying the preclinical stage of AD have come with longitudinal studies of aging and the study of individuals genetically susceptible to AD (Bondi et al., 1994; Hom et al., 1994; Newman et al., 1994). The study of individuals with mutations on the Apo E4 allele on chromosome 19 that appears to be a susceptibility gene for AD offers the opportunity to study people at risk by virtue of being carriers of this genotype (Corder et al., 1993). Study of elderly population is another useful procedure because of the particularly high risk of very late onset AD associated with advanced aging.

MCI is typically defined by the following criteria: complaints of memory problems, memory performance below age-based reference norms, normal performance in other cognitive domains, absence of impairment in instrumental and basic activities of daily living and no diagnosis of dementia (Peterson et al., 1997; Peterson, 2000). These criteria differ from those proposed for a variety of other approaches to MCI such as AAMI or AACD. AAMI requires defective memory performance relative to people under age 50 (Crook et al., 1986; Feher et al., 1994). AACD requires defective performance in any cognitive domain, with norms based on the perfor-

Table 7.1: Diagnostic and descriptive terminology for MCI in older people.

Terminology	Reference
Aging Associated Cognitive Decline (AACD)	Levy (1994)
Age-Associated Memory Impairment (AAMI)	Crook et al. (1986)
Age-Consistent Memory Decline (ACMD)	Crook (1993)
Age-Consistent Memory Impairment (ACMI)	Blackford and La Rue (1989)
Age-Related Cognitive Decline (ARCD)	American Psychiatric Association (1994)
Age-Related Memory Decline (ARMD)	Blesa et al. (1996)
Benign Senescent Forgetfulness (BSF)	Kral (1962)
Isolated Memory Decline (IMD)	Small et al. (1999)
Isolated Memory Impairment (IMI)	Berent et al. (1999)
Isolated Memory Loss (IML)	Bowen et al. (1997)
Limited Cognitive Disturbance (LCD)	Gurland et al. (1982)
Late Life Forgetfulness (LLF)	Blackford and La Rue (1989)
Mild Cognitive Impairment (MCI)	Reisberg et al. (1982)
Minimal Dementia (MD)	Roth et al. (1986)
Mild Neurocognitive Disorder (MND)	American Psychiatric Association (1994)
Questionable Dementia (QD)	Morris et al. (1993)

mance of age-matched elders (Levy, 1994; Richards et al., 1999). Yet another definition for MCI is "questionable dementia", which involves mild deficits in cognitive and functional status. This state is recognized in the 0.5 category of the CDR (Hughes et al., 1982). In spite of intense research, the boundary area between aging and AD remains poorly delineated. Accurate characterization of the limit between normal aging and the very early stages of dementia of the Alzheimer type has been difficult particularly because many of the early cognitive signs of dementia overlap with modifications observed in normal aging (La Rue et al., 1992). The best future opportunities for treatment of AD may exist in the presymptomatic, earliest stage of the disease before further, irreversible brain damage has occurred. However, there are no definitive early diagnostic markers for the disease. The early diagnosis of AD in the elderly is complicated by the need to differentiate common age-related cognitive decline from pathological deficits. A study reported that 30% of individuals over the age of 65 years were found to have cognitive deterioration without meeting diagnostic criteria for dementia (Ebly et al. 1995). The average age of these "cognitively impaired not demented" subjects was 80.5 years. It is unknown how many of these individuals will progress to dementia. In addition, large numbers of elderly persons fear that occasional lapses of memory presage impending cognitive decline. Therefore, the ability to identify individuals at very low risk for dementia could provide important reassurance to these persons.

There is some skepticism about the diagnosis of preclinical AD in a clinical setting. Some clinicians feel that preclinical AD cannot be distinguished from mild dementia. In this view, subjects with a diagnosis of preclinical AD are in fact demented subjects. In addition, it may be difficult to distinguish subjects with preclinical AD from subjects who have MCI that is nonprogressive and that is, for example, caused by normal aging or mild affective disorders. A major question therefore is whether preclinical AD is a diagnostic entity in a clinical setting. A study designed to investigate whether the preclinical stage of AD can be diagnosed in a clinical setting assessed the memory of 23 subjects with preclinical AD, 44 subjects with nonprogressive MCI and 25 subjects with mild AD-type dementia (Visser et al., 2001). Age and delayed recall performance were the most useful variables for distinguishing subjects with preclinical AD from subjects with nonprogressive MCI. The overall accuracy was 87%. The score on the Global Deterioration Scale and a measure of intelligence could best discriminate between subjects with preclinical AD and subjects with very mild AD-type dementia. The overall accuracy was 85%. Therefore, subjects with preclinical AD can be distinguished from subjects with nonprogressive MCI and from subjects with very mild AD-type dementia. These data suggest that preclinical AD is a diagnostic entity for which clinical criteria should be developed.

Follow-up of MCI subjects has demonstrated that progression toward a clearly demented state occurred in about two to three subjects (Rubin, 1989). Those who deteriorated, in retrospect, were considered to be in the "amnestic plateau" phase of AD (Morris et al., 1991). About 1/3 of AD subjects in fact show such an isolated mild memory deficit for the first few years and then manifest a more global cognitive decline (Haxby et al., 1992). Such decline is a predisposition to dementia since it has been reported that about half of MCI subjects develop dementia of the Alzheimer's type within 3 to 5 years of diagnosis (Chertkow, 1998).

7.2 Boundary between normal aging and MCI

The boundary between normal aging and very early AD is becoming a central focus of research. The transition from normal cognition to AD is gradual (Petersen, 2000; Ritchie & Touchon, 2000). There is no simple way to define the upper boundary of MCI in very old age. Usually, it is acceptable to define a limit by statistical means. If the mean test results for a patient are to be found between 1 and 2 standard deviations (SD) below the mean or below 1 or 1.5 SD (without a diagnosis of dementia), then MCI is diagnosed. However, it is a complex question whether it is correct to distinguish on one hand, a normal deterioration which advances with age and, on the other hand, degenerative diseases and lesion-related brain impairments which account also for a performance reduction.

The term MCI may, and often does, involve other cognitive domains besides memory including deficits in language, attention, motivation, affect and executive function. A number of studies have noted motor deficits in older people who meet criteria for MCI. Older adults meeting criteria for MCI performed worse than normal elderly patients on tasks involving fine and complex motor skills (mainly tests of manual dexterity) (Kluger et al., 1997). AD patients, by contrast, scored more poorly than normal elderly patients on tests of more rudimentary motor control. These findings suggest a gradient of motor as well as cognitive performance in which MCI patients again fall between normal elderly patients and people who meet criteria for AD.

MCI is considered to be a transitional stage between normal aging and dementia. MCI patients have a higher rate of progression to AD (12–15% per year) than elderly persons without memory disorders (1–2% per year) (Petersen et al., 1999; 2001). Some studies show that people with MCI who ultimately progress to AD show mild functional deficits (such as occasional need for help or need for cuing and supervision in activity) and reductions in physical activity before AD (Touchon & Ritchie, 1999; Friedland et al., 2001). The significance of MCI lies in the identification of patients at high risk for developing dementia and in the potential for treating these patients so as to prevent further decline. Subjects with MCI followed for up to 4 years have a high risk of progressing to dementia with an annual conversion rate ranging from 6% to 25% (Petersen et al., 2001). Therefore, these data strongly suggest that MCI, at least when it is restricted to people who meet criteria for CDR 0.5 is a dementia prodrome rather than a benign variant of aging.

Brain potentials and reaction time were examined in elderly controls and MCI using a target detection paradigm (Golob et al., 2002). Subjects listened to a sequence of tones and responded to high-pitched target tones that were randomly mixed with low-pitched tones. Measures were a pre-stimulus readiness (RP), post-stimulus potentials (P50, N100, P200, N200, P300) and reaction time. In the target detection task, MCI subjects were distinguished from healthy elderly controls by significantly larger P50 amplitudes and longer P50 latencies. Significantly longer P300 latencies and somewhat longer reaction times were also observed.

The differences in brain electrical activity between MCI and AD during rest condition (eyes closed) were analyzed in two studies (Jelic et al., 1996; Zappoli et al., 1995). They showed that there are clear differences in spectral power (relative power) of the theta band (4–8 Hz) between AD patients and subjects with MCI as well as between AD patients and healthy subjects (Jelic et al., 1996). Differences between subjects with MCI and healthy controls could not be observed in this study. However, Zappoli et al. (1995) found no differences in theta power (4–7.5 Hz) between AD patients and patients with MCI. Grunwald et al. (2002) suggest that neuropsychological tests are sensitive to early perceptive-cognitive and functional deficits in subjects with MCI.

According to Bozoki et al. (2001), elderly subjects with MCI involving domains other than memory are two or more times as likely to develop AD within 2-5 years compared to those with memory impairment alone. Recent clinical research has established that MCI, particularly when associated with verbal memory impairment, is a risk factor or precursor for AD (Bozoki et al., 2001; Petersen et al., 1999).

7.3 Neuropsychological markers

One means of clarifying this distinction is by identifying those cognitive changes that enable the prediction of which elderly persons will develop dementia before they come to clinical attention. One possible way of detecting highly sensitive markers of AD with tests such as the California Verbal Learning Test (CVLT) involves the administration of neuropsychological tasks to populations of apparently intact elderly individuals who are at high risk for the disease. A number of studies have reported that memory decline is one of the earliest changes in persons who become demented (Hanninen et al., 1995; Masur et al., 1994; Petersen et al., 1994). A prospective, longitudinal study of healthy individuals examined at yearly intervals with sensitive neuropsychological tests indicated that individuals who subsequently developed dementia showed evidence of verbal memory impairment at their initial examination which was of 2.8 years before clinical evidence of dementia (Howieson et al., 1997). Performance of individuals who did not develop dementia remained relatively stable during follow-up for up to 5 years. This study suggests that AD has a preclinical stage in which verbal memory decline is the earliest sign. The measures that showed significant group differences were the Boston Naming Test, CERAD Word-List Delayed Recall, WAIS-R Picture Completion and Block Design, and WMS-R Logical Memory I and II and Visual Reproduction I. Of these, the measure that contributed the most to the classification of subjects who subsequently became demented from those who did not was the story recall task (WMS-R Logical Memory II). Performance on this task was better at classifying who would subsequently become demented than age. The other cognitive variable that distinguished the subjects who subsequently became demented from those who maintained their cognitive status was confrontational naming (Boston Naming Test).

Normative data were collected in a study population of 150 randomly selected elderly subjects (Zaudig, 1992). Using the SIDAM (Structured Interview for the Diagnosis of Dementia of the Alzheimer Type, multi-infarct dementia, and dementias of other etiology according to DSM-III-R and ICD-10), both the dimensional and the categorical aspects of dementia and MCI were considered. With the SIDAM score (SISCO) (range 0–55, maximum, no cognitive impairment) and the SIDAM Mini Mental State Examination (MMSE) (range 0–30), appropriate cutoffs for the category of DSM-III-R and ICD-10 dementia and MCI were defined. MMSE scores of 0-22 were found to be indicative of DSM-III-R and ICD-10 dementia. For MCI, MMSE score of ≤ 22 was found to differentiate between DSM-III-R/ICD-10 dementia and MCI with a specificity of 92% (ICD-10, 95.6) and a sensitivity of 96% (ICD-10: 96%). For MCI, a SISCO between 34–51 was found. The SISCO covers a broader range of cognitive functions than the MMSE and is more useful in detecting even very mild cognitive decline. Furthermore, the newly defined category of MCI could be validated successfully by means of GDS stages 2–3 and CDR stage 0.5. These findings confirm the value of the SIDAM as a short diagnostic instrument for measurement and diagnosis of dementia and MCI.

Recent longitudinal studies have demonstrated that complaints of word-finding difficulty and reduced memory of recent events correlated with development of severe dementia to a greater degree than did early changes in affect (Persson & Skoog, 1992). The development of dementia in the elderly can be predicted with some success on the basis of changes in specific neuropsychological test scores well before the appearance of clinically significant cognitive change. Evaluation of a cohort of normal elderly persons with the Blessed Information-Memory-Concentration Test (BIMC) has been useful in identifying those at high risk for the development of SDAT (Blessed et al., 1968; Fuld, 1978, Katzman et al., 1989). Recall scores from both the Fuld Object Memory Evaluation (FOME) and the Buschke Selective Reminding Test (SRT) have predicted the development of dementia 1 year before clinical onset at a level of about four times the base rate of the disease for this population (Fuld, 1981; Buschke, 1973; Buschke & Fuld, 1974; Fuld et al., 1990; Masur et al., 1990). A preliminary study using the WAIS demonstrated that the Information and Digit symbol subtests were the most robust preclinical predictors of dementia (Masur et al., 1990a). Clinicopathologic evidence for the preclinical detection of dementia has been tentatively demonstrated by moderate negative correlations between recall scores on the FOME and the presence of cortical plaques in subjects with some degree of cognitive change but who did not develop dementia (Crystal et al., 1988; Fuld et al., 1987). In a study that assessed neuropsychological performance in 317 initially nondemented elderly persons and followed them for at least 4 years as part of the Bronx Aging Study, it was reported that it was possible to predict which patients would develop AD (Masur et al., 1994). Four measures of cognitive function from the baseline assessment (delayed recall from the Buschke SRT, recall from the FOME, the Digit symbol subtest from the WAIS and a verbal fluency score) can identify one subgroup with an 85% probability of developing dementia over 4 years and another with a 95% probability of remaining free of dementia. This study clearly demonstrates that baseline measures of cognitive function, often performed many years before the actual diagnosis of dementia, can provide important information about demential risk. The group likely to develop dementia becomes a target for preventive or early therapeutic interventions, and the group unlikely to develop dementia can be reassured.

In a study designed to determine the association between cognitive dysfunction and motor behavior in older adults, 41 cognitively normal elderly, 25 nondemented patients exhibiting MCI and at risk for future decline to dementia and 25 patients with mild AD were examined using a wide array of motor/psychomotor and cognitive assessments (Kluger et al., 1997). The outcome measures included 16 motor/psychomotor tests categorized a priori into gross, fine, and complex, as well as eight cognitive tests of memory and language. Relative to the normal group, MCI subjects performed poorly on cognitive, fine and complex motor measures but not on gross motor tests. AD patients performed worse on cognitive and all motor domains. Differences in complex motor function persisted after adjustment for performance on cognitive and on less complex motor tests. Motor/psychomotor assessments were observed to be comparably sensitive to traditional tests of cognitive function in identifying persons affected by the earliest stages of AD pathology and may improve identification of at-risk nondemented elderly.

A study performed in 1159 elderly residents around Bordeaux in south-western France, which used the logistic regression with backward stepwise approach, showed that the clinical diagnosis of dementia or AD was preceded by a preclinical stage of at least 2 years detectable by psychometric testing (Fabrigoule et al., 1998). It also showed that preclinical deficits in dementia and AD reflect the deterioration of a general cognitive factor, which may be interpreted as the disturbance of central, control processes. The digit symbol substitution test (DSST) and Isaacs Set Test (IST) scores characterize the principal component analysis (PCA) factor 1 better than the other tests test scores, since their loadings on this factor were at least twice as large as the loadings of any other factor. The Benton Visual Retention Test (BVRT) and Zazzo's Cancellation Task (ZCT) timescores are very close to fulfilling these criteria. From a neuropsychological perspective, this stage seems to be characterized by a homogeneous deterioration of cognition that could be interpreted as an impairment of control processes.

A longitudinal study of asymptomatic individuals at risk of autosomal dominant FAD was performed to assess clinical and neuropsychological features of the disease (Fox et al., 1998). Over a 6-year period, 63 subjects underwent serial assessments. Individuals who later became clinically affected already had significantly lower verbal memory ($p = 0.003$) and performance IQ ($p = 0.030$) scores at their first assessment when they were ostensibly unaffected. Subsequent assessments showed progressive decline in multiple cognitive domains. These findings imply that in FAD cognitive decline predates symptoms by several years and that verbal memory deficits precede more widespread deterioration.

A recent study that investigated the retrieval and encoding of episodic memory in normal aging and patients with MCI suggests that encoding of episodic memory may be impaired in MCI patients (Wang & Zhou, 2002). There was significant decline in the function of orientation, language and praxis besides memory impairment in the MCI group. Impairment of encoding and retrieval of episodic memory was also observed in the MCI group. It was suggested that MCI might be a transitional state from normal aging to AD.

Intellectual decline in certain domains has been described as an inevitable consequence of normal aging but the severity of these changes varies widely among individuals. Up to the age of 67, there is a 90% overlap in many cognitive dimensions between younger and older subjects. Studies using the WAIS suggest that the verbal IQ remains relatively stable over time while the performance IQ declines (Storandt, 1977). There is some decline of inductive reasoning test scores but performance remains fairly stable until the late 70s. Even by the age of 81, between 60–70% of people maintain their previous abilities over time.

7.4 Eyeblink classical conditioning

Eyeblink classical conditioning is a useful paradigm for the study of the neurobiology of learning, memory and aging which also has application in the differential diagnosis of neurodegenerative diseases expressed in advancing age.

Investigators who first reported adult age effects in eyeblink conditioning compared young and older adults in the delay paradigm in which the tone conditioned stimulus (CS) begins first and is then overlapped about half a second later with the corneal airpuff unconditioned stimulus (US). The main important finding was the relative inability of the older participants to acquire conditioned responses (CRs). In eyeblink classical conditioning, the learned response or CR is a blink response to the

CS before the onset of the US. The inability of older adults to acquire CRs in the delay eyeblink classical conditioning procedure is somewhat unexpected because classical conditioning is considered to be a relatively simple form of learning. When distinctions are made between declarative (explicit) and nondeclarative (implicit) forms of learning and memory with classical conditioning in the delay procedure labeled as a nondeclarative form, large age differences in eyeblink conditioning contradict the perspective that nondeclarative forms of learning and memory are relatively spared in old age (Woodruff-Pak & Finkbiner, 1995; Cohen & Squire, 1980; Squire & Zola-Morgan, 1991). The conclusion that nondeclarative forms of learning and memory were stable was based on studies of repetition priming (Graf, 1990; Light & La Voie, 1993). Data on eyeblink classical conditioning assessed over the entire adult age span indicated that the age differences appeared as early as middle age. Using the delay eyeblink conditioning paradigm with a 400-ms interval between a tone CS and corneal airpuff US in adults ranging from 18 to 83 years of age, it was found that age differences in acquisition in eyeblink conditioning emerged in the age decade of the 40s (Woodruff-Pak & Thompson, 1988; Solomon et al., 1989).

A case study was presented in which eyeblink conditioning detected impending dementia six years before changes on other screening tests indicated impairment (Woodruff-Pak, 2001). A participant who was tested on eyeblink classical conditioning over a 10-year provided evidence that performance in the 400-ms delay procedure detects AD early in its course before declarative memory measures show impairment. Because eyeblink conditioning is simple, non-threatening and non-invasive, it may become a useful addition to test batteries designed to differentiate normal aging from MCI that progresses to AD and AD from other types of dementia.

7.5 A standardized clinical assessment?

To identify aspects of a standardized clinical assessment that can predict which individuals within the category of questionable AD have a high likelihood of converting to AD over time, detailed semi-structural interviews were performed at baseline and annually for three years (Daly et al., 2000). Likelihood of progression to AD during the follow-up period was strongly related to the Total Box score. For example, more than 50% of individuals with a Total Box score of 2 or higher at baseline developed AD during the follow-up interval, whereas about 10% of individuals with a Total Box score of 1 or lower developed AD during this same period. Selected questions from the standardized clinical interview also were highly predictive of subsequent conversion to AD among the study population. Of the 32 questions examined, only 8 questions (see Tables 7.2 and 7.3) differed significantly between the normal, questionable and converter groups. These eight selected questions from the clinical interview at baseline, combined with the CDR Total Box Score, identified 88.6% of such individuals accurately. Five of the questions that significantly discriminated the three groups from one another pertained to the category of judgment and problem solving, two to home and hobbies and one to personal care. None of the questions that

Table 7.2: Some studies showing cognitive markers preceding AD

Cognition affected	Experimental design	Results	References
Verbal memory	Prospective, longitudinal study of healthy 139 individuals (65–106 years) examined at yearly intervals with neuropsychological tests selected to be sensitive to the early detection of dementia.	Individuals who subsequently developed dementia showed evidence of verbal memory impairment at their initial examination, which was a mean of 2.8 years before clinical evidence of dementia.	Howieson et al., 1997
General cognitive factor	The study of cognitive performances of 1159 elderly subjects (all > 65 years old) in the PAQUID (Personnes Agées quid) cohort, at a fixed lag time of 2 years before the clinical diagnosis of dementia.	The clinical diagnosis of dementia or AD was preceded by a preclinical stage of at least 2 years detectable by psychometric testing.	Fabrigoule et al., 1998

differentiated the three groups from one another pertained to the category of memory, orientation, or community affairs. For example, questions from the category of memory significantly differentiated the normal from the questionable group ($p < 0.001$) and the normal from the converter group ($p < 0.001$) but not the questionable from the converter group ($p = 0.08$). Therefore, a standardized clinical assessment can be used to identify the subgroup of individuals within the category of questionable AD who have a high likelihood of converting to AD over time. After a 3-year follow-up, 15% improved, 29% remained the same, and 55% had progressive difficulty with memory, including 19% who converted to AD. The CDR Total Box score at baseline, combined with eight selected questions from the clinical interview at baseline, identified 88.6% of such individuals accurately. This accuracy was based on a semi-structured evaluation that can be applied by a skilled clinician in any clinical setting.

7.6 Discussion

A reason to investigate MCI, especially in adulthood, is that job demands in modern life are much more cognitively demanding than in the past. MCI mostly leads to job loss and probably to early retirement (Reischies, 1999). Early detection of AD

Table 7.3: Questions that differed significantly between the normal, questionable and converter groups (Daly et al., 2000)

Groups	Questions
Judgment and problem solving	1. Does the subject have increased difficulty handling problems (e.g., an increased reliance on others to help solve problems or make plans)? 2. Is there a change in the pattern of driving not secondary to visual difficulty (e.g., increased cautiousness, trouble making decisions)? 3. Is the subject's judgment as good as before or is there a change? 4. Is the subject having increased difficulty managing finances (e.g., maintaining a checkbook, making complicated financial decisions, paying bills)? 5. Does the subject have more difficulty handling emergencies (e.g., makes unsafe decisions, needs increased cueing)?
Home and hobbies	1. Is the subject having increased difficulty performing household tasks (e.g., cooking, learning how to use new appliances)? 2. Has there been any change in the subject's ability to perform hobbies (e.g., decreased participation in complex hobbies, increased difficulty following rules of games, reading less or needing to reread more)?
Personal care	1. Does the subject now need prompting to shave or shower?

has multiple advantages. When new treatments are found for AD, it will be good to provide these to individuals at risk of developing the disorder to prevent that irreparable damage ensues; it may be possible in the future to prevent this disease in persons at risk. Identifying dementia in its preclinical stages will increase in importance as early pharmacological treatments of dementia emerge. The detection of subthreshold diseases as in the development of AD may be improved by the integrated information gained from neuropsychological examination and follow-up data such as the speed of cognitive or mnestic decline, brain imaging techniques and neurochemical markers. It is still unclear which is the earliest symptom of a deteriorating time course in MCI. Various parameters may be considered as follows: certain neuropsychological dysfunctions, the speed of cognitive decline or indicators of dysfunction of certain CNS subsystems, assessed in electrophysiological or functional magnetic resonance research or markers of neuronal dysfunction or neuronal damage, as a final common pathway of degenerating diseases.

Given the considerable differences between the inclusion and exclusion criteria of published classification systems, there is surprising consensus regarding the cognitive, genetic and cortical correlates of MCI. Episodic memory impairment, hippocampal atrophy and the Apo E4 allele are all consistently shown to be associated with cognitive impairment in older people. Although there is substantial variability, the rate of expression of these outcome measures in older people with MCI is broadly similar to that observed in patients with clinically diagnosed AD. For example, criteria that require the individual to display moderate deficits report a higher rate of progression to AD, and a lower estimated rate of prevalence than classification systems that require only mild impairment (Almkvist et al., 1998). A large proportion of older individuals with MCI do not progress to develop clinically recognizable AD, regardless of the severity of their deficit (Collie & Maruff, 2000). Two reasons may explain such findings. First, the period of time between classification of impairment and determination of clinical outcome in many studies may not have been sufficient for all incipient cases of AD to be expressed clinically. Second, the systems for classifying cognitive impairment may not be specific to the preclinical stages of AD. As a result of these factors, accurate and consistent estimates of the outcome of MCI in otherwise healthy older people are yet to be obtained.

Changes in frontally mediated behaviors are common in very early and mild stages of cognitive impairment, even before functional decline in daily living is evident (Ready et al., 2003). These behaviors deserve more study in MCI because they may have implications for prognosis, treatment adherence, family distress, and patient quality of life. People with MCI are not sure what to make of their conditions, whether it is part of normal aging or the first signs of AD. The label "MCI" gives them a medical condition without a clear guidance on how to interpret or treat it. Patients are left with a disease label with unclear clinical significance. People with MCI then are placed in the difficult position of knowing that their memory ability is not what it should be but not knowing whether these deficits matter for daily activity. They do not perceive major changes in activity but report some concern that they may already have begun to alter daily patterns of activity to avoid challenges and situations that might expose memory problems.

Numerous studies have demonstrated that a large proportion of patients diagnosed with mild to moderate AD lack full awareness of their progressive cognitive decline (DeBettignies et al., 1990; Kotler-Cope & Camp, 1995; Kiyak et al., 1994). A study designed to assess the predictive utility of self-reported and informant-reported functional deficits in patients with MCI for the follow-up diagnosis of probable AD showed that in patients with MCI, the patient's lack of awareness of functional deficits identified by informants strongly predicts a future diagnosis of AD (Tabert et al., 2002). In practice, it is often difficult to recognize when a patient with memory complaints has a process that is likely to progress to further intellectual and functional decline. Because there is no diagnostic test to identify dementia, the clinician must rely on a careful and accurate history from the patient and a reliable informant as well as a good mental status examination. Clinicians who assess

MCI patients should obtain both self-reports and informant reports of functional deficits to help in prediction of long-term outcome. Another interesting longitudinal study directly compared change in different cognitive abilities and other key clinical milestones in individuals with MCI to those without cognitive impairment (Bennett et al., 2002). The results showed that on average, subjects with MCI had significantly lower scores at baseline in all cognitive domains. Over an average of 4.5 years of follow-up, 30% of subjects with MCI had died, a rate 1.7 times higher than those without cognitive impairment. The results also indicated that 34% of subjects with MCI developed AD, a rate 3.1 times higher than those without cognitive impairment. Furthermore, subjects with MCI declined significantly faster on measures of episodic memory, semantic memory and perceptual speed but not on measures of working memory or visuospatial ability as compared with normal control elderly subjects. These data suggest that MCI is associated with an increased risk of death and of AD and a greater rate of decline in selected cognitive functions. MCI subjects tended to be younger, to have lower levels of education, more frequent depressive symptoms and slightly higher family histories of AD (Meyer et al., 2002).

Chapter 8
Cognitive impairment in Alzheimer disease

8.1 Introduction

The most important cognitive deficit of AD is the progressive loss of memory that is manifested according to the traditional classification in both STM and LTM tasks (see Tables 8.1, 8.2 and 8.3). The earliest and most prominent symptom in AD is a profound impairment in the ability to acquire and remember new information whether tested by recall or recognition (Grady et al., 1988; Welsh et al., 1991). Many patients have purely episodic memory impairment for a number of years (Perry & Hodges, 2000; Perry et al., 2000). Semantic memory, the database of conceptual knowledge that gives meaning to sensory experience, is eventually affected in AD, but early in the disease patients show mild and variable impairment of semantic memory. Written language deficits are frequently noted and occur at early beginning of the disease (Eustache & Lambert, 1996). They initially take the form of a lexical agraphia: the characteristic symptom is the production of errors so called "de régulation" (of regulation) occurring during the writing of words with irregular orthography (e.g. femme > fame). On the contrary, deficits of phonological type often appear with the progression of the dementia.

A number of cross-sectional studies suggest that the pattern of cognitive impairment differs in early- *versus* late-onset AD. A recent study examined the relation of age at onset and visuocognitive disturbances in AD using a large sample of patients, quantitative neuropsychological measures and multivariate statistics controlling for gender, education, stage of dementia and disease duration (Fujimori et al., 1998). The results indicated that the patients with early-onset AD performed worse than late-onset patients on digit span (forward and backward), visual memory span (forward 0–14, backward 0–12), visual counting (0–14), block design of WAIS-R (0–61) and Rey-Osterreith figure-copy (0–36) tests. These findings confirm the greater attentional and visuospatial impairments in early onset patients when all confounding factors are controlled. Neuropsychological studies of patients with AD have also revealed selective and heterogeneous patterns of cognitive impairments. For example, some patients present severe language dysfunction along with milder

Table 8.1: Distinguishing features of early dementias.

	Executive function	Visuospatial	Declarative memory	Language	Behavior
Alzheimer disease	Preceded by memory loss	Early topographic disorientation	Early, STM loss > LTM loss (–) cueing	Poor word list generation	Socially appropriate; late agitation; misidentification
Vascular dementia	Variable	Variable	Variable; (+) cueing	Aphasia if cortex involved	Apathy or depression
Dementia with Lewy bodies	Impaired	Impaired	Fluctuating alertness; memory spared	Slower	Hallucinations, delusions
Fronto-temporal dementia	Early decline	Spared	Decreased concentration > STM loss; (+) cueing	Unrestrained but empty. Aphasia may precede dementia.	Early: disinhibition, hypochondriasis, affective disorders, mania

visuoconstructional dysfunction, whereas other patients present with the opposite pattern (Haxby, 1990). Such difference in the pattern of cognitive deficits suggests a variable distribution of neocortical abnormalities that can be examined with the magnetic resonance spectroscopy technique. In fact, several studies using PET have shown that early-onset patients demonstrated significant parietal hypometabolism (Grady et al., 1987; Mielke et al.,1991; Small et al., 1989).

8.2 Cognitive impairments

Cognitive disorders characteristic of AD include impairment of declarative memory, semantic aspects of language (naming and comprehension), visuospatial skills, arithmetic abilities and executive function. Patients with AD have been shown to have intact functioning of the articulatory rehearsal and phonological storage components of working memory, but impairments in the functioning of the central executive (Baddeley et al., 1991). Although speech may be fluent, it is rather empty of meaning or the patient may present with some difficulty finding words. Visuospatial deficits are often manifest by impairment of topographical memory, when patients easily gets lost. A number of studies have indicated that careful neuropsychological testing can accurately identify individuals who are experiencing mild (and even unrecognized) cognitive impairment (Rubin et al., 1989; Masur et al., 1994; Green et al., 1995; Jacobs et al., 1995; Linn et al., 1995).

AD patients almost invariably show severe deficits when delayed recall is assessed, but immediate recall is less typically affected (Welsh et al., 1991). LTM impairments, reflecting pathological changes in structures of the medial temporal lobes, are believed to be the primary determinant of poor delayed recall in these patients. Immediate free recall is thought to depend only in part on LTM and immediate recall decrements may involve additional deficits in STM or other cognitive domains (Pepin & Eslinger, 1989; Diesfeldt, 1978). Severe deterioration of episodic and semantic memory occurs very early in the AD process while working memory shows a gradual deterioration over time. Some aspects of working and implicit memory can be spared in the mild to moderate stages of AD.

Difficulty with the acquisition of new information is generally the first and most salient symptom to emerge in patients with AD. When clinical neuropsychological tests are used to evaluate memory in AD patients, it is clear that recall and recognition performance are impaired in both the verbal an non-verbal domain (Wilson et al. 1983; Storandt & Hill, 1989). Experimental studies have examined AD patients to determine whether the manner in which information is lost over brief delays is unique in any way to this patient group. Patients with AD were compared to a group of amnesic patients who had alcoholic Korsakoff's syndrome (KS), a group of dementing patients with Huntington's disease (HD), and a group of normal controls (NC) (Moss et al., 1986). All of the subjects were administered the delayed recognition span test (DRST). All of the patient groups were impaired in their recognition performance with respect to controls, but there was overlap among the patient groups. There was no significant difference among the three patient groups in their ability to recognize new spatial, color, pattern or facial stimuli; patients with HD performed significantly better than the other two groups when verbal stimuli were used. However, when fifteen seconds and 2 minutes after completion of the last verbal recognition trial, the subjects were asked to recall words that had been on the disks, the AD patients differed considerably from the other patients. They recalled significantly fewer words over this brief delay interval (2 min) than either HD or KS patients. Although all three patient groups were equally impaired relative to normal controls at the 15-s interval, patients with AD recalled significantly fewer words than either the HD or KS groups at the 2-min interval. In fact, only the AD group performed significantly worse at the longer, as compared with the shorter interval. This pattern of recall performance demonstrated that patients with AD lose more information over a brief delay than other patients with amnesic or dementing disorders.

A comparison of patients with AD and patients with FTD also demonstrates the severe recall deficits of the AD patients (Moss & Albert, 1988). Patients with AD and patients with FTD, equated for overall level of cognitive impairment, were administered the DRST. As in the previous study, the difference in total recall between the 15-s and the 2-min delay interval (i.e. the savings score) differentiated the groups. The retention of the FTD patients over this delay interval approached normality, whereas the AD patients lost a substantial amount of information. In

Table 8.2: Summary of memory systems in AD.

Memory system	Tests or paradigms	Changes in minimal AD	Changes in mild AD	Changes in moderate AD	References
Working memory					
Central executive system	Dual task	↓	↓	Not tested	Chen et al. 2002
	Digit span	↓	↓	↓	Reid et al., 1996
Visuospatial sketchpad	Corsi's blocks test	↓	↓	⇓	Perry & Hodges, 2000
Articulatory loop	Recency effect	Preserved	Preserved	↓	
	Phonological similarity effect	Preserved	Preserved	Not tested	
	Word length effect	Preserved	Preserved	Not tested	
Long-term memory					
Explicit memory					
Episodic	Encoding	⇓	⇓	⇓	Albert, 1996
	Free recall	⇓	⇓	⇓	
	Cued recall	↓	⇓	⇓	
	Recognition	↓	↓	⇓	
Source	Recall of contexts of learning	preserved	preserved	preserved	
Semantic	Category fluency tasks	↓	↓	⇓	Brandt et al., 1988
	Pyramids and palm trees test	↓	↓	⇓	Lekeu et al., 2003
Autobiographical	Autobiographical incidents schedule:				
	Childhood	↓	↓	Not tested	Fromholt & Larsen, 1991
	Early adult life	↓	↓	Not tested	
	Late adult life	⇓	⇓	Not tested	Greene et al., 1995

Table 8.2 (continued)

Memory system	**Tests or paradigms**	**Changes in minimal AD**	**Changes in mild AD**	**Changes in moderate AD**	**References**
Personal semantic	Personal semantic memory schedule:				
	Childhood	↓	↓	Not tested	Kazui et al., 2003
	Early adult life	↓	↓	Not tested	
	Late adult life	⇓	⇓	Not tested	
Prospective	Subtests of the Rivermead behavioral test	↓	↓	Not tested	
Implicit memory					
Skill learning	Pursuit-rotor task	preserved	preserved	preserved	Jelicic et al., 1994
Verbal priming	Mirror reading task	preserved	preserved	preserved	
Data-driven processes	Word identification tasks	preserved	preserved	preserved	Ergis et al. 1998
	Incomplete picture tasks	preserved	preserved	preserved	
	Word-stem completion tasks	preserved	preserved	preserved	
Conceptually-driven processes	Word-stem completion tasks	↓	↓	↓	

↓ = *mild or moderate deficit;* ⇓ = *severe deficits.*

Table 8.3: Suggested neuropsychological assessment of dementia at early stage.

Cognitive function assessed	Potential tests	Comments	References
Overall severity of dementia	Mattis Dementia Rating scale	AD patients performed significantly worse than patients with Lewy bodies pathology on the Mattis Dementia Rating scale	Mattis, 1976, Connor et al., 1998
Short-term memory	Digit span (Wechsler Memory scale) Block tapping test	With an earlier age at onset, significantly more impairment on tests of digit span and praxis was observed.	Reid et al., 1996 Spinnler & Tognoni, 1987
Verbal LTM Organized information	FCSRT Logical memory (Wechsler memory scale)	Tests of immediate and delayed recall, administered over brief delays can be used to differentiate AD patients from controls, and from patients with a wide variety of dementing disorders.	Grober et al., 1988, 1997 Albert, 1996
Visual LTM	Subtest of the Wechsler memory scale revised		
Intelligence	Subtest of the WAIS		
Frontal lobe test	Stroop test Digit cancellation test Go-no Go test	Left-handed patients with AD do not differ from right-handed patients in the severity or pattern of neuropsychological deficits.	Stroop, 1935, Spinnler & Tognoni, 1987 Doody et al., 1999, Muller et al., 1993
Motor speed	Finger tapping		
Constructional abilities			
Spontaneous speech			
Confrontation naming			
Comprehension	Token test, short version		De Renzi & Faglioni, 1978

general, these findings suggest that the nature and severity of the AD patients' memory disturbance, in relation to delays spanning the first 10 min after encoding, is likely to be the result of a unique pattern of neuropathological and/or neurochemical dysfunction.

8.2.1 Executive function

Executive function associates the capacity to perform the elements of a complex task with the orchestration of the task and its actual execution. Patients with impaired executive function behave like 6-year-olds: they can zip their zippers, button their buttons and tie their shoes but they cannot get dressed by themselves. Executive dysfunction is present in the very old and in patients with disease involving the frontal lobe and its subcortical circuits including uncontrolled hypertension.

8.2.2 Working memory deficit

The deterioration of working memory in AD has been observed repeatedly and studies carried out using a dual task paradigm have shown a selective deterioration of the Alzheimer group under these conditions (Baddeley et al., 1986; Spinnler et al., 1988).

AD patients exhibit both a decrease in their span capacity and a lower performance on the classical Brown-Peterson procedure than that of normal aged-matched controls (Sullivan et al., 1986; Wilson et al., 1983). These anomalies in short-term recall have been attributed to a dysfunction of the attentional component of working memory.

8.2.3 Episodic memory deficit

Episodic memory loss is thought to be followed by progressive cognitive decline in other domains including semantic memory, attention, language, central executive function, visuospatial and perceptual abilities and abstract thinking (Perry et al., 2000; Elias et al., 2000).

8.2.4 Semantic memory deficit

Although the characteristics of the memory deficits that accompany AD are comparable in many respects to those of other amnesic disorders, patients with AD may exhibit a disproportionate disruption of semantic memory processes, as evidenced by poor performance on confrontation naming and word fluency tasks (Martin & Fedio, 1983; Nebes, 1989). A priming task involving a word-stem completion paradigm was administered to patients with AD, patients with HD and normal control subjects (Randolph, 1991). The task was done under conditions of both implicit and explicit recall. Explicit and implicit recall were positively correlated in all three groups. After controlling for explicit recall ability through statistical procedures, AD patients were found to be normally susceptible to the effects of priming on implicit recall. HD patients, however, exhibited significantly increased susceptibility to priming, suggesting that they may have carried out the implicit task in a manner different from that of normals and AD patients. AD patients were also found to supply words of significantly lower association strength than the other two groups in a "free association" task using words from a published list of word association norms. This degradation of semantic memory was strongly correlated with explicit performance, suggesting that explicit, implicit, and semantic memory functions decline in parallel in AD.

One current controversy revolves around the presence and specificity of semantic memory impairments in AD. Semantic priming paradigms used in recent investigations of AD and other memory-impaired patients generally fall into two classes, those that assess the influence of priming on subsequent response speed and those that assess the influence of priming on the subsequent retrieval of words. The majority of studies that have looked at the influence of semantic priming on subsequent processing speed have concluded that AD patients and controls benefit in a similar fashion from priming manipulations (Nebes et al., 1986, 1989). On the other hand, studies investigating the influence of semantic priming on subsequent measures involving the retrieval of words have concluded that AD patients exhibit a deficit relative to controls, patients with HD and patients with KS (Heindel et al., 1989; Salmon et al., 1988).

8.2.5 Naming deficit

Although it is now well established that progressive naming difficulties are one of the main characteristics of AD, the underlying functional deficit causing the naming problem is not well understood. Hypothetically, the naming deficit in AD may reflect deficient visual channels or a loss of semantic knowledge. A study designed to assess the source of the naming deficit in AD showed that AD subjects display more overall naming errors than vascular dementia subjects (Lukatela et al., 1997). The AD group made significantly more errors than the vascular dementia group across the most error types (i.e. omissions, visual, semantic). However, the pattern of errors was similar between the two groups. If it is assumed that semantic knowledge is hierarchically organized, these data suggest that in early AD access to lower and more detailed nodes is more impaired than access to higher-level concepts. Both visual and semantic factors seem to contribute to the naming deficit in AD.

A study designed to examine a possible naming improvement in mild AD patients using a computer-aided therapy method based on episodic and semantic lexical training was performed (Ousset et al., 2002). It was shown that lexical therapy may help treating anomia in mild AD patients and its efficacy relies on episodic linking between objects and lexical labels.

8.2.6 Intrusion errors

Not only are people with AD unlikely to recall much information, they are also prone to making errors. Clinical studies indicate that people with AD make many more intrusion errors than controls (Jacobs et al., 1990). There is also evidence that intrusion errors are particularly diagnostic of AD. A correlation between the number of intrusion errors and the biological features of the disease (SP counts and levels of the enzyme ChAT) was demonstrated (Fuld et al., 1982). Research has also shown that even when matched for recall levels, AD patients make more intrusion errors than patients with a sub-cortical dementia, HD (Kramer et al., 1988).

Two prominent explanations of intrusion errors have been exposed. Because the intrusion errors in AD are predominantly for semantically related items (e.g. recalling diamond instead of sapphire), the propensity to make intrusion errors is believed to reflect a disruption in semantic memory processing (Brandt & Rich, 1995). Intrusion errors have also been proposed to be due to an inability to monitor output during retrieval (Correa et al., 1996). Either account could arise from a deficit in inhibitory processes in AD. AD patients may fail to inhibit other semantic activations during retrieval or fail to inhibit output. However, a recent study that used the retrieval-induced forgetting paradigm that is ideal to assess inhibition in episodic memory showed that AD patients have normal levels of inhibition (Moulin et al., 2002). Therefore, a deficit in inhibitory processes during retrieval is not the reason for the high levels of intrusion errors made in recall in AD.

8.2.7 Visuo-spatial deficit

Studies that have examined visuospatial dysfunction find it to be common sequelae of AD although it has received relatively little attention (Cogan, 1985; Mendez et al., 1990). Early in the disease, visual disturbance is considered to be minimal or even absent. In the later stages of the disease, impairments have been described in visual acuity, visual fields, color vision and depth perception, all of which contribute to measurable visual perceptual impairment that in turn affects object recognition and word-reading ability. Neuropathological studies of regional and laminar changes in AD indicate significant involvement of secondary and association cortices with relative sparing of primary sensory areas (Lewis et al., 1987; Pearson et al., 1985). These areas are believed to mediate higher-order perceptual functions including visual perceptual threshold, color discrimination and contrast sensitivity function (Turvey, 1973; Bulens et al., 1989).

Visuospatial skills are easily tested by having patient copy a three-dimensional cube or with the less challenging skill of copying the intersecting pentagons that appear in the MMSE.

8.2.8 Language

Undisturbed language production and verbal communication require multiple and complex functions such as capabilities of perception, articulation, memory or concentration. In diseases such as AD that may affect a large number of cortical structures, language impairment presumably results from a disturbance of several mental functions and their interaction rather than from an impairment of one or few defined cortical regions. Disturbance of language is an intrinsic part of AD and appears to be among the earliest of symptoms, showing progression with severity of language decrement often correlated with severity of illness (Kertesz, 1994; Kirshner, 1994). A study that discussed the potential of various assessments of language function in the diagnosis of AD showed that AD patients have distinct semantic speech disturbances whereas they are not impaired in the amount of produced speech (Bschor et al., 2001).

The speech of patients with AD is often described as "empty" as it contains a high proportion of words and utterances that convey little or no information. For example, AD patients produce a low ratio of propositions to words and the thoughts they express are often left incomplete (Hier et al., 1985). However, perhaps the single most obvious characteristics of AD patients" so-called empty speech is the overuse of "empty words" ("thing", "it" etc.) (Kempler, 1995). Semantic impairments could cause the empty speech associated with AD.

Word-finding difficulties represent one of the prominent deficits in AD, even at an early stage and frequently become a major problem with progression of the disease. This word finding difficulty is not merely the result of general forgetfulness that may make patients forget what they wanted to refer to because patients often exhibit word finding difficulty even in the presence of a referent or its picture (Bayles et al., 1989). Patients often have more difficulty with referents from only specific semantic categories (e.g., living things) (Gonnerman et al., 1997). Rather, the word finding difficulty is likely caused by a genuine semantic impairment in processing lexical information from semantic LTM. Although a debate still exists regarding the nature of the underlying anomic deficit, there is a general agreement emerging from numerous works that the naming problem in AD can be related to impaired semantic processing or to impaired binding between the semantic and lexical level, the phonological level being spared (Hodges et al., 1991; Astell & Harley, 1996). Looking at future directions for research, language in AD as it varies according to stage of illness is as yet poorly defined. Understanding the progression of language impairment in AD can be best achieved through follow-up or longitudinal investigations of which there are at present very few (Appell et al., 1982; Kertesz et al., 1986; Bayles & Kaszniak, 1987).

Although normal women are known to have a modest superiority in language compared with men, several investigators have reported that AD has a greater negative impact on language skills of women patients (Henderson & Buckwalter, 1994). Ripich et al. (1995) reported that naming and word recognition skills of women are more affected by the neurodegenerative disease processes than those of men, after making gender performance comparisons on the Boston Naming Test,

Peabody Picture Vocabulary Test-Revised, Word Fluency Test and shortened Token Test (Spellacy & Spreen, 1969; Dunn & Dunn, 1981). However, Bayles et al. (1999) questioned the conclusion that language is more affected in women with AD. Their studies showed that that there is no evidence of significant gender effects in either cross-sectional or longitudinal analyses.

8.2.9 Memory distortions

Patients with probable AD suffer from distortions of memory in addition to their failure to retrieve desired information (Förstl et al. 1994). Because patients may believe that they turned off the stove or took their medications when they have only thought about performing these activities, memory distortions may impair the ability of AD patients to live independently. Memory distortions in AD patients are an important clinical problem. The etiology and treatment of such distortions remain largely unexplored.

AD patients have been examined using paradigms that allow measurement of a type of memory distortion known as false recognition. False recognition occurs when people incorrectly claim to have previously encountered a novel word or event. These studies have shown that AD patients show either greater or lesser levels of false recognition and recall than healthy older adults depending on the particular paradigm and analysis used. For example, using corrected recognition scores to control for unrelated false alarms, AD patients were noted to exhibit lower levels of false recognition of perceptually related novel objects compared with older adults (Budson et al., 2001). Watson et al. (2001) demonstrated that AD patients and older adults showed similar rates of false recall of semantic associates, phonological associates and hybrid lists combining semantic and phonological associates. More impaired AD patients, however, showed higher levels of false recall relative to their true recall performance when compared with the less impaired AD patients.

Budson et al. (2002) reported the use of a false recognition memory test in a clinical trial of patients with AD. Tests of false recognition allow assessment of two components of memory: the specific details of a prior encounter with a particular item (item-specific recollection) and the general meaning, idea or gist conveyed by a collection of items (gist memory). Because cognitive enhancing medications used to treat AD may improve gist memory or item-specific recollection preferentially, use of this type of paradigm may allow more sensitive detection of drug effects than standard memory tests.

8.2.10 Implicit memory performance

Individuals with AD often perform normally on perceptual implicit memory tests (see Fleischman & Gabrieli, 1998). For perceptual implicit memory tests, participants are engaged in processing perceptual information at test (e.g., identifying words or pictures) and memory is measured either as a reduction in response time or as an increase in accuracy that is associated with items that have previously been presented. This facilitation in performance is referred to as priming. AD patients have shown normal priming in the following perceptual implicit memory tasks: word pronunciation, lexical decision, perceptual identification and stem completion (Keane et al., 1994; Fleischman et al., 1997).

In contrast to their generally normal performance on perceptual implicit memory tests, AD patients often exhibit pronounced priming deficits on conceptual implicit memory tests (see Fleischman & Gabrieli, 1998). Conceptual tests are tests that provide a retrieval cue that is related to previously studied words by meaning, rather than by perceptual form. The demonstration of mnemonic effects on implicit conceptual tests is referred to as the “conceptual priming effect”. This effect is defined as the more frequent occurrence of responses for studied than for unstudied words on implicit conceptual tests of memory where no reference is made to the study episode. For example, Salmon et al. (1988) used a free-association task in which participants were asked to generate associates to target words. They observed that elderly control participants were more likely to produce items that were presented in an earlier study list but that AD patients failed to show such an effect. Similar conceptual priming deficits have been reported in other studies with this task (Carlesimo et al., 1995; Vaidya et al., 1999). Conceptual priming deficits in AD patients have also been reported for the exemplar generation tasks in which participants are asked to produce examples of semantic categories and priming on this task is measured as the likelihood of generating a studied compared with a non-studied target item in response to the category cue (Vaidya et al., 1999).

It is important to distinguish between “conceptual priming tasks” and “semantic priming tasks” as the effects of AD on these two types of memory tasks may be different (Ober & Shenaut, 1995; Shenaut & Ober, 1996). Semantic priming tasks involve facilitation between two different, semantically related items that is measured over a very short interval. For example, on semantic priming task the word “doctor” is identified more rapidly if it is immediately preceded by the word “nurse”. Results suggest that semantic priming remains intact in early AD (Ober & Shenaut, 1995).

8.3 Cognitive differences in neurodegenerative disorders

In diagnosing the cause of dementia, it is important to distinguish between failures of storage (or retention), associated with damage to the limbic and especially hippocampal structures, retrieval associated with frontal-subcortical dysfunctions and STM associated with temporo-parietal lesions. Storage disorders are characterized on testing by deficits in both recall and recognition and rapid loss of information. Retrieval disorders are characterized by a difficulty in accessing information and short-term memory disorders are characterized on testing by reduced memory span and rapid loss of information measured by the Brown-Petersen paradigm (Petersen & Petersen, 1959).

Some studies have found that declarative LTM functions are more intact in FTD patients compared with patients diagnosed clinically with AD (Binetti et al., 2000; Rahman et al., 1999). It has been proposed that because MTL involvement is an early and almost universal feature of the pathology in AD but is less common in FTD, then FTD patients should retain greater capacities for encoding new information into LTM, at least early in the disease (Lavenu et al., 1998). FTD patients' retained capacity for encoding new information into long-term declarative memory is likely due to relatively spared medial temporal lobe involvement.

The most likely explanation for the abnormalities in memory that characterize the early stage of AD pertains to the damage to the hippocampal formation seen in these patients (Hyman et al., 1985; 1986). It has been suggested that abnormalities in these regions produce a functional isolation of the hippocampus. These findings suggest that neuropathological damage to these medial temporal lobe structures may be responsible for the marked STM impairment evident in the early stages of AD.

Although episodic memory deficits occur in both cortical and subcortical dementia syndromes, these deficits result from damage to different memory systems. The deficit in AD is a true impairment in the episodic memory system due to direct damage within the medial temporal lobes. Consequently, AD is both quantitatively and qualitatively similar to the consolidation deficit that occurs in patients with circumscribed amnesia. Thus, like amnesic patients, AD patients show a little improvement in acquiring information over repeated learning trials, a tendency to recall only the most recently presented information in free recall tasks, a failure to demonstrate normal improvement in performance when memory is tested with a recognition rather than a free recall format and a rapid forgetting of information over time. AD patients also display a severe retrograde amnesia that is temporally graded, with memories from the distant past being better remembered than more recent memories.

8.3.1 Typical AD

In general, AD has an insidious onset, followed by gradually progressive cognitive deterioration with little or no focal neurological symptoms and signs, and it has typical degenerative features. It can be separated into predictable clinical stages ranging from prodromal MCI to mild, moderate and profound dementia. On the other hand, vascular dementia is supposed to have an abrupt onset of dementia, followed by stepwise deterioration of cognitive performance associated with neurological signs and symptoms reflecting focal brain lesions (Hachinski et al., 1974).

The first symptom of typical AD is impairment of recent memory: poor learning and retention of information over time (McKhann et al., 1984) (see Table 8.4). Patients with AD show poor learning over repeated trials and may make intrusion errors (Butters et al., 1995). This is a disorder of storage, retrieval and later of STM. The test for delayed recall has been found to be the best overall discriminatory measure to differentiate patients with early AD from cognitively normal elderly controls

Table 8.4: Distinguishing features of early dementias

	AD	FTD	DLB
Memory	Early, short term loss > long term; (–) cueing	Decreased concentration > short term memory loss; (+) cueing	Fluctuating alertness; memory spared
Executive function	Preceded by memory loss	Early decline	Impaired
Language	Poor word list generation	Aphasia may precede dementia	Slower
Behavior	Socially appropriate; late agitation; misidentification	Early disinhibition, hypochondriasis, affective disorders, mania	Hallucinations, bizarre, delusions

with the Committee of the Consortium to establish a registry for Alzheimer disease (CERAD) battery (Albert, 1996; Locascio et al., 1995; Welsh et al., 1991, 1992).

8.3.2 Frontotemporal dementia

The paradigm of frontal lobe dementia is that described by Pick in 1892, which was associated with circumscribed atrophy of both the frontal and temporal lobes. This form of dementia is much less common than AD. It is more frequent in women. It may be inherited through a single autosomal dominant gene, although most cases are sporadic. The diagnosis of FTD requires the presence of frontally predominant atrophy on MRI, frontal hypoperfusion on SPECT, or frontal hypometabolism on PET.

There are distinguishing features that reflect the underlying pathologic changes of Pick's disease and separate it from AD. In particular, abnormalities of behavior, emotional changes and aphasia are frequent features. Some authors have noted elements of the Kluver-Bucy syndrome at one stage or another in the disease (Cummings & Benson, 1983). Interpersonal relationships deteriorate, insight is lost early, and the jocularity of frontal lobe damage may even suggest a manic picture. The aphasia is reflected in word-finding difficulties, empty, flat, nonfluent speech and aphasia. With progression, the cognitive changes become apparent: these include memory disturbance but also impairment of frontal lobe tasks. Ultimately, extrapyramidal signs, incontinence, and widespread cognitive decline are seen. Considering the possibility to discriminate further FTD from AD, recent studies indicate that, whereas personality features and neuroimaging findings distinguish these dementias, neuropsychological measures including conventional tests of frontal functions fail to distinguish between FTD and AD (Mendez et al., 1996; Pachana et al., 1996).

It has been hypothesized that the cognitive estimations test (CET) could help in the differential diagnosis of FTD from AD (Mendez et al., 1998). CET is a measure of judgment and reasoning, frontal functions that are disproportionately affected in FTD (Lund & Manchester Groups, 1994). The CET was administered to 31 FTD and 31 AD patients of comparable dementia severity plus 31 normal elderly controls. Both dementia groups gave significantly more extreme estimates on the CET than did controls and AD patients gave more extreme estimates than did FTD patients. The CET may be particularly impaired in AD because it reflects impaired memory and impaired numerical ability as well as disturbed judgment and reasoning. Therefore, the CET, in conjunction with other clinical and mental status measures, may be helpful in discriminating some patients in moderate stages of AD from those with FTD. At a cut-off score of 10, the CET correctly discriminated all patients with AD from the normal elderly controls.

8.3.2.1 Semantic dementia

In semantic dementia, the presenting feature is a profound breakdown in semantic memory but in contrast to AD, episodic memory is relatively spared as judged by the patient's preserved orientation, recall of recent autobiographical events and recognition-based tests of anterograde memory for pictures (Hodges et al., 1992; Graham & Hodges, 1997; Snowden et al., 1996). The predominant cognitive feature of semantic dementia is a progressive deterioration of semantic knowledge about people, objects, facts and word meanings. Patients perform poorly on neuropsychological tests dependent upon conceptual information such as picture naming, category fluency (i.e. generating as many exemplars from a semantic category as possible in one minute), word-picture matching, defining concepts in response to their names or pictures and sorting words according to pre-specified criteria. The deficit is also observed on non-verbal tests of semantic knowledge such as selecting the correct color for a black-and-white line drawing, drawing animals or objects from memory. The selective nature of the semantic deficit in these patients has been confirmed by their good performance on assessments of current day-to-day memory, short-term verbal memory, visuo-spatial skills, non-verbal reasoning, phonology and syntax until very late in the course of the disease (Hodges et al., 1992).

In semantic dementia, Warrington (1975) found that knowledge of subordinate categories (e.g., the name of a specific animal) was more impaired than knowledge of superordinate categories (e.g., animals or birds). Similar findings were obtained by other researchers (Snowden et al., 1989; Hodges et al., 1992). Hodges et al. (1995) monitored the deterioration of a single patient over the course of 2 years, finding that the pattern of relative preservation of superordinate knowledge was demonstrable both during a particular test session and across test sessions longitudinally. However, over the course of time, there was a progression from subordinate to superordinate errors. Warrington (1975) had interpreted this pattern of deterioration as reflecting the hierarchical manner in which knowledge is organized but Hodges et al. (1995) argued that superordinate words have multiple connections within the semantic network and consequently are less vulnerable to degradation, a view that does not necessarily imply a hierarchical structure of the organization of knowledge. A different view proposed by Funnell (1995) is that both superordinate and subordinate knowledge are affected, after controlling for word frequency and familiarity effects, suggesting that the apparent sparing of superordinate knowledge is an artifact of factors such as word frequency and familiarity. Other classes of knowledge, such as number and the ability to perform mental calculations, may be surprisingly well preserved despite the severe disruption of verbal and visual semantics in this disorder (Butterworth et al., 2001; Cappaletti et al., 2001).

Unlike amnesic patients who often show better recall of distant memories compared with more recent personal events, patients with semantic dementia showed the opposite pattern (Snowden et al., 1996; Graham & Hodges, 1997). They have better recall of recent memories compared with those from childhood and early adulthood. Several studies have highlighted the patients" progressive loss of knowledge about the meanings of non-verbal as well as verbal stimuli (Lambon Ralph & Howard, 2000; Bozeat et al., 2000).

The syndrome of semantic dementia provides an ideal testing ground for investigating the role of conceptual knowledge in various cognitive domains. These include the issue of whether the ability to use familiar objects is reliant, in whole or in part, upon intact conceptual knowledge about these items. It has been observed on the basis of clinical observation that patients with semantic dementia, despite sometimes severe semantic deficits, are relatively competent in their daily activities (Snowden et al., 1996). For instance, a number of patients with notable loss of conceptual knowledge have continued to be engaged in hobbies and to cook and complete various domestic activities (Lauro-Grotto et al., 1997). These cases have been reported to use a number of objects correctly, even the same objects for which they cannot provide names, descriptions or correct associative semantic judgments.

8.3.3 Lewy body dementia

According to several studies, dementia associated with Lewy bodies now ranks second to only AD as a cause of dementia (Hansen et al., 1990; Perry et al., 1990). Vascular dementia, once considered the second most common cause of dementia, is now ranked third. Numerous terms have been used to describe the same or similar entities. Some of the more frequent terms encountered include diffuse Lewy body disease and senile dementia of the Lewy body type (Perry et al., 1990; Lewy, 1912). The neuropathological hallmark for these disorders is the Lewy body, a spherical intraneuronal cytoplasmic inclusion originally described in brainstem nuclei in Parkinson's disease (Lewy, 1912). In Lewy body dementia, Lewy bodies are found in subcortical nuclei such as the substantia nigra as well as diffusely in the neocortex (Kosaka et al., 1984, Kosaka, 1990). With the advent of more sensitive techniques to identify Lewy bodies, it is now known that neocortical Lewy bodies are also commonly seen in idiopathic Parkinson's disease, a finding that has created some controversy. Given the frequent clinical overlap in Parkinson's disease and Lewy body dementia and the similar neuropathological features, some investigators feel that Parkinson's disease and Lewy body dementia represent a phenotypic spectrum of the same disease process (Kosaka, 1990). Further adding to the confusion is the fact that concomitant neuropathological AD changes, i.e. SPs and NFTs, may also be present in Lewy body cases (Hansen et al., 1990). The true boundaries of Lewy body dementia remain to be clearly defined given the significant clinical and pathological overlap with both Parkinson's disease and AD.

The cognitive impairment in Lewy body dementia in many ways resembles the dementia of AD although there appear to be some key symptoms that are suggestive of Lewy body dementia rather than AD (Hansen et al., 1990; Perry et al., 1990). As with AD, memory deficits are the initial complaint in about 70% of cases. A study has shown severe but similar degrees of impaired performances in tests of attention/STM (digit span) frontal lobe function (verbal fluency, category, and Nelson card-sort test) and motor sequencing in both Lewy body dementia and AD groups as in Parkinson's disease patients and controls (Gnanalingham et al., 1997). However, Lewy body dementia patients often experience fluctuations in their cognitive ability and performance (Perry et al., 1990). They also may have fluctuations in their alertness in the form of a confusional state or transient reduction in the level of consciousness (Perry et al., 1990). Loss of consciousness has also been reported. These fluctuations are important to recognize because they may be misinterpreted as vascular events or as sundowning. However, neuropsychological testings have indicated that patients with Lewy body dementia display disproportionately severe deficits in attention, fluency, visuospatial and constructional abilities and psychomotor speed (Hansen et al., 1990; Salmon et al., 1996). These feaures may help to distinguish Lewy body dementia cases from typical AD cases.

8.4 Discussion

The emergence of functional neuroimaging techniques offers unprecedented opportunities to discover how the brain learns and remembers. Combining information from brain imaging with that regarding genetic predisposition or risk through Apo E genotyping promises to be a powerful tool for early AD detection. Brain imaging technologies, such as MRI and PET, are not required diagnostic tools but enable investigators to study the brains of living patients and follow these patients longitudinally to evaluate disease progressions and the effects of therapeutic interventions (Fox et al., 1996; Reiman et al., 1996; Schuff et al., 1997).

The potential desirability of using both types of tests when assessing patients at risk of dementia is reinforced when the complementary shortcomings of each type of test are considered. Cognitive tests may be affected by education and premorbid ability. Poorly educated patients may be misclassified as demented, and dementia in well-educated patients may be missed (Tombaugh & McIntyre, 1992). Cognitive testing also requires intact sensorimotor and language faculties. Important report questionnaires have been shown to be uncontaminated by premorbid intelligence and education, and they are unaffected by patients" physical disabilities. However, scores on these tests may be influenced by noncognitive factors, such as the affective state of the patient and the informant, the personality of the patient, and the quality of the relationship between the patient and informant (Jorm, 1996). It has been suggested that the use of the informant report and cognitive testing are complementary approaches to the assessment of dementia (Mackinnon & Mulligan, 1998). Used together, these tests provide more information than either test alone. This information can be used to improve the accuracy of screening and diagnosis. From a clinical perspective, the combination of cognitive testing and informant report adds little to the clinician's burden. The short form of the Informant Questionnaire on cognitive decline in the elderly comprises only 16 items and is self-administered. In many cases, it is possible for an informant to complete this questionnaire while waiting during the time the patient is examined.

The International Consensus Conference on Dementia and Driving has suggested that before a decision about driving can be made, the etiology of the presumed cognitive impairment or the diagnosis must be clear. A conclusion related to driving fitness requires an adequate examination of multiple cognitive domains and a solid understanding of the individual's functional abilities (Lundberg & Johansson, 1997). Clinical guidelines for driving-related recommendations have been proposed by the American Academy of Neurology based on a review and meta-analysis of epidemiological studies and driving performance studies investigating crash risk among drivers with a diagnosis of AD (Dubinsky et al., 2000). The Academy guidelines recommend that patients with presumed Alzheimer's dementia with a CDR of 1 be instructed not to drive because of an increased risk of an automobile accident. These guidelines also recommend that physicians inform patients with possible AD and a CDR of 0.5 that they are at greater risk for driving difficulty than other drivers in their age range and that these patients be referred for a formal driving assessment by a qualified examiner. Because of the relatively high probability that these patients may progress over time to a CDR rating of 1, it is also proposed that dementia severity and ability to continue driving be reassessed every 6 months.

Chapter 9
Behavioral and psychological impairments

9.1 Introduction

AD and other types of dementias are associated with many problematic behaviors. Noncognitive feature is a terminology that draws the distinction between cognitive loss and the heterogeneous group of symptoms that include psychiatric symptoms such as depression, hallucinations, delusions and misidentifications. This term also includes behavioral disturbances such as aggression, agitation, pacing and eating disorders (Burns et al., 1990). An alternative expression is neuropsychiatric features (Cummings et al., 1994). The term "Behavioral and Psychological Symptoms of Dementia" (BPSD) has also been defined as "symptoms of disturbed perception, thought content, mood and behavior that frequently occur in patients with dementia" (Consensus Conference on the Behavioral Disturbances of Dementia, 1996). Noncognitive disorders can cause distress both to patients and carers. Such changes are often a determinant for institutional care. A study reported that psychotic symptoms (e.g., hallucinations and delusions) were present in about 50% of patients with dementia and disruptive behavior (e.g., physical violence, hiding things, wandering, demanding or critical behavior) in up to 70% (Rabins et al., 1982). These include hitting, yelling, grabbing people, wandering, hoarding, and repetitive gestures and questioning (Moak & Fisher, 1990; Reisberg et al., 1987). All of these problematic behaviors are often grouped together as "agitation" or "agitated behavior".

Recent studies found a relatively high incidence of delusions and hallucinations in AD patients early in the course of the disease (Paulsen et al., 2000). Common psychiatric phenomena in AD include disorders of thought content, such as simplex and complex delusions, persecutory ideation and disorders of perception such as hallucinations and misidentification syndrome (Ballard et al., 1999). Overall, the number and severity of behavioral problems increase with worsening of cognitive impairment (Finkel, 1998). BPSD can be mainly grouped into behavioral and psychological symptoms of dementia (see Table 9.1). The former are usually identified on the basis of patient observations while the latter are mainly assessed on the basis of interviews with patients and relatives.

Table 9.1: Behavioral and psychological symptoms of dementia

Group I **Most common/** **most distressing**	**Group II** **Moderately common/** **moderately distressing**	**Group III** **Less common/** **manageable**
Psychological Delusion Hallucinations Depressed mood Sleeplessness	*Psychological* Misidentifications	*Behavioral* Crying Cursing Lack of drive Repetitive questioning, anxiety Shadowing behavior and disinhibition
Behavioral Physical aggression Restlessness	*Behavioral* Agitation Culturally inappropriate Screaming, wandering	

9.2 Psychological impairment

9.2.1 Anxiety

Anxiety is a multidimensional concept, having physiologic, psychological and emotional dimensions. It triggers the autonomic nervous system, resulting in symptoms of increased heart rate and blood pressure, trembling, increased respiration, dry mouth and queasy stomach. It is often expressed as agitation in demented patients, and has been described as probably the most prevalent psychiatric condition affecting individuals with dementia, since approximately 60% of them present with agitation during the course of their illness (Mintzer & Brawman-Mintzer, 1996). Patients who have insight into their memory problems and those with coexisting physical illness may also have higher levels of anxiety (Shankar et al., 1999). Anxiety can also exacerbate poor performance on cognitive tests, leading to an overestimation of the severity of dementia (Yesavage, 1984). Recently, a study found a 36% prevalence of clinical anxiety in patients with dementia receiving psychiatric services (Shankar et al., 1999).

The Rating Anxiety in Dementia Scale has been developed to measure specifically anxiety in patients with dementia (Shankar et al., 1999). This scale had good inter-rater reliability, test/retest reliability and internal consistency. The validity of the scale is also good because the scale significantly correlated with other anxiety scales and independent ratings the physician and the caregiver using visual analogue scales.

9.2.2 Agitation and aggression

Agitation is a broad concept and is often used interchangeably with aggression. Agitation can be divided into four subtypes, based on the Cohen-Mansfield Agitation Inventory (CMAI): physically non-aggressive behaviors (restlessness, pacing, mannerisms, hiding things, inappropriate dressing or undressing); physically aggressive behaviors (hitting, pushing, scratching, biting, kicking and grabbing); verbally non-aggressive behaviors (negativism, repetitions, interruptions, constant requests for attention); and verbally aggressive behaviors (screaming, making strange noises, cursing and temper outbursts) (Cohen-Mansfield et al., 1989). Agitation has been identified as a significant problem in dementia patients, occurring in 70–90% of individuals with dementing illness at some point during the course of the disease (Cohen-Mansfield & Billig, 1986). An overall rate of aggression for patients satisfying NINCDS/ADRDA criteria for AD is reported at 19.7% (Burns et al., 1990). In AD, agitation has been found to increase as the disease progresses from mild to severe stages and in many patients, agitated behaviors occur on a daily basis (Teri et al., 1992). Thus, agitation is a primary source of stress and burden to family caregivers. Because of its prevalence and impact on both patients and caregivers, the assessment of agitation is an important area of investigation.

Physically aggressive behaviors are more common in men, in severe dementia and are associated with carer stress while verbally non-aggressive behaviors are more common in women, in mild to moderate dementia and are associated with depression and poor social relationships (Cohen-Mansfield, 1996).

9.2.3 Depression

Evidence suggests that there is a depression specific to AD and that the features and course often differ from those of major depressive disorder in the DSM (APA, 2000). Depression in individuals with dementia is associated with wandering, aggression, constant requests for help, complaining and negativism, and impaired activities of daily living (Lyketsos et al., 1997). Depression can be present at all stages of dementia and it also appears to accelerate loss of functioning. A study reported that major depression occurs in 5–15% of AD patients and in 20–25% of patients with multi-infarct dementia; minor depression in 25%; depressive features occur in 50% of dementia patients; and family-attributed depression in their relatives who have dementia in 50–85% (Greenwald, 1995). In fact, memory complaints in the elderly may be more related to depression than to objective memory impairment. The essential features of major depression are sadness that is persistent or anhedonia (loss of interest in usual activities). In addition to memory impairment and poor concentration, depressed patients experience sleep and appetite disturbances, loss of energy, psychomotor retardation, feelings of worthlessness or guilt or recurrent thoughts of death. It is important for the clinician to be aware of the risk of suicide in patients with AD. Suicide risk is likely to be greatest in those with severe depression, preserved awareness and the ability to perform planned actions (absence of apathy).

The differentiation between depression and dementia can be difficult because loss of interest, decreased energy, apathy, weight loss, sleep disturbance and agitation are common to both (Burke et al., 1988). As dementia progresses, the diagnosis of depression becomes more difficult due to increasing difficulties with both verbal and non-verbal communication. When depression symptoms occur just before the development of AD, they may be early symptoms of the dementing process. The majority of studies concur that dysphoria and loss of interest are the most common symptoms of depression in AD (Ballard et al., 1996; Devanand et al., 1997). A recent study shows that depression symptoms before the onset of AD may be associated with the development of AD even in families where first depression symptoms occurred more than 25 years before the onset of AD (Green et al., 2003). These data suggest that depression symptoms are a risk factor for later development of AD.

A number of scales have been specifically developed to assess depression in dementia patients. The Depressive Signs Scale relies on direct observation of an individual's behavior by trained raters and has been found to be highly reliable, but its internal consistency and validity requires further evaluation (Katona & Aldridge, 1985). The Cornell Scale for Depression in Dementia is a 19-item instrument that uses information gathered from an interview with both the patient and a carer (Alexopolous et al., 1988). The scale has been shown to be more reliable than both the Hamilton Rating Scale for Depression and the Dementia Mood Assessment Scale in elderly patients with dementia (Camus et al., 1995). The signs and symptoms of depression are categorized into five areas: mood-related signs, behavioral disturbances, physical signs, cyclic function and ideational disturbances.

The determination of depression of AD is predicated primarily on careful clinical assessment, rather than symptom rating scales. The clinician must establish the diagnosis of dementia of the Alzheimer type and identify that there are clinically significant depressive symptoms. This will involve thorough clinical assessment, with consideration of the temporal associations between the onset and course of the depression and the dementia. Finally, the clinician must judge that the depression is not better accounted for by a primary depression, other primary mental disorders, other medical conditions or adverse effects of medication. Thus, considerable clinical judgment is involved in making this diagnosis.

9.2.4 Psychosis

Psychotic symptoms such as hallucinations and delusions occur in individuals with AD. Most studies report these symptoms occurring in 20–40% of patients with AD (Wragg & Jeste, 1989). A recent study reported the cumulative incidence of psychosis to be 51% (Paulsen et al., 2000). The actual figures vary from one study to another depending on the rating scales, study population, severity of the dementia and the diagnostic criteria. AD with psychosis has been associated with more severe cognitive deficits in patients with AD matched on other clinical characteristics (Stern et al., 1987; Jeste et al., 1992). The presence of psychotic symptoms also is an independent predictor of aggressive behavior, more rapid functional decline and premature institutionalization (Gilley et al., 1997; Lopez et al., 1999; Mortimer et al., 1992). Studies conflict regarding whether AD with psychosis is associated with more rapid cognitive dysfunctions although most have found an association (Paulsen et al., 2000; Lopez et al., 1999; Levy et al., 1996). Visual hallucinations are the most common hallucinations in AD but their relation to other cognitive and noncognitive symptoms is unclear. It has been observed that AD patients with visual hallucinations have more behavioral disturbances and perform worse on mental status testing than those without visual hallucinations, despite similar disease duration (Lerner et al., 1992). Auditory hallucinations occur in up to 17% of patients with dementia but olfactory and tactile hallucinations are rare (Gilley et al., 1991).

A wide variety of delusions have been reported in AD, including delusions of persecution, delusions of infidelity, delusions of abandonment, or delusions that deceased individuals (e.g. parents) are still living (Tariot et al., 1995). Delusions have been reported in 10–73% of patients with dementia (Cohen et al., 1993; Flynn et al., 1991). Persecutory or paranoid delusions are the most common delusions in dementia, with delusions of theft occurring in 9–28% (Allen & Burns, 1998). Reports of misidentifications vary between 6.5 and 49% (Allen & Burns, 1998). Misindentifications delusions include the belief that one's home is not one's home, belief that a family member is someone else, has been reduplicated or is an imposter and the belief that images on the television are actually people present in the house. The underlying mechanism for psychosis in AD is not well understood. A recent hypothesis suggests that genetic risk factors for other psychiatric syndromes or variants in neurotransmitter systems are risk factors for psychopathology in the course of AD.

Determining the presence of hallucinations in AD patients presents different challenges. Since many AD subjects have expressive language difficulties, it may not be clear whether they are indeed hallucinating or experiencing illusions. Nocturnal hallucinations present another challenge. Hallucinations transiently occurring upon falling asleep or awakening can be normal phenomena, but it may not be possible to make this distinction in an AD patient. Finally, care must be taken to exclude the possibility that hallucinations (and delusions) were not solely limited to a period of acute delirium, or a transient, medication-induced state (Jeste & Finkel, 2000).

The association between depression and psychosis in AD is relatively unstudied. In individuals that develop psychosis, the prevalence of depression in AD ranges from 40–50% (Wragg & Jeste, 1989). A population-based study using psychiatric rating scales found frequent presence of hallucinations (13%) and delusions (22%) (Lyketsos et al., 2000). Whereas criteria for psychosis of AD require the presence of delusions or hallucinations, delusions and hallucinations are not exclusionary criteria for depression of AD (Jeste & Finkel, 2000). Therefore, when depression and psychosis co-exist in an AD patient, both should be diagnosed.

9.2.5 Sexual impairment

When AD strikes, a spouse with these commitments is faced with a situation that, in terms of sexuality at least, is fraught with problematic contradictions. The marriage plays a vital role in the commitment to caregiving, and part of marriage is sexual intimacy. Yet, emotional intimacy, lost in AD, is also a condition for acceptable sexuality. Some caregivers are no longer interested in sexuality or physical intimacy. Common reasons given for this are the patient's inability to identify the spouse or remember the spouse's name, the patient's incontinence or poor hygiene leading to physical distaste on the part of the spouse, and the change in relationship roles such that the caregiver feels more like a parent than a spouse. Still others fear sexual arousal that will later leave them with feelings that have no respectable outlet.

Other spousal caregivers express an interest in maintaining sexual activity or physical intimacy but feel frustrated by problems that arise. For example, the patient may forget the sequence of events involved in intercourse and be unable to participate as a sexual partner. The spousal caregiver may also be too physically tired to have energy for sexuality, or the spouse may feel guilty or embarrassed about initiating sex with a partner who cannot clearly consent or refuse.

In women, there is a close linkage between physical and emotional intimacy. Sex is fine while the husband with AD still recognizes his wife. When that ends, sex can become conflictual and women may want to avoid sexual activity.

Few studies exist to date concerning the effect of AD on the patient's sexual interest and performance or the spouse's willingness to continue to engage in sexual relations with the patient. A study that interviewed the wives of 26 male AD patients, inquiring specifically about changes in their sexual relationships, noted that 26 wives reported that the onset of their husbands' disease had affected the sexual intimacy with their spouses (Shapira & Cummings, 1989). Their answers revealed three types of changes as follows: less interest in sex by eight of these AD patients after the onset of their disease with sexual contact rarely initiated, greater libido shown by four AD patients who sought more frequent sexual activity than before onset of the illness and inability of ten of the wives to engage in sexual relations with their husbands after their disease became evident, even though the husbands continued to want sex.

Recent surveys indicate that sexual behavior problems in the afflicted are more common than suggested. A study that investigated the impact of AD from a human developmental perspective to ascertain perceptions of marital quality and coping from a caregiver and afflicted spouses and to compare their answers with those of relatively healthy couples showed that indeed sexual behavior problems in the afflicted are common (Wright, 1991). Only 27% of the AD couples *versus* 82% of the well couples were still sexually active. High sexual activity occurred in 14% (n = 4) of male afflicted spouses (50% of sexually active couples), and of these, two were age ≤ 60. High sexual activity was problematic to three-fourths of the female caregiver spouses. Caregivers evidenced adaptation and control. Afflicted spouses tended to deny problems and had distorted perceptions of interactions with their caregiver spouses.

Erectile failure is a frequent problem in men with AD and it was reported in 53% of 55 male AD patients with a mean age of 70.25 (Zeiss et al., 1990). Loss of erection is not related to degree of cognition impairment, age or depression. A man with early AD may feel that he is no longer a true husband or appropriate sexual partner. The wife may share this view. The couple may focus their attention on making other salient adaptations to the disease such as dealing with its impact on the extended family and their own financial resources. The stress of these adaptations may preclude having the emotional energy to continue an active sexual relationship. Future research might follow a group of patients from early onset to get a more precise assessment of the time of onset of erectile failure, as well as other sexual complaints.

9.2.6 Delirium

Delirium or acute confusional state is another condition that is common in the elderly. According to the DSM-IV, delirium requires the disturbance of consciousness (i.e., reduced clarity of awareness of the environment) with reduced ability to focus, sustain or shift attention (APA, 1994). In addition, a change in cognition (e.g., memory deficit, disorientation, language disturbance) or the development of a perceptual disturbance that is not better accounted for by a pre-existing, established or evolving dementia is noted. The disturbance develops over a short period (usually hours to days) and tends to fluctuate during the course of the day. Patients with dementia are at increased risk for delirium. Both delirium and dementia may coexist (Tune & Ross, 1994).

9.2.7 Sleep disturbances

Diurnal rhythm changes and sleep disturbances are common in dementia (Rabins et al., 1982). In AD, approximately 22% of patients have some sleep disturbance (Allen & Burns, 1995).

9.2.8 Wandering

Wandering is one of the most troublesome behaviors and is a frequent cause of referral to psychiatric services. It can be of different types, e.g. checking, trailing and pottering (Hope & Fairburn, 1990).

9.2.9 Apathy

Apathy which denotes lack of motivation without dysphoria or vegetative symptoms of depression can occur in up to half of patients in mild to moderate stages of AD and can be difficult to differentiate from depression.

9.2.10 Alterations in dietary habit

Approximately 40% of AD patients have alterations in dietary habit and binge eating or excessive appetite is noted in 10-20% (Allen & Burns, 1995). This may be an underestimation as more than half of the families do not see eating behavior as a problem even when it is present (Rabins et al., 1982).

9.3 Discussion

Lawlor et al. (1994) suggested that AD is a clinically heterogeneous illness. Younger age at onset may be related to more prominent language and praxis impairment and the development of depression during the course of the illness. Manic symptoms are reported less often in dementia, the rate varying between 2.5 and 17% (Allen & Burns, 1995). Apart from being less common, they may be under-reported and difficult to differentiate from other neuropsychiatric symptoms. Further research may help in clarifying these issues and may lead to identification of specific subtypes of AD with an associated pattern of BPSD.

It is suggested that various neurotransmitters, including ACh, norepinephrine, dopamine, serotonin and glutamate may be involved in the causation of neuropsychiatric symptoms in AD but the evidence is often indirect. Förstl et al. (1994) found that AD patients with psychotic symptoms had lower neuronal counts in parahippocampal gyrus, CA1 region of hippocampus, dorsal raphe nuclei and locus ceruleus. Delusions and hallucinations were found to be related to higher SP counts in presubiculum and delusional misidentifications were associated with lower neuronal counts in the CA1 region of the hippocampus (Förstl et al., 1994; Zubenko et al., 1991). Neuroimaging evidence suggests that psychiatric and behavioral symptoms in dementia are not random consequences of diffused brain disease but are fundamental expressions in the regional cerebral pathology (Sultzer, 1996). Research into biological factors responsible for the neuropsychiatric symptoms of AD is expanding and may further the understanding between brain and behavior.

Chapter 10
Assessment of memory

10.1 Introduction

Carrying out the neuropsychological assessment at an early stage of dementia has the goal of revealing memory disorders, which are not always associated with memory complaints. It also characterizes the memory disorder in the context of cognitive neuropsychology, thus permitting other cognitive functions to be integrated with the memory disorder into a broader syndrome.

As memory dysfunction is generally the first and most severe cognitive impairment in AD, the choice of memory testing to be used in these studies is of great importance. It should reflect an understanding of memory systems being assessed with neuropsychological tests and the fact that some tests can be more appropriate than others to show benefit with certain classes of cognition-enhancing medications. Various consensus statements recommend diagnostic assessments that include a complete history from someone who knows the patient well, physical and neurological examination, and a careful mental status examination. Cognitive deficits can be detected several years before the clinical diagnosis of dementia and establishing the neuropsychological profile often indicates the underlying neuropathology (Linn et al., 1995). The importance of well-designed clinical trials using valid, sensitive and reliable tools of neuropsychological evaluation methods to measure cognitive and behavioral effects of drugs to demonstrate efficacy is crucial. Neuropsychological variables remain the most informative approach to detect and analyze the activity of the drugs on the different components of cognition. Assessment of cognitive function is the essential co-primary measure of efficacy in clinical trials. Treatments that improve, or reduce the decline of, cognition would be expected to provide important benefits to patients with AD and their caregivers (Wimo et al., 1998).

Ratings can be self-reported, observer-rated or based on information from an informant. The choice of instrument is often based on a combination of the user's knowledge of the scale, the time available for its application and the presence and reliability of an informant. Subjective ratings are highly dependent on the cooperation of patients and their ability to comprehend either written or verbal instructions. Informant-based assessments are often used for patients with dementia who may not be reliable observers of their own functioning or behavior. Such ratings may be subject to bias as it may be influenced by the informant's mood state or perceptions. Often, using both a proxy reporting with a direct patient interview gives the best result.

Table 10.1: Cognitive portion of the ADAS-cog

Item	Possible points
Spoken language ability	5
Comprehension of spoken language	5
Recall of test instructions	5
Word-finding difficulty	5
Following commands	5
Naming objects, fingers	5
Construction (drawing)	5
Ideational praxis	5
Orientation (time, day, date, month, year, season, person, place)	8
Word recall	10
Word recognition	12
Total	70

10.2 Memory assessment

10.2.1 The Alzheimer Disease Assessment Scale – Cognitive Subscale

The ADAS-cog measures memory, orientation, reasoning, language and other areas of cognition (see Table 10.1; Rosen et al., 1984). Scores range from 0 to 70. The higher the score, the greater is the cognitive impairment. In general, a 3-point to 4-point improvement is needed to see a clinical difference in symptoms. It has become a *de facto* standard, and most recent clinical trials have used ADAS-cog as the primary index of cognitive change. Its reliability and validity have been established in the original version. The test-retest and inter-rater reliability of the original English ADAS scale, ADAS-cog and non-cognitive subscales are good. The inter-rater reliability of the ADAS-cog is 0.989. Its test-retest reliability is 0.915 which is much better than that of the non-cognitive subscale of 0.588 (Rosen et al., 1984; Mohs & Cohen, 1988). Versions in Japanese, Greek, German, French, Spanish, Dutch, Swedish, Chinese, Icelandic, Finnish, Danish and Hebrew have also been developed. For the Chinese version, pictures instead of words for the assessment of recall and recognition were used in order to accommodate illiteracy (Liu et al., 2002). The Chinese ADAS-Cog has high internal consistency (Cronbach's alpha = 0.87) and very high interrater reliability and test-retest reliability. The National Institute of Aging AD Cooperative Study group has also been developing an expanded ADAS-cog to assess attention and executive function more directly (Mohs et al., 1997).

Compared with other cognitive scales, ADAS-cog is more sensitive to a wider range of disease severity and has greater specificity to the major dysfunctions experienced by patients with AD. Regulatory authorities often define a four-point improvement on ADAS-cog relative to baseline as a clinical “responder”. The ADAS, which is used principally for longitudinal studies, was designed to provide a composite assessment of longitudinal investigations and clinical trials including patients with AD (Rosen et al., 1984).

The modified ADAS, another important neuropsychological test, was constructed as a severity scale applicable to patients diagnosed as suffering from AD (Rosen et al., 1984; Mohs & Cohen, 1988). It is a good representative of the sort of scale containing both cognitive and non-cognitive items. The scale consists of 21 items – 11 items pertaining to the cognitive section (ADAS-cog) and 10 items to the interview-based ratings of behavioral symptoms (ADAS-Ncog). Due to its facility to use, ADAS has become one of the most widely used research tools for the assessment of treatment efficacy in large clinical trials of pharmacological treatment of AD. This scale has also been adapted to several languages. In a large-scale multicenter clinical trial, ADAS clearly demonstrated reliability and validity in the assessment of symptom severity in AD patients (Weyer et al., 1997).

10.2.2 The Mini-Mental State Examination

The ADAS-cog is primarily a research tool. Practitioners in the clinical setting may be more likely to use other cognitive tools such as the MMSE. The MMSE was developed as a bedside tool to evaluate the cognitive status of elderly persons in clinical settings (Table 10.2). It was not intended for use as a clinical screening instrument but has nonetheless been incorporated into clinical trials as a staging criterion for entry and has been used as a secondary measure of cognitive functions (Tombaugh & McIntyre, 1992; Feher et al., 1992). It is brief and easy to administer, and has shown good reliability. Validity as a screening test is generally acceptable although certain limitations have been identified (Feher et al., 1992). Validity may be weak in samples that include psychiatric patients, and the MMSE may not detect focal brain dysfunction or mild dementia (Feher et al., 1992; Faustman et al., 1990; Naugle & Kawczak, 1989). This may be due to limited coverage of right hemisphere functions, and to the use of overly simple items (Feher et al, 1992; Mazzoni et al., 1992). An MMSE score of 27 or higher is usually taken as excluding mental impairment, while one of 23 or lower generally indicates sufficient cognitive decline for the diagnosis of dementia to be made (Folstein et al., 1985). Published normative data allow interpretation of scores according to the patient’s age and education (Crum et al., 1993).

MMSE is the most frequently used scale worldwide for screening procedures for dementias (Folstein et al., 1975). This very short examination contains seven areas of cognitive evaluation: orientation (for time and place), attention and calculation, language, ideomotor praxis, constructional praxis and memory. The MMSE is based exclusively on the cognitive components of dementia and is usually considered to have high inter-rater reliability and validity. It is commonly used to assess cognition in clinical practice. A score of less than 23 is most commonly used to classify

Table 10.2: The Mini-Mental State Examination

Question	Maximum Score	Patient Score
Orientation		
Give 1 point for each correct answer.		
What is the (year) (season) (date) (day) (month)?	5	–
Where are we: (state) (county) (town) (hospital) (floor)?	5	–
Registration		
Name three common objects (such as an apple, table, penny). Take one second to say each. Then ask the patient to repeat all threeafter you have said them. Give 1 point for each correct answer. Repeat them until he or she learns all three. Count trials and record. Trials: –	3	–
Attention and calculation		
Ask the patient to count backward from 100 by sevens. Stop after five answers. Give 1 point for each correct answer. Alternatively, spell "world" backward; score 1 point for each letter in the correct order.	5	–
Recall		
Ask the patient to recall the three objects previously stated. Give 1 point for each correct answer.	3	–
Language		
Show the patient a wristwatch; ask the patient what it is. Repeat with a pencil. (2 points) Ask patient to repeat the following: "No ifs, and, or buts." (1 point) Ask the patient to follow a three-stage command: "Take a paper in your right hand, fold it in half, and put it on the floor." (3 points) Ask the patient to read and obey the following sentence, which you have written on a piece of paper: "Close your eyes." (1 point) Ask the patient to write a sentence. (1 point) Ask the patient to copy a design. (1 point)	9	–
Total score:		–

patients as cases (Tombaugh & McIntyre, 1992). A MMSE score of 26 or less should be considered abnormal in an individual with a high school education. A score of 27 to 30 with evidence of decline of intellectual function (usually manifested as difficulty with work or household tasks) requires more extensive neuropsychological testing to document objective cognitive impairment. However, one has to be careful not to over-interpret MMSE scores, because low levels of education and speech impairment will reduce its validity. Highly educated individuals may have high MMSE scores despite an obvious decline in their prior abilities. As a cognitive assessment, the MMSE is very limited; the majority of the measures can be qualified as cerebral posterior ones, neglecting executive functions and working memory alterations. The evaluation of episodic memory, severely altered in AD, is almost inexistent. Therefore, in minimal and mild AD, the probability of having ceiling effects is strong. Nevertheless, the MMSE is useful for confirming the impression from history taking that there is cognitive loss, and the test can be repeated over time to assess deterioration and response to treatment.

The MMSE is a nonspecific screen for cognitive function and has some limitations. It is not sensitive for detecting cognitive impairment in individuals with higher levels of education or high levels of premorbid functioning. Conversely, those with low levels of education or minority cultural backgrounds may score low on the test without impairment. Patients with high education levels may show normal cognitive function on the MMSE despite the presence of functional decline, while conversely, some patients with low education levels may show MMSE scores yet no functional decline. MMSE has disadvantages for the screening of vascular dementia; it emphasizes language and verbal memory, it lacks the recognition part of memory, it has no timed elements and it is not sensitive to impairment in executive functions or mental slowing (Roman et al., 1993). However, the MMSE is especially useful when repeated regularly to follow illness progression. Over 12 months, untreated patients would be expected to decline by 2–4 points on the MMSE (Aisen et al., 2000; Mulnard et al., 2000).

10.2.3 The Blessed dementia scale

The Blessed dementia scale (BDS) is, to our knowledge, the oldest scale to be used in trials of cognition-enhancing drugs (Blessed et al., 1968). It contains two main subdivisions: BDS-part 1 relates to ADL with a negative scoring system (0 to 28), and a BDS-part 2 relates to cognition with a positive scoring system (0 to +37). BDS-part 2 is also called the information-memory-concentration test.

10.2.4 The Gottfries-Brane-Steen scale

The Gottfries-Brane-Steen scale (GBS) contains four areas of evaluation such as motor functions, emotional functions, intellectual functions and different symptoms common in dementia (e.g. confusion or restlessness) (Gottfries et al., 1982). Clinical experience with the GBS scale suggests that an approximate 10-point or 10–15% treatment in the total score is clinically relevant in studies conducted in patient populations with a baseline score of approximately 80 points (moderate dementia) (Waldemar et al., 2000).

10.2.5 The Cambridge Mental Disorders of the Elderly Examination

Some interview schedules explore the cognitive functioning of the patients, their daily living adaptation and the presence of psychiatrically relevant symptoms. Among the few standardized interview schedules currently available, the CAMDEX which contains a cognitive section (CAMCOG) offers high psychometric quality due to its sensitivity to different levels of severity of dementia, rated on the basis of internationally established criteria (O'Connor et al., 1989, 1990; Roth et al., 1986, 1988). In particular, the CAMDEX reliably detects cases of minimal and mild dementia and is independent of cultural factors (Linas et al., 1990; Neri et al., 1992). The design of a short version of this interview (administration of initial version requires 60–90 min) which requires 30 min to administer and consists of 106 items of the 340 items of the full form make it more feasible (O'Connor, 1990). The CAMDEX is another interesting scale to be used in AD research (Roth et al., 1986). It is a structured instrument made up of eight sections – an interview with the subject, a cognitive section (the CAMCOG), the interviewer's observations of the subject, a physical examination, a note of medication, any additional information and an interview with an informant. The resulting information provides a formal diagnosis in a number of categories: four types of dementia, delirium, depression, anxiety, paranoid disorder and other psychiatric disorders. Interrater reliability is excellent. The CAMDEX has been used extensively in research studies. The European Community guidelines state this scale as the preferred scale to use in AD research. These three scales appear as the "golden standard" scales to be used in pharmacological research.

10.2.6 The 7-minute neurocognitive screening battery

The 7-minute neurocognitive screening battery known as the "7 Minute Screen" is another test that has been demonstrated to be very reliable and can be more rapidly administered. It is more neurocognitive in nature and demonstrates less bias than the MMSE. In contrast to the MMSE, which takes approximately 30 min to administer, the 7 Minute Screen is reported to take an average of 7–8 min to administer and can be used with minimal training in primary care offices as well as long-term facilities (Solomon et al., 1998). It has four sections, each measuring a different aspect of cognition often compromised in those with AD that take advantage of the evolving understanding of the cognitive differences between AD and the normal aging process. The tests are an orientation test, a memory test, clock drawing and verbal fluency. The test yields high degree of sensitivity (92%) and high specificity (96%) (Scinto & Daffner, 2000).

10.2.7 The Mattis dementia rating scale

The Mattis dementia rating scale (DRS) was designed as a screening instrument to detect the presence of brain pathology in impaired geriatric patients (Mattis, 1976). It evaluates a broad array of cognitive functions and includes subtests for attention, initiation, construction, conceptualization, verbal and nonverbal memory. Thus, it is sensitive to frontal and fronto-subcortical dysfunctions. High test-retest reliability has been reported and normative data have been published (Mattis, 1976; Schmidt et al., 1994).

The DRS is widely used in clinical and research settings as a global measure of cognitive functioning in patients with AD. The DRS has greater sensitivity to longitudinal cognitive decline in AD than either of two other commonly used assessments tools, the BIMC test and the MMSE (Salmon et al., 1990). The DRS has reasonably high power to discriminate AD from other dementing illnesses (Salmon et al., 1989; Rosser & Hodges, 1994). The indices thought to best discriminate patients with AD from healthy older individuals are the memory and initiation/perseveration subscales (Monsch et al., 1995).

10.2.8 The Brown-Peterson task

The Brown-Peterson task has been used to determine whether the rate of forgetting is similar to that of the control group or whether the AD patients forget more quickly. Some initial reports indicate that the loss of information was indeed quicker in the patient group (Sullivan et al., 1986). Nevertheless, subsequent studies that were able to equate the initial level of acquisition demonstrated that although AD patients clearly performed less well than the controls, they did not forget more quickly (Dannenbaum et al., 1988; Koppelman, 1985, 1991). The forgetting curves were parallel for the two groups with verbal and non-verbal items.

Among the most frequently used STM tasks are verbal memory span tasks and the Brown-Peterson task (Brown, 1958; Peterson & Peterson, 1959). In the former, some investigators have reported significant differences between AD patients and normal elderly whereas others have not found such differences (Belleville et al., 1996; Collette et al., 1999; Morris, 1984, 1986, 1987; Lines et al., 1991; Weingartner et al., 1983). A recent study designed to determine whether the poor performance of AD patients in the Brown-Peterson paradigm might be accompanied by a pattern of errors that is qualitatively different from that of the control group, or whether the errors are of the same type but only in large numbers, showed that the rate of forgetfulness was similar in AD and control groups (Sebastian et al., 2001). Contrary to other studies where an AD forgetfulness tendency of omissions was observed, this study showed an excess of perseveration indicating problems in the central executive (Sebastian et al., 2001; Dannenbaum et al., 1988; Kopelman, 1985). Thus, these data suggest that the performance deficit shown by AD patients in the Brown-Peterson task is primarily due to a difficulty in renewing or updating the contents of working memory.

10.2.9 The Cognitive Drug Research Computerized Assessment System

The Cognitive Drug Research Computerized Assessment System (COGDRAS) was originally designed to evaluate the cognitive effects of drugs (Wesnes et al.,

1987). A further version of the system was developed to examine aspects of cognitive performance in people with dementia. Thus, the COGDRAS has established sensitivity in assessing cognitive changes associated with drug therapies in many populations from young to elderly. In one study, the performance of 23 elderly patients with established dementia was compared with that of control subjects (Simpson et al., 1991). The demented patients showed large and significant impairments in the speed of choice reaction time and the accuracy and speed of all memory tasks. There was good correlation with standard diagnostic instruments and good test-retest reliability, showing that the assessment package was a valid reliable tool in dementia. Using the COGDRAS to assess memory in patients (range: 28–83 years) that were divided into five groups (worried well, depressed, demented, minimally cognitively impaired and other brain disorders), the performance of the demented group was observed to be significantly impaired in comparison with the worried well group (Nicholl et al., 1995). The depressed group tended to perform slightly less well that the worried well and the "other" group showed a wide range of scores consistent with its diversity. The minimally impaired had scores intermediate between the demented and worried well. Demented elderly patients and controls performed a version of the COGDRAS comprising cognitive tasks measuring choice reaction time, vigilance, and the sensitivity and speed of digit, word and picture recognition (Simpson et al., 1991). The demented patients showed large and highly significant impairments in the speed of choice reaction and in the sensitivity and speed of all memory tasks. These findings, together with evidence of good test-retest reliability, indicate that for demented patients, the utility of this system to assess cognitive change should be comparable to that of previous versions with the young and with non-demented elderly.

The relationship between speed and accuracy on computer testing requires further work. It appears that in the early states of cognitive impairment an individual may be able to obtain the correct response at the expense of taking a longer time to perform the task.

10.2.10 Informant Report Questionnaire

Alternative approaches to screening for dementia have been proposed and of these, informant report is the most promising (Mulligan et al., 1996). This approach involves the administration of a questionnaire to a person who knows the patient well. Questionnaires of this type inquire about aspects of memory and intellectual function in everyday situations compared to earlier in life.

While clinical practice in the assessment of dementia often includes an informal interview with a relative of the patient, only the administration of an informant report questionnaire permits a quantitative rating of the changes observed. The Informant Questionnaire on Cognitive Decline in the Elderly is a self-administered questionnaire that asks an informant about changes in an elderly patient's cognitive performance over the previous 10 years. Studies of this instrument with a variety of patients have shown high reliability and a good ability to differentiate demented from undemented patients (Jorm et al., 1991; 1996).

10.3 Choice of an appropriate sensitive cognitive test

Choosing an appropriate test to measure cognitive decline in AD in clinical trials is not an easy task. A good approach to use these assessment methods is to know how and when to use which of the existing scales. The qualities of these methods to consider would be their reliability, especially the inter-rater reliability and the validity. A good cognitive test must not only show adequate reliability and validity in detecting the presence and severity of cognitive dysfunction, but must also include a wide range of items with increasing difficulty. Ideally, it should include both easy and difficult items in each of the cognitive domains such as memory, spatial skills and language that is evaluated. If difficult items are missing in the test, this would limit the sensitivity of the test to milder and more subtle cognitive dysfunction, making the test susceptible to "ceiling" effects. On the other hand, if easier items are not included in the test, this would cause the test to lack sensitivity at the more severe end of the range of cognitive dysfunction, leading to "floor" effects. Both "ceiling" and "floor" effects affect the sensitivity or power of a test to detect significant differences in cognitive impairment that produced a higher rate of false-negative results (Nunnally, 1978).

The best instrument for assessing memory disorders in early dementia is probably the Free and Cued Selective Reminding Test (FCSRT) (Grober & Buschke, 1987; Grober et al., 1997). Unlike most clinical memory tests that do not control cognitive processing, this test includes a study procedure in which subjects search for items (e.g. grapes) in response to cues (e.g. fruit) that are later used to elicit recall of items not retrieved by free recall. Moreover, this test provides a characterization of the memory impairment that distinguishes AD from subcortical dementia and from FTD (Pasquier, 1996; Pillon et al., 1993). This test is sensitive to early neuropathological changes in AD, in comparison to global status tests (Grober et al., 1997).

For the assessment of the effects of drugs on working memory and measuring the CES and the articulatory loop, the DS of the subtest of the WAIS has often been used. For the functioning of the visuospatial sketchpad, many other tests which include the Corsi's blocks test, the complex visual search test, various cancellation tasks, the trail making test (TMT-A), the Barbiset-Cany's 7/24 immediate recall, a STM test for figures, and a sequential visual test memory test have been used. The DS subtest of the WAIS has been the test of working memory most frequently utilized in drug trials.

Several tests of episodic memory have been used in drug trials. The most common ones were, in order of frequency: (1) the logical memory subtests of Wechsler memory scale (WMS, immediate and delayed free recall of orally presented paragraphs or stories), (2) the Buschke selective reminding test (learning of a list of words or drawings of objects visually presented with recognition, free and cued recall paradigms), (3) the paired associates learning subtest of the WMS (learning a list of semantically related and non-related paired words orally presented with a cued recall paradigm, (4) the Rey auditory verbal learning test (RAVLT, learning of a list of words orally presented with free recall and recognition paradigms), (5) the 5-item test (acquisition of learning phase, 10 min and 24 h delayed recall), (6) the names-learning test and (7) various tests of verbal and visual recall and recognition. The paired associates learning subtest of the WMS has not been very successful in showing any change in performance in groups treated with different drugs. The RAVLT has been comparatively more useful than the paired associates test in demonstrating change in episodic memory performance in patients treated with cognition enhancer drugs (Taylor, 1959; Rey, 1964).

Three tests have been used to assess semantic memory: verbal fluency tasks, the Boston naming test and the vocabulary subtest of the WAIS (Wilson et al., 1981; Goodglas & Kaplan, 1972; Borod et al., 1980). Verbal fluency tasks have been the semantic tests most extensively employed.

The clinical and cognitive scales most frequently used in trials of cognition-enhancing drugs are the MMSE, the ADAS, the Sandoz Clinical Assessment-Geriatric (SCAG), the Self-Assessment Scale-Geriatric (SASG), the BDS, the GBS, the GDS, the WAIS and the WMS (Folstein et al., 1975; Rosen et al., 1984; Shader et al., 1974; Yesavage et al., 1981; Blessed et al., 1968; Gottfries et al., 1982; Reisberg et al., 1982). The clinical global impression of change (CGIC) and the clinical interview based impression of change (CIBIC) are the most frequently used general observational evaluations.

Earnst et al. (2000) attempted to identify cognitive predictors of legal standard competency judgments in patients with AD. Tests of basic and simple verbal communication such as the Boston Naming Test were predictive of patients' capacity to make a reasonable treatment choice. Subsets of the Dementia Rating Scale and the Wechsler Memory Scale were predictive of the more complex abilities needed to meet standards for appreciation of treatment choice, rational reasoning and understanding of treatment situation and choice. Overall, neuropsychological tests do not provide automatic answers to competency but by providing objective measures of capacity relative to other known populations, they can provide a more substantial basis for making judgments.

10.4 Discussion

All cognitive tests are not equally suited for longitudinal clinical trial. Comparing the sensitivity of three cognitive tests (MMSE, BIMC and DRS) in AD, it was found that the DRS showed greater annual rates of change scores in the moderately to severely impaired patients than in the mildly impaired patients (Stern et al., 1994; Salmon et al., 1990; Mattis, 1989). This sensitivity difference may be explained by the fact that the DRS includes more easy items than the other two instruments. It was recommended to use the DRS when following more severely impaired patients. Cognitive or mental status testing should include assessment of attention, level of arousal, orientation, recent and remote memory, language, visuospatial function, calculations, and judgment. Techniques used to assess these domains are discretionary and may include tests and screening instruments combined with clinical impressions. Clinicians are advised to take into account educational and cultural factors when interpreting test results. The digit span test, the RAVLT, the Buschke selective reminding test and the verbal fluency tasks are the most sensitive memory tests, whereas the most sensitive scales are the Sandoz clinical assessment geriatric, the Gottfries-Brane-Steen and the BDS.

The lack of sensitivity of the BIMC in mild and severe cases of AD may have led to failures to detect cognitive changes over time in clinical trials (Stern et al., 1994). Also the Blessed rating instruments do not seem ideal for evaluating the severity of dementia in AD (Reisberg et al., 1988). More research examining sensitivity issues of cognitive instruments in AD needs to be done to help investigators select the best appropriate tests to analyze adequately the cognitive decline in AD patients according to the experimental designs performed.

In a recent study, the interaction of AD severity and visual stimulus complexity in relation to brain function was studied in controls, mildly and moderately to severely affected patients with AD (Mentis et al., 1998). Functional failure was observed in patients with mild dementia when large neural responses were required and in patients with moderate/severe dementia when large and intermediate responses were required. Functional abnormalities in AD in striate and immediate extrastriate association areas (Brodmann's areas 18 and 19) suggest a dysfunction in perceptual processes. It was suggested that as accurate perception facilitates accurate memory and visuospatial and executive functions, perceptual anomaly may likely contribute to the dementia of AD and that drug treatment should also be directed toward modulating perceptual networks in addition to those underlying the usual cognitive and attentional processes. Such visual information processing approach may provide important information to identify functional anomalies at an early stage in mild dementia in the presence of minimal pathology and for drug treatment evaluation. It may also be used as objective psychophysiological measures of drug efficacy.

The ADAS-cog is more elaborate than the MMSE, but many of the criticisms aimed at the MMSE are also appropriate for the ADAS, in particular those concerning the use of posterior measures and the probability of ceiling effects in minimal and mild AD. Unlike the MMSE, the ADAS assesses to a limited extent the capacity of episodic memory. However, the cueing paradigm, of interest for minimal and mild AD, is not employed and there is no evaluation of the consolidation capacity (with the delayed recall paradigm). Furthermore, the ADAS has no quantitative and thorough evaluation of semantic memory, and no evaluation of working memory, autobiographical memory or personal semantic memory. This lack of measures sensitive to the first stages of the disease, and specifically, the lack of memory system measures in the ADAS has already been reported (Mohr et al., 1995). The FDA now recommends the ADAS as the scale of choice in psychopharmacological research.

Commonly used global cognitive screening tests such as the MMSE and the original version of the CAMCOG have been shown to be insensitive to early cognitive deficits associated with AD but are still widely used to aid clinical diagnosis. The CAMCOG has a number of subscores for different cognitive domains with limited demand per domain. Highly educated and high-IQ subjects may show ceiling effects on these tests (Huppert et al., 1995; Cullum et al., 2000). The MMSE has been criticized for its lack of specificity if used with low-education groups (Jacqmin-Gadda et al., 2000). Domain-specific cognitive measures have been shown to be more effective at predicting global cognitive decline than global screening measures (Swainson et al., 2001). In addition to age and education, impairment in hearing and vision as well as cultural and language background may affect performance on cognitive testing. Nonnative English speakers may have difficulty with cognitive function tests given in English. Individuals relatively fluent in English may still perform at a higher level in their native language given the stress of cognitive function tests. Cognitive screening tests translated into the native language should be used when possible. Even when assessing cognition in the language most comfortable for the patient, culturally biased items on the screening test used may still affect performance, requiring interpretation of the test results within culture-specific norms. Reliable translations and adaptations of screening scales developed for Western cultures such as the MMSE are now available (Katzman et al., 1988; Ganguli et al., 1995; Kabir et al., 2000; Teng et al., 1994; Glosser et al., 1993). However, the level of the education of the subjects influences these cognitive screening scales. To overcome this problem, the Community Screening Interview for Dementia was proposed for use in transcultural studies (Hall et al., 2000). This scale combines a cognitive test for nonliterate and literate populations and an informant interview concerning performance in everyday living. This combination produces better sensitivity and specificity.

The development and use of simple short tests may be interesting in dementia screening. Recently, the visual association test (VAT) to detect early dementia of the Alzheimer type was proposed (Lindeboom et al., 2002). This test is a brief learning test based on imagery mnemonics. The materials consist of six line drawings of pairs of interacting objects or animals, for example, an ape holding an umbrella. The person is asked to name each object and later is shown one object from the pair. He is requested to name the other. The VAT detects with high specificity a high proportion of patients with AD a year before the diagnosis (Lindeboom et al., 2002). Sensitivity was 79% and specificity was 69%. Therefore, the VAT appears a promising tool for the assessment of AD.

The ability of psychometric tests to identify AD is heavily influenced by variables such as the patient's educational attainment, age and social background. Therefore, the importance of culture in cognitive assessment has been emphasized (Ardilla, 1995). A recent study designed to validate a concise neuropsychologic assessment battery to identify early AD in an elderly Chinese population studied how the level of literacy influenced the psychometric performance and determined the best individual and combination of tests from the battery to detect AD (Sahadevan et al., 2002). The results indicate that the majority of tests in the neuropsychologic battery validly identify early AD in the elderly Chinese subjects. However, for all the psychometric tests the influence of education was significant such that the optimal cutoff scores for each test became higher in the more literate group. Thus, correct interpretation of test scores required the adjustment for patients' educational level. Verbal memory, verbal fluency and visuospatial functioning were the most important diagnostic variables. This neuropsychological pattern has broad similarities with findings from the Western countries, supporting the transcultural cognitive manifestation of AD.

The widespread use of standardized ratings such as the MMSE and more advanced variations such as the CAMDEX have greatly improved doctors' ability to screen cognitive function. Despite the challenges, memory tests can differentiate individuals in the earliest phase of AD from those with memory impairment who will not progress to AD over a few years of follow-up (Tuokko et al., 1991; Jacobs et al., 1995; Rubin et al., 1998). Additional tests of cognition that assess, for example, the domains of language and praxis have also been shown to differentiate these two groups, although these additional measures have provided only modest precision (Masur et al., 1994; Small et al., 1997). Neuropsychological tests were highly accurate in identifying individuals with memory problems who were destined to develop AD in one study in which researchers found that testing delayed recall of word list and mental manipulation of well-learned sequences predicted onset of AD 2 years later (Tierney et al., 1996). A recent study identified measures of delayed recall and recognition as significant early predictors of subsequent cognitive decline in high-functioning older adults (Chodosh et al., 2002). Future efforts to identify those at greater risk of cognitive impairment may benefit by including these measures. Overall, the balance of research findings suggests that memory tests, particularly those involving delayed recall, may be useful in determining which individuals are most likely to develop cognitive impairment over time.

Future investigations should use sensitive memory tests, together with behavioral and psychiatric scales, rather than general observational evaluations. Future studies of cognition-enhancing drugs should include tests that assess memory systems known to be affected by AD in its minimal, mild and moderate stages as well as memory tests that have been shown to be responsible to specific neurotransmitters or have been able to detect improvement in several previous trials. Working memory, episodic and semantic memory, autobiographical and personal semantic memory, prospective memory should therefore be evaluation targets in drug trials.

Chapter 11
Functional abilities and behavioral symptom assessments

11.1 Introduction

In clinical practice, the behavioral disturbances seen in patients with dementia are helpful in determining disease severity and the need for support care. In patients with AD, the early appearance of behavioral symptoms is associated with faster disease progression. The first neuropsychological dysfunction in AD is described as a loss of attention capacity, followed by loss of language and visuospatial function (Grady et al., 1988). Behavioral complications of dementing illness are common, adversely affecting the quality of lives of patients and caregivers. They create stress for caregivers, complicate patient management and often precipitate institutionalization. About 70–90% of patients with dementia exhibit behavioral and psychological symptoms at some time during the course of the disease. There has been growing interest in the detection and quantification of the emotional, behavioral, ideational, perceptual and vegetative symptoms that accompany and complicate the lives of persons with AD and their caregivers. The availability of scales to detect and quantify these aspects of the disease is important to clinical investigators seeking effective means of treating these symptoms. Reliable quantification of behavior is desirable to facilitate documentation of symptoms, provide a tool for follow-up of behavioral changes over time and enable researchers to measure the effects of behavioral and pharmacological intervention.

Treatment of noncognitive features alleviates suffering in patients and carers. Treatments that delay the emergence of the behavioral symptoms or reduce existing symptoms would be expected to have beneficial effects on patients and their families (Knopman et al., 1996). The relationship of these early neurologic deficits to the patient's ability to perform independently in ADLs is not fully understood. There are neuropsychological tests that can identify areas of function and dysfunction.

11.2 Functional abilities and activities of daily living assessment

The maintenance of baseline levels of functional abilities compared with placebo treatment should mean an improvement in quality of life for patients and carers. Also, the maintenance of ADL may allow a patient to live in an independent fashion for longer and may delay costly institutionalization (Wimo et al., 1998).

11.2.1 Blessed-Roth dementia scale

The Blessed-Roth Dementia Scale has been widely used in dementia research in senile dementia (Blessed et al., 1988). The scale consists of 22 items, half of which measure changes in functional abilities. Eight items measure changes in performance of everyday activities, three measure changes in performance of everyday activities, three measure changes in self-care habits and 11 assess behavioral changes. The scale has been shown to be a valid measure of the severity of dementia in that it is correlated with cognitive performance and aspects of the pathology of AD (Blessed et al., 1988).

11.2.2 Progressive deterioration scale

The PDS was used in clinical trails of rivastigmine, donepezil and tacrine (Rösler et al., 1999; Knapp et al., 1994; Winblad et al., 1999). It is composed of 29 items rated by a caregiver, largely addressing ADL but also including items reflecting behavior and quality of life (DeJong et al., 1989). Scores range from 0 to 100. Studies using this scale as a measure of ADL have been criticized because the scale measures a range of domains and the resulting score does not necessarily assess changes in ADL (Bentham et al., 1999).

11.2.3 Interview for deterioration in daily living activities in dementia

The IDDD was designed as a severity instrument for quantifying impairment of patients' ADL. It is a 33-item structured interview of caregivers consisting of self-care items (e.g., washing, dressing, eating) and complex activity items (e.g., shopping, writing, answering the telephone) performed by men and women (Teunisse et al., 1991). Descriptions are elicited and frequency of assistance is rated on a three-point scale for each item. Both the initiative to perform activities and the performance itself are evaluated.

11.2.4 AD cooperative study – Activities of daily living inventory

The AD Cooperative Study – Activities of Daily Living (ADCS-ADL) inventory was developed by the ADCS to assess functional performance in patients with AD (Galasko et al., 1997). In a structured interview format that takes less than 15 min to administer, informants are asked whether patients attempted each of 24 items in the inventory during the previous 4 weeks and if so, to comment on their levels of performance. The ADCS-ADL scale has good test-retest reliability and clearly distinguishes the stages of severity of AD patients from very mild to severely impaired.

11.2.5 Alzheimer disease functional assessment and change scale

The Alzheimer Disease Functional Assessment and Change Scale (ADFACS) is used for the assessment of ADL in patients with AD with particular reference to outcomes in clinical trials (Galasko et al., 1997). It is informant-based and takes 20 min. The scale has been used in clinical trials and consists of ten items for instrumental ADL: ability to use the telephone, performing household tasks, using household appliances, handling money, shopping, preparing food, ability to get around both inside and outside the home, pursuing hobbies and leisure activities, handling personal mail and grasping situations or explanations. These are rated from no impairment to severe impairment. The scale was developed from 45 activities of daily living items. Changes on some of the items have been shown to be sensitive to change over 12 months and to correlate with the MMSE with excellent test-retest reliability (Galasko et al., 1997).

11.2.6 Disability assessment for dementia

The DAD scale was developed specifically for use in patients with AD (Gelinas & Auer, 1996; Gelinas et al., 1999). The DAD scale includes 17 items related to basic self-care and 23 to instrumental ADL. The instrument is completed either by the caregiver as a questionnaire or is administered as a structured interview (Gelinas et al., 1999).

11.3 Behavioral assessment

It is generally agreed that behavioral abnormalities are common in dementia and that most behavioral change is not thought to be solely a consequence of cognitive impairment. Some altered behavior may be manifested before there is much change in cognition and may be the presenting feature indicating a high risk for subsequent development of dementia. Therefore, it is important to adequately assess behavioral disturbances in AD. The best approach for behavioral assessment appears to be directly observable behavior documented by a trained clinician with a well-validated rating instrument (Patterson & Bolger, 1994; Hope & Fairburn, 1992).

The use of behavioral scales is an important component in determining efficacy of new drugs in clinical trials for AD. Behavioral assessment in clinical trials must be sensitive to disease heterogeneity, disease progression and drug modification of behavior. Three such scales, the behavior pathology in AD rating scale (BEHAVE-D), the consortium to establish a registry for AD behavior rating scale for dementia (C-BRSD) and the CMAI are useful in clinical trials. The BEHAVE-AD, C-BRSD and CMAI scales are valid, reliable, rapid to administer, cover relevant behaviors occurring during the course of the disease and are appropriate for use in AD clinical trials.

11.3.1 The dementia behavior disturbance

The Dementia Behavior Disturbance (DBD) scale was developed in the late 1980s, a time when the importance of behavioral symptoms in dementia was increasingly being recognized. The DBD scale was designed to quantify specific observable behaviors usually associated with dementia through an interview with the patient's primary caregiver (Baumgarten et al., 1990).

11.3.2 The eating behavior scale

Eating is complex behavior that requires attention, initiation, conceptualization, visuospatial abilities and planning. The deficits, identified through the neuropsychological testing support the observations of mealtime difficulties in AD patients.

The Eating Behavior Scale (EBS) was developed to measure AD patients' functional ability during eating (Tully et al., 1998) (see Table 11.1). The EBS incorporates the cognitive components of functional abilities providing a greater depth of information for assessment purposes. The EBS is an easy-to-use observational tool that does not place an additional burden on the patient. Accurate assessment of the impact of interventions can be made by using such a standardized tool. The significant correlations between the EBS and other neuropsychological tests support the validity of this measurement tool (e.g., correlation coefficients between EBS scores and MMSE = 0.8210, Mattis Dementia Rating Scale = 0.8865, Raven Colored Matrix sub-test = 0.9912 and Ravens Colored Progressive = 0.9912). Use of the EBS can provide the family and other care providers with a realistic, objective appraisal of the patient's current functional status so that appropriate interventions can be implemented.

11.3.3 Neuropsychiatric inventory

The Neuropsychiatric Inventory (NPI) was developed to assess psychopathology in dementia patients (Cummings, 1997). It is a structured interview of a caregiver or informant addressing 10 (or 12 in later versions) behavioral domains common in dementia: agitation, irritability, anxiety, dysphoria, hallucinations, delusions, apathy, euphoria, disinhibition, aberrant motor behavior, night-time disturbances and appetite and eating abnormalities (Cummings, 1997). In addition to assessing the severity and frequency of each symptom, the NPI also assesses the amount of caregiver distress engendered by each of the disorders.

11.3.4 Behavioral pathology in Alzheimer disease rating scale

The Behavioral Pathology in Alzheimer disease Rating Scale (BEHAVE-AD), a 25-item assessment scale, is designed for assessment of outpatients with mild to severe AD (Reisberg et al., 1987). The BEHAVE-AD measures a wide range of behavioral disturbances with assessment based solely on caregiver report. The BEHAVE-AD takes 20 min to administer by a clinician and was designed particularly to be useful in prospective studies of behavioral symptoms and in pharmacological studies to

Table 11.1: National Institutes of Health Warren G. Magnuson Clinical Center Nursing Department Eating Behavior Scale (EBS).

Patient:	Admit date	
Observation date	Observer initials	
Meal start	Finished	
Patient room	Day room	Time-minutes ....
Circle only one answer:	Maximal score = 18	Total = ...

Observed behavior	I.*	V.**	P.***	D.****
Was the patient:				
1. Able to initiate eating?	3	2	1	0
2. Able to maintain attention to meal?	3	2	1	0
3. Able to locate all food?	3	2	1	0
4. Appropriately use utensils?	3	2	1	0
5. Able to bite, chew and swallow without choking?	3	2	1	0
6. Able to terminate meal?	3	2	1	0

Comments:

**I = Independent, **V = Verbal prompts, ***P = Physical assistance, ****D = Dependent*

document behavioral symptoms in patients with AD. The BEHAVE-AD is a good choice of scale in clinical trials in which cognitive function and behavioral symptoms are the primary target of treatment (Kluger & Ferris, 1991). It is in two parts: the first part concentrates on symptomatology and the second requires a global rating of the symptoms on a four-point scale of severity. It focuses on seven behavioral domains: (1) paranoid and delusional ideation, (2) hallucinations, (3) activity disturbances (e.g., wandering or purposeless activity), (4) aggressiveness, (5) diurnal rhythm disturbances, (6) affective disturbances (e.g., tearfulness and depressed mood), and (7) anxieties and phobias. Each symptom is scored on a 4-point scale of severity, where 0 = not present, 1 = present, 2 = present, generally with an emotional component and 3 = present, generally with an emotional and physical component. The BEHAVE-AD also contains a 4-point global assessment of the overall magnitude of the behavioral symptoms in terms of disturbance to the caregiver and/or dangerousness to the patient.

The instrument had adequate inter-rater reliability for individual and total scale scores (Mack & Patterson, 1994; Sclan et al., 1996). Studies also support the validity of clinical information obtained from proxy report utilizing the BEHAVE-AD (Mack & Patterson, 1994). If present, behavioral symptoms are rated as mild, moderate or severe. The BEHAVE-AD and its subscales have been used in AD clinical trials assessing the efficacy of antipsychotics and cholinesterase inhibitors (Katz et al., 1999; Corey-Bloom et al., 1998; Rösler et al., 1999).

The BEHAVE-AD may help to differentiate FTD from AD and to characterize the behavioral features of FTD (Mendez et al., 1998). Patients presenting for an initial dementia evaluation who score high on the BEHAVE-AD are more likely to have FTD than AD. On the BEHAVE-AD, FTD patients tended to demonstrate verbal outbursts such as disinhibited or abrasive comments and inappropriate activity such as unacceptable behaviors and immodest acts. The BEHAVE-AD subscale scores also show that FTD patients have more aggressiveness and less anxiety or depression than AD patients.

Scores on the BEHAVE-AD are based on caregiver reports of symptoms occurring generally within the preceding 2 weeks. However, for various reasons, caregivers may overreport or underreport symptoms. Family caregivers may exaggerate symptoms because of the stress they cause. Professional caregivers, on the other hand, may downplay symptoms in an attempt to demonstrate their competence in caring for the patient. To overcome the possible bias of caregivers, an observer-rated instrument called the Empirical Behavioral Pathology in Alzheimer's Disease Rating Scale (E-BEHAVE-AD) was proposed (Auer et al., 1996).

The E-BEHAVE-AD consists of 12 items in the following six categories: paranoid and delusional ideation, hallucinations, activity disturbances, agitation, affective disturbances, and anxieties and phobias. Consequently, six of the seven BEHAVE-AD categories are assessed with the E-BEHAVE-AD. The only BEHAVE-AD category not assessed by the E-BEHAVE-AD is diurnal rhythm (sleep) disturbance. An observer rates each behavior during a clinical interview with the patient apart from the caregiver. Reliability of the E-BEHAVE is similar to that for the BEHAVE-AD. The utility, construct validity and responsivity of this rating scale in evaluations of psychotropic treatment intervention are supported by several studies.

11.3.5 The committee of the consortium to establish a registry for Alzheimer disease

The CERAD behavior rating scale for dementia was developed in 1986 to provide a standardized method of evaluating a wide range of behaviors seen in patients with varying degrees of dementia. CERAD has developed valid and reliable methods and assessment forms for the standardized evaluation of AD (Morris, 1997).

Items on the scale are well anchored and homogeneously scaled according to frequency, not severity. The scale can be administered by interviewers with limited clinical experience in approximately 20–45 min. An instruction manual, demonstration videotape, scoring manual and normative data are available. The scale was developed to correlate changes in behavior with selected characteristics of dementia, such as age at onset, duration, severity, sex, rate of cognitive decline and neuropathologic or neurochemical features. It is also being used as an outcome measure in intervention trials conducted by the Alzheimer disease Cooperative Study and other investigators. The scale has been translated into several languages, permitting cross-national and cross-cultural studies. Given the length of the behavior rating scale for dementia, a shortened version is available for clinicians interested in screening only. The scale is also being modified for application to patients with severe dementia. To maximize respondent participation in a study of the prevalence, incidence and natural history of dementia, carefully selected CERAD-certified nurses could independently carry out assessments for the most common types of dementia either in the clinic or at home (Trapp-Moen et al., 2001). Use of nurses for such activity need not be limited to epidemiological studies but is relevant also in clinical practice.

11.3.6 The MOUSEPAD

The Manchester and Oxford Universities Scale for the Psychopathological Assessment of Dementia (MOUSEPAD) is a standardized assessment scale that includes both observed behavioral and psychiatric symptoms (Allen et al., 1996). The main indication for use of the scale is the assessment of psychiatric symptoms and behavioral alterations in patients with dementia. The MOUSEPAD is administered to carers by an experienced clinician and it takes 15–30 min, most items being given a 3-point severity score. It is uniquely designed to allow simultaneous evaluation of the month before interview and the whole of the dementia syndrome. It does not include sections on depression because these symptoms are satisfactorily covered in brief tests that are already available.

The test-retest and interrater reliability are both high for this instrument. For the interrater reliability, the values ranged from 0.56 to 1.0 excluding reduplications and severity scores ranged from 0.61 to 1.0. Four measures of change can be derived from the MOUSEPAD: change in symptom count ever, change in symptom count in the last month, change in symptom severity ever and change in severity in the last month. The MOUSEPAD is likely to be a useful outcome measure because the test is relatively brief and yet covers the noncognitive features associated with dementia in some depth. The MOUSEPAD distinguishes between symptoms and behavior and achieves a satisfactory compromise between brevity and breadth.

11.3.7 The Cohen-Mansfield agitation inventory

The CMAI was developed to evaluate agitated behaviors in nursing home residents with AD but has been adapted for use in outpatient in clinical trials (Cohen-Mansfield, 1989). It was designed to measure agitation, which is defined as “inappropriate verbal, vocal or motor activities not explained by apparent needs or confusion” in demented and non-demented nursing home residents (Cohen-Mansfield et al., 1989). The 7-point rating system of the CMAI assesses 29 different agitated behaviors in patients with cognitive impairment. It takes 10–15 min and is carried out by carers. Training is necessary. The scale rates the frequency of each behavior in the prior 2 weeks, using a 7-point scale with a range from 0 (never) to 6 (several times an hour). An additional “not applicable” category is used to rate those behaviors that could not possibly be performed by the patient (e.g., kicking for a paraplegic person).

The agitated behaviors include wandering, aggression, inappropriate vocalization, hoarding items, sexual disinhibition and negativism and are rated on a 7-point scale of frequency. In attempts to understand the interrelationships of behavioral symptoms, a number of factor-analytic studies of the CMAI have been performed. The instrument was conceptualized as measuring agitation in two dimensions, verbal and physical, each of which has two poles, aggressive or non-aggressive. Studies based on this conceptualization found behavioral abnormalities of elderly subjects to be comprised of three main factors: verbally aggressive behavior, verbally non-aggressive behavior and physically non-aggressive behavior (Cohen-Mansfield, 1986; Cohen-Mansfield et al., 1995). Weiner et al. (2002) explored the applicability of the standard scoring of the CMAI in community-dwelling persons with AD. Analysis of the data suggested that conventional CMAI subscoring did not adequately describe the responses of these patients. Overall, the CMAI seemed best suited to describe the overall level rather than the specific subtypes of behavioral dyscontrol in community-dwelling persons with AD.

Recently, a French version of the test was developed. The results demonstrate that the inter-rater reliability (r = 0.72), test-retest reliability (r = 0.72), internal consistency (Cronbach's alpha from 0.75 to 0.77), concurrent validity (r = 0.74), and construct validity of this French version of the CMAI to be significant (Deslauriers et al., 2001).

The CMAI reliably detects behaviors that occur frequently in AD patients and normal controls (Koss et al., 1997). An important feature of the CMAI is that behaviors are operationally defined and it offers a focused description of behaviors commonly observed in demented persons. The CMAI represents a promising way to characterize agitated behaviors in AD patients and it can be used as an instrument to assess agitation in clinical trials designed to test the efficacy of pharmacological treatment for cognitive and behavioral problems.

11.3.8 The geriatric mental state schedule

The Geriatric Mental State Schedule (GMSS) is one of the most widely used instruments for measuring a wide range of psychopathology in older people in all settings (Copeland et al., 1976). The GMSS can be administered *via* a laptop computer, and has been translated into a number of different languages. It has to be administered by a trained interviewer. The use of the GMSS is limited to research.

11.3.9 The geriatric depression scale

The Geriatric Depression scale (GDS) is a self-report scale designed to be simple to administer and not to require the skills of a trained interviewer (Yesavage et al., 1983). Each of the 30 questions has a yes/no answer with the scoring dependent on the answer given. A sensitivity of 84% and specificity of 95% have been documented with a cutoff score of 11/12. A cutoff of 14/15 decreased the sensitivity rate to 80% but increased specificity to 100%.

11.3.10 The Cornell scale

The Cornell Scale is specifically for the assessment of depression in dementia and is administered by a clinician (Alexopoulos et al., 1988). It takes 20 min with the carer and 10 min with the patient. The 19-item scale is rated on a 3-point score of "absent", "mild or intermittent" and "severe" symptoms with a note when the score is unavailable. A score of 8 or more suggests significant depressive symptoms. It is the best scale available to assess mood in the presence of cognitive impairment.

It differs from other depression scales in the method of administration rather than in analysis of any different symptoms profile seen in depression with dementia compared with depression alone (Purandare et al., 2001).

11.3.11 The consortium to establish a registry for AD behavior rating scale for dementia

The consortium to establish a registry for AD Behavior Rating Scale for Dementia (C-BRSD) takes about 30 min to administer, evaluates the frequency of behavioral symptoms in AD and provides a psychopathologic scale of clinically relevant items (Tariot et al., 1995). Unlike the BEHAVE-AD, which assesses severity of symptoms, the C-BRSD evaluates the frequency of behavioral symptoms in AD. The C-BRSD was developed as a survey instrument for the assessment of community-dwelling persons with AD with mild to moderate cognitive impairment. It is administered in an interview with the patient's primary caregiver and consists of 51 items rated according to frequency of occurrence on a scale of 1 to 4 (1 = occurred on only 1 or 2 days during the past month; 4 = occurred on 16 or more days in the past month). The original version was divided into eight factors: depressive features, psychotic features, defective self-regulation, irritability/agitation, vegetative features, apathy, aggression and affective lability (Tariot et al., 1995). A second version was developed with six subscales: depressive symptoms, psychotic symptoms, inertia, vegetative symptoms, irritability/aggression and behavioral dysregulation (Patterson & Mack, 1996). Scoring of this instrument is by subscale and total score.

Although the C-BRSD is quite new, there are already considerable data indicating its validity in terms of identifying behavioral problems that are associated with AD. Tariot et al. (1995) who examined results for individual items and Patterson et al. (1997) who reported total C-BRSD scores both found the C-BRSD to pick up behavioral problems in AD. Studies of C-BRSD reliability have been generally encouraging (Tariot et al., 1995). Test-retest reliability for a total C-BRSD score over a 1-month interval was also found to be satisfactory (Patterson et al., 1997).

11.4 Global function assessment

Global measures fall into two broad categories: those that assess disease severity and those that assess change. Severity measures help to provide a structure by which the clinician can assess a patient's dementia although staging dementia on the basis of severity alone does not necessarily assist management or predict future course of the disease. Change scales differ from severity scales in that they only assess change from a patient's baseline at the beginning of a trial.

In order to measure the severity or progression of established dementia, several global clinical scales have been developed to monitor changes in function brought about by an individual's intellectual decline in the context of his/her past performance. These scales are based on a multidimensional assessment of cognitive, functional and behavioral aspects of the disorder using a semi-structured interview method. This approach bypasses many of the well-known confounding factors such as education, age, culture and practice effects that affect standardized cognitive and neuropsychological testing.

11.4.1 The clinical dementia rating scale

The CDR scale is the simplest of the global assessment scales. It is to be completed by a clinician after an examination of the patient and an interview with an informant (Hughes et al., 1982; Berg, 1984). Used in this way, the CDR is known to be a reliable and valid instrument for both staging and screening (Berg et al.,1988; Juva et al., 1995). Both the patient and an informant participate to rate performance in six domains: memory, orientation, judgment and problem solving, community affairs, home and hobbies and personal care. A value of 0 in the current expanded version indicates no impairment, 0.5 indicates questionable impairment (see Table 11.2). The values of 1, 2, 3, 4 and 5 indicate mild, moderate, severe, profound and terminal cognitive impairment, respectively. The scale is ordinal, but not necessarily equal interval. For personal care, no distinction is made between levels 0 and 0.5. Dooneief et al. (1996) who validated the extended components of the CDR scale reported that the "profound" and "terminal stages" predicted shortened survival and that the scale should be useful in studies of the later stages of dementia. Subjects with CDR = 4 or 5 are measurably more demented than subjects with lower CDR ratings and have poorer prognosis for survival. The CDR is commonly used as a staging measure of severity but can be utilized as a diagnostic tool because it does discriminate "no dementia" from "dementia" (Morris, 1993). Although it was designed for use in clinical situations, the scale has also been used in epidemiological studies (Forsell et al., 1992; Juva et al., 1995). Interrater reliability on the CDR has been good, as detailed by Burke et al. (1988) with Kendall's tau B = 0.91, kappa = 0.74, weighted kappa = 0.87. A study that assessed the interrater reliability of the CDR in a multicenter clinical trial showed that the overall interrater reliability was 0.62 (Rockwood et al., 2000). Within the CDR domains, the global kappas ranged from 0.33 ± 0.06 to 0.88 ± 0.06.

The performance of the patient in each area is scored on the basis of the patient's cognitive loss only. Training in the CDR has been shown to result in moderate to high inter-rater reliability (Sano et al., 1993; Burke et al., 1988; McCullar et al., 1989). Because the six cognitive domains rated by the CDR are linked to validated clinical diagnostic criteria for AD, the CDR demonstrates content validity (Berg et al., 1982; Morris et al., 1988; Fillenbaum et al., 1996). This scale also proved to be a good screening tool for dementia (Juva et al., 1992). With the cutpoint 1 (mild dementia), the sensitivity was 95% and the specificity 94%. The information obtained in these six areas can be summarized in two ways. The scores in each area can be summed yielding a "Sum of Boxes" score with a potential scoring range of 0-30 (Berg et al., 1988). Alternatively, with memory as the principal component, the number of higher or lower scores among the other five cognitive domains is assessed and a prescribed algorithm is applied to identify the global CDR score (Morris, 1993). The range of possible global CDR scores is 0 through 5, the same as that of the component scores. Their meanings are 0, not demented; 0.5, dementia questionable (used for very early cases or those with incipient dementia such as persons with impairment in memory but in no other area). The integers 1 through 5 represent mild, moderate, severe, profound and terminal dementia. Both hand- and computer-based scoring are feasible (Morris, 1993; Gelb & St. Laurent, 1993; 1994).

The algorithm for computing the global CDR is straightforward as long as all the secondary scores cluster fairly tightly around the value of memory rated category. When the scores are not distributed uniformly, the algorithm sometimes produces a global CDR that seems inconsistent with the goal of the scoring system. In such situations, a small change in one of the category scores may result in an inordinately large change in the global CDR or even a global CDR change in the opposite direction. However, an alternative algorithm that eliminates these incongruities while retaining the fundamental features of the original method has been proposed (Gelb & Laurent, 1993).

A study designed to assess the validity of the CDR when the rating is based solely on informant data showed moderate agreement with the clinician CDR (Grayson et al., 1999). This suggests that it would be a valid substitute in situations in which the subject could not be examined. Using a CDR score of 1+ to indicate dementia, the informant CDR had a sensitivity of 80% and a specificity of 98% against the clinician's score. Tractenberg et al. (2001) reported that raters-in-training experienced the most difficulty with rating normal and questionable dementia. Memory appears as the most difficult domain to score for both naïve and experienced raters. Hence, they recommended greater emphasis on the memory, home and hobbies, and orientation domains during CDR training. Increasing the information provided for the judgment and problem solving domain on the standardized CDR worksheets could improve the consistency of raters and increase the efficiency with which they are trained to use the CDR.

11.4.2 The global interview-based change scales

This broad category of clinical assessments of change includes the CGIC, the CIBIC and the NYU-CIBIC-plus (Guy, 1976; Knopman et al., 1994; Corey-Bloom et al., 1998). These are different from other scales in that they assess only change from a specified baseline. Global change scales were used extensively as a primary outcome

Table 11.2: CDR Scale.

	Memory	Orientation	Judgment and problem solving	Community affairs	Home and hobbies	Personal care
None (0)	No memory loss or slight, inconsistent forgetfulness	Fully oriented	Solves everyday problems well, including financial and business affairs; judgment good in relation to past performance	Independent function at usual level in job, shopping, volunteer and social groups	Life at home, hobbies, intellectual interests well maintained	Fully capable of self-care
Questionable (0.5)	Consistent slight forgetfulness: partial recollection of events; "benign" forgetfulness	Fully oriented except for slight difficulty with time relationships	Slight impairment in solving problems, similarities, differences	Slight impairment in these activities	Life at home, hobbies, intellectual interests slightly impaired	Fully capable of self-care
Mild (1)	Moderate memory loss, more marked for recent events; defect interferes with everyday activities	Moderate difficulty with time relationships; oriented for place at examination; may have geographic disorientation elsewhere	Moderate difficulty in handling problems, similarities, differences; social judgment usually maintained	Unable to function independently at these activities though may still be engaged in some; appears normal to casual inspection	Mild but definite impairment of function at home; more difficult chores and more complicated hobbies and interests abandoned	Needs prompting
Moderate (2)	Severe memory loss; only highly learned material retained; new material rapidly lost	Severe difficulty with time relationships: usually disoriented in time, often in place	Severely impaired in handling problems,similarities, differences; social judgment usually impaired	No pretense of independent function outside home; appears well enough to be taken to functions outside home	Only simple chores preserved; very restricted interests, poorly sustained	Requires assistance in dressing, hygiene, keeping of personal effects

Table 11.2 (continued)

	Memory	Orientation	Judgment and problem solving	Community affairs	Home and hobbies	Personal care
Severe (3)	Severe memory loss; only fragments remain	Oriented to person only	Unable to make judgments or solve problems	No pretense of independent function outside home; appears too ill to attend functions outside home	No significant function in home	Requires much help with personal care; frequent incontinence
Profound (4)	Even fragments generally are lost; often unable to test memory because of unintelligible or irrelevant speech	Occasionally responds to own name	Unable to follow even simple instructions or commands	Unable to participate meaningfully in any social setting	Unable to engage meaningfully in any hobby or home activity	May attempt to dress or feed self; nonambulatory without assistance
Terminal (5)	No meaningful memory function; often uncomprehending or obtunded	No recognition of self	No awareness of problems or comprehension of surroundings	Completely unable to engage in any activity	Completely unable to engage in any activity	Needs to be fed; bed-ridden
Box Score						

0: No dementia, 0.5: Uncertain or deferred diagnosis; 1: Mild dementia; 2: Moderate dementia; 3: Severe dementia; 4: Profound dementia: 5: Terminal dementia criterion in early clinical trials for antidementia drugs, usually with the completely unstructured CGIC (Guy, 1996; Schneider & Olin, 1996).

The CIBIC was designed by researchers in consultation with the FDA in response to growing research activity concerning antidementia drug treatment. The CIBIC has since been developed to take account of information provided by a carer (CIBIC-plus). A widely used version of the CIBIC-plus includes textual data on the patients' history, general appearance, mental cognitive state, behavior, functional ability and 7-point Likert scales recording disease severity and changes during treatment (Olin et al., 1996). These scales have been shown to have face validity and predictive validity (Schneider et al., 1997; Knapp et al., 1994). The CIBIC-plus scales have been shown to be sensitive to longitudinal change in 24- and 30-week studies (Knapp et al., 1994; Schneider et al., 1997). At 12 months, the change scores of patients' global scales were significantly associated with the change scores of the CDR, GDS, MMSE and Functional Assessment Staging (Schneider et al., 1997). However, the criteria underlying global (numeric) scores of patients' changes have nevertheless had little formal evaluation. A study that assessed the reliability of the CIBIC-plus for a range of patients showed that levels of interrater agreement on a global rating of change that included carer information varied widely from patient to patient (Boothby et al., 1995). An important contributor to this variation was carer information. However, if carer information was presented before patient information, considerably better levels of agreement were obtained on the global rating than if the reverse order was used. Agreement on the CIBIC-plus can be improved by presenting carer data first and making sure that the carer is "believable". Therefore, carer information selected and presented properly may contribute to a more valid and acceptable clinical decision.

Studies that have used global measures of change have focused on the numeric scores to the exclusion of the written data. A study designed to explore both sets of data of the CIBIC-plus to assess efficacy in drug trials showed that there were no consistent effects (Joffres et al., 2000). The global (numeric) scales were inconsistently supported by the textual data. Improved standardization of note-taking in the CIBIC-plus textual data may allow for a better understanding of the typical profiles and clinical importance of changes seen in the course of dementia treatment. It is important that written clinical assessments need to capture those aspects of the clinical interview process that best ensure validity. The file format used should be consistent across participating investigators. Clinicians should use similar forms and understand how and when they have to be filled in while at the same time individualizing and giving context to changes in patients' AD.

11.4.3 The global deterioration scale

The GDS is a global cognitive deterioration scale for primary degenerative dementia. It defines six stages of dementia: stage 1, no cognitive decline; stage 2, very mild cognitive decline; stage 3, mild cognitive decline; stage 4, moderate cognitive decline; stage 5, moderately severe cognitive decline; stage 6, severe cognitive decline

(Reisberg et al., 1982). The use of this scale also requires the administration of a detailed psychometric evaluation. The GDS provides caregivers and practitioners an idea of where a patient is in the disease process by comparing the individual's behavior to the GDS. Thus, it is useful for describing and staging dementia. This test takes 2 min to score once the information has been collected. Inter-rater and test/retest reliability have been demonstrated regularly at over 0.90 for both (Resisberg et al., 1982).

11.4.4 AD cooperative study clinical global impression of change

The ADCS-CGIC is used by clinicians to address clinically relevant overall changes. It includes 15 areas under the domains of cognition, behavior and social and daily functioning (Schneider & Olin, 1996). For each area there is a list of sample probes, taken in part from existing rating rates, which the clinician uses as a guideline. It relies on both direct examination of the patient and interview of informants.

11.4.5 Clinician's global impression of change

The Clinician's Global Impression of Change scale is administered by a trained rater and takes 10–40 min (Guy, 1976). The ratings depend on the ability of the clinician to detect change, and any change that is clinically detectable is significant. By definition, these measures are global ratings of a patient's clinical condition and draw information from several sources. The scale has been used in many clinical trials of antidementia drugs. It can usefully assess change from a specified baseline (Knopman et al., 1994; Schneider & Olin, 1996).

11.5 Quality of life

Health-related quality of life (QOL) is regarded as an important element in the overall evaluation of health. It is increasingly used as an outcome measure in clinical trials. These measures are used to determine whether individuals consider themselves fundamentally better or worse off because of an intervention. This issue has become more important as people faced with choices among alternative treatment or therapies that have varying benefits and risks. There is no consensus on a definition of QOL in dementia, although there are some areas of agreement. The definition of QOL by the International Working Group on Harmonization of Dementia Drug Guidelines may be considered as it is quite broad (Whitehouse et al., 1997). It includes cognitive function and the ability to perform ADL, in addition to mood and affect. Logsdon et al. (2000) similarly include the evaluation of physical health, mood, interpersonal relationships, ability to participate in meaningful activities and financial status. Despite some difficulties, there seems to be an agreement that mood and affect are critical components of QOL. Patients' preserved abilities to experience positive emotions, feelings of belonging and enjoyment are universally recognized as important components of QOL. Relative absence of negative emotional experiences such as depressed mood and anxiety are also considered important.

Measuring QOL in patients diagnosed with dementia presents challenges. Lack of insight refers to the decreased capacity of patients with cognitive impairment to accurately describe their deficits. This lack of insight complicates the assessment of QOL in AD because it may affect patients" self report. Compared to caregiver perceptions, AD patients tend to overestimate their abilities and report less impairment (Ott et al., 1996; Vasterling et al., 1997). Since the cardinal feature of dementia is decline in cognitive abilities, and language impairment is an early symptom of all dementing conditions, it is assumed that self-reports of health status by patients with dementia are not valid. These patients are often unaware of their decreasing abilities, a feature known as anosognosia (Rossor, 1993). When interviewed, patients often describe themselves as "very well" and even after interviewer prompting may admit to no problems. Although patients in the early stages of dementia can answer questions about their QOL, it has been suggested that these patients are likely to give overly optimistic ratings (Lawton, 1994). Despite these challenges, reasons for measuring QOL in people with cognitive impairment are compelling. QOL assessments provide a format for individuals and their caregivers to express whether an intervention made an important difference in the patient's life. Such assessments allow researchers to draw conclusions about the extent to which treatments provide intended and clinically meaningful benefits. Furthermore, monitoring changes in QOL in individuals with progressive cognitive impairment may suggest new areas of intervention to maintain or improve life quality.

Although a number of conceptual, practical and ethical issues remain the subject of some debate, several scales to assess QOL in dementia have been developed. Researchers have used a number of methods including patient self-reports, semi-structured interviews, proxy reports and observational methods (Selai & Trimble, 1999; Albert & Logsdon, 1999). The DQOL instrument is an example of a valid and reliable dementia-specific scale that is based on direct interviews with patients suffering from mild to moderate dementia who have been shown to be able to respond to the questions appropriately (Brod et al., 1999). It represents five domains: self-esteem, positive effect and humor, negative effect, feeling of belonging and sense of esthetics. Assessments of QOL vary depending on how much patients are asked to participate in the evaluation. For example, assessment of QOL can involve a joint interview with the patient and caregiver (Ready et al., 2002) After the interview, the clinician incorporates information from both sources into QOL ratings. In another approach to measuring QOL in dementia, a clinician conducts separate interviews with patients and caregivers (Logsdon et al., 2000). The two sets of QOL ratings can then either be merged into a single rating or kept separate. This approach allows for systematic evaluation of differences and similarities in patient and caregiver perceptions. It is generally acknowledged that, in the later stages of dementia, proxy ratings of QOL are required since patients are no longer capable of making an evaluation (Stewart et al., 1996). However, proxies tend to underestimate the patient's QOL (Pierre et al., 1998).

A study that examined the agreement between self-reports and proxy reports of QOL in patients with mild to moderate AD and their proxies, the proportion of exact agreement between patients and proxies on the 17-item Duke Health profile ranged from 26.3 to 52.6% (Novella et al., 2001). These data reveal poor to moderate agreement (intraclass correlation coefficients from 0 to 0.61 for 10 subscales) between patients' and proxies' reports. Agreement was higher for measures of function that are directly observable (physical health, disability) and relatively poor for more subjective measures. Further, these data suggest that patients with mild to moderate AD are aware of, and able to report, their own health-related QOL than do family proxies. Agreement varies by QOL subscales and this discrepancy increases as the level of patients' cognitive impairment increases. Thus, assessments by proxies should be used with caution for patients with mild to moderate AD. The use of a proxy is useful for severe AD because the rate of nonresponse is much higher in that case. Patients with MMSE scores of 10 or greater can usually participate in an interview about their QOL to some degree, providing data that are as reliable as caregiver reports (Logsdon et al., Selai et al., 2000).

A promising method to measure QOL in dementia is an approach that incorporates patient and caregiver reports. Joint assessment in dementia has been recommended for completion of an activity schedule, where caregivers are helpful for rating characteristics of activities and patients may provide input about enjoyment (Teri & Logsdon, 1991). This approach has advantages over methods that rely solely on either patient or caregiver report because subjective patient perception is incorporated into the assessment and caregivers assist in recall of relevant events and symptoms that the patient may forget. Ready et al. (2002) described the development of a QOL scale (the Cornell-Brown Scale) based on the conceptualization that high QOL typically is indicated by presence of positive affect, satisfaction, self-esteem and the relative absence of negative affect. A clinician completes the new QOL scale after a joint interview with patients and caregivers. Preliminary data indicate that the Cornell-Brown Scale for QOL is a brief, easily administered, reliable and valid measure of QOL.

Research on QOL in AD has confirmed the intuitive notion that QOL declines as the disease progresses and is adversely influenced by associated features of dementia such as psychiatric problems and functional decline (Albert et al., 1996; Karlawish et al., 2001). Participation in some activities hypothesized to be associated with QOL were not found to decrease as dementia severity increased, such as spending time with the family, while participation in other activities such as reading declined significantly. Strategies to improve patient QOL would be to provide better education and training to family caregivers of AD patients.

An important distinction has been emerging between "health-related quality of life" (HRQOL) and "individual quality of life" (IQOL) (Joyce et al., 1999). IQOL is based on a growing consensus that the individual concerned should, where possible, be the primary source of information regarding his/her QOL. To date there has been inadequate consideration of the individual nature of each caregiver's situation, perceptions and needs (Zarit, 1994). A need for more time away from the patient is a major QOL concern for highly burdened caregivers and a perceived lack of adequate informal support and financial constraints are contributory factors (Coen et al., 2001).

11.5.1 Concept of utilities

Utilities are ratings of preferences for a particular state of health. Perfect health receives the highest rating of 1 with poorer health states associated with lower ratings. This concept assumes that lower ratings (less preferred states) are associated with poorer quality of life. A low utility assigned to a health state indicates that the health state is less desirable than other states and it is ranked lower than a state of optimal health. The utility rating expresses the rater's perception of the degree to which the health state is worse than optimal health (Patrick & Erickson, 1993). The utility rating for a particular state of health can be considered an indicator of health-related QOL. Utilities may be established for any type of outcome, allowing the comparison of QOLs across different types of efficacy assessments.

Utility-based quality of life measures have played an important role in economic analyses because they allow for quality of life comparisons across a wide range of diseases. Generic, preference-based health status indexes such as the European Quality of Life Instrument (EQ-5D), Health Utilities Index (HUI) and Quality of Well Being (QWB) Scale are commonly used to measure health status and to generate utilities for use in cost-utility analysis. Of the three utility-based measures, the QWB scale included the largest number of quality of life attributes.

11.6 Discussion

The CGIC and the CIBIC are the general observational evaluations of cognition and behaviors most frequently used in drug trials. They are subjective in nature and not psychometrically structured or validated. The CIBIC is a global evaluation of change used to determine whether the effects of a drug are important enough to be detected by a skilled and experienced clinical observer during a clinical interview. In this kind of evaluation, every CIBIC rater involved in a drug trial uses his own instruments and questions to evaluate the clinical change, even though guidelines with general areas of cognition that need to be observed are generally provided by the trial sponsors. The CGIC has the same characteristics as the CIBIC except that it involves the presence of the caregiver at every assessment to confirm the accuracy of the patient's statements.

Despite their lack of formal structure, there is a general opinion that CGIC ratings are sensitive to clinically meaningful effects. CGICs are not intended as sensitive measures of small changes that are unlikely to be clinically meaningful. In principle, a clinician rating a patient as changed on a global change scale is determining "clinically meaningful and distinct change."

The CMAI and C-BRSD appear to be interchangeable as measures of agitation with the CMAI possibly more useful for patients who lack language and the C-BRSD more sensitive to apathy and depression (Weiner et al., 1998). Recently, the Agitated Behavior in Dementia (ABID) scale, a new measure for agitation in AD outpatients, was proposed (Logsdon et al., 1999). The ABID provides a reliable and valid assessment of both frequency and caregiver reaction to common agitated behaviors experienced by individuals with dementia. It is brief, easy to administer and focuses on observable behaviors. However, further research is required to examine the use of ABID in population-based samples of dementia subjects and assess its sensitivity to change over disease progression and in treatment outcome studies.

Informant ratings of both cognitive and functional decline may be free of educational bias and may supplement the results of testing. Because of the special requirements and the nature of a dementing illness, it is particularly important that clinicians should take both their own perception of the patient's behavioral problems and the caregiver's perception into account before arriving at a decision about a specific treatment or dosage adjustment and before coming to a precise determination of the treatment goal.

An important research goal should be the development and evaluation of new instruments in relevant domains that are sensitive, reliable and valid for detecting change in normal ageing and early AD. Such tools should be self-administered and do not require significant involvement of professional staff, and this can minimize requirements for data monitoring and data entry. New uses of technology such as computerized assessments and telephonic methods are required in this area.

The assessment of QOL in patients with cognitive impairment is an important and fruitful area for future investigation. It has been assumed in the past that patients with dementia would not be able to rate their own QOL because of the nature of their disease. However, recent investigations suggest that this is not the case and that individuals can rate their own life quality well into the progression of the disease (Logsdon et al., 2002). Caregiver ratings do not substitute for patient ratings. Future research should address factors that affect both patient and caregiver ratings and identify strategies to improve the accuracy and reliability of evaluations of QOL and other subjective states in cognitively impaired older adults.

Part III
Treatment of Alzheimer disease

Chapter 12
Pharmacological treatment of cognitive deficits

12.1 Introduction

Since the ACh deficit associated with AD was recognized in the mid-1970s, there has been a systematic attempt to increase CNS cholinergic activity by pharmacological manipulations. The main classes of drugs tested so far include ACh precursors, releasing agents, cholinesterase inhibitors (CI) and muscarinic agonists. The expectations were that cholinergic enhancement would improve symptoms of AD in a way similar to dopamine enhancement in Parkinson's disease.

Aβ amyloid accumulation is one core pathological feature of AD. How does this accumulation relate to the neuronal degeneration that is manifest as a progressive cognitive impairment with widespread neurologic and neuropsychiatric disturbances? An answer is slowly emerging in which the Aβ amyloid induces a variety of neurotoxic phenomena, including reactive oxygen species. To date, however, only the secondary degenerative effects have been amenable to therapy, as seen in the beneficial effects of cholinergic-enhancing strategies. CI are currently being used for the symptomatic treatment of patients with AD, based on the consistent finding of a cholinergic deficit in the brain of these patients. Their clinical effectiveness is best described as a stabilization of cognitive and functional symptoms over one year, with a slower decline thereafter (Winblad et al, 2001). The recent addition of memantine, acting on glutamatergic NMDA receptors, adds a new dimension to the treatment of AD.

One of the limitations of conventional CI is that efficacy may not be sustained long-term because of the development of drug tolerance and disease progression. This has led to interest in therapies with more than one mode of action. In addition to the four licensed CI tacrine, donepezil, rivastigmine and galantamine, and the NMDA receptor agonist memantine, there are many drugs awaiting approval or in phase III trials (Table 12.1). While drugs specifically targeting the Aβ amyloidogenic pathway are only now beginning to emerge in a preclinical setting, most others are directed at the cholinergic system. There are many psychotropic agents available to treat the behavioral manifestations of AD, including antipsychotics, agitation-reducing drugs, antidepressants, anxiolytics and sedative-hypnotics. Interventions in AD include treatment of the underlying disease process and amelioration of neurochemical deficits produced by the cellular changes. This chapter discusses current perspectives in the pharmacotherapy of AD and examines how the different disciplinary approaches are being incorporated into clinical research for effective drug

Table 12.1: AD-relevant drugs that are awaiting approval or are in phase III trials.

Status	Drug names	Mechanisms of action
Awaiting approval	Idebenone	Antioxidant
	Nebracetam	M1-muscarinic receptor agonist
	Nefiracetam	M1-muscarinic receptor agonist
Phase III	Amridin	AChE inhibitor
	Eptastigmine	AChE inhibitor
	Cevimeline	M1 agonist
	Milameline	M1 agonist
	Talsaclidine	M1 agonist
	Xanomeline	M1 agonist
	Dehydroepiandrosterone	Neurosteroid
	Montirelin	ACh release stimulator, TRH agonist
	NS	Nootropic agent, ACh/GABA modulator
	Selegiline	Monoamine oxidase-B inhibitor
	Taltirelin hydrate	Thyrotropin-releasing hormone agonist

treatments. Attention is drawn to new compounds with novel mechanisms of action that could have a tremendous impact in the future treatment of AD.

12.2 Modulation of the cholinergic system

The cholinergic hypothesis claims that low levels of ACh lead to cognitive decline. Different strategies have been developed to boost the cholinergic system, including increased ACh production with cholinergic precursors (choline and lecithin), prevention of synaptic ACh destruction with CI (physostigmine, tacrine (9-amino-1,2, 3,4-tetrahydroacridine; Cognex), donepezil (E2020; Aricept), rivastigmine (SDZ ENA 713; Exelon), galantamine (Reminyl)), or direct stimulation of postsynaptic muscarinic receptors with receptor agonists (for example, Xanomeline, which is an M1 muscarinic agonist). At the present time, CI are the only cholinergic drugs that have been shown to produce significant improvements on cognitive, behavioral and global performance indices in large controlled trials of patients with AD.

Evidence now indicates that some CI may also provide neuroprotective effects, perhaps through the activation of nicotinic receptors, and may even enhance neurotrophic regeneration. Other possible actions include the effect of cholinergic agonists on the processing and secretion of APP and Aβ (Jope et al., 1997; Nitsch et al., 1992).

12.2.1 Tacrine

Tacrine hydrochloride is a reversible inhibitor of the enzyme AChE. After the initial positive and overly optimistic reports in 1986 on the efficacy of tacrine, it was subsequently noted to be an even stronger inhibitor of the butyrylcholinesterase family of enzymes. More recently, apart from AChE inhibition, tacrine has been shown to possess a much broader pharmacological profile such as blockage of potassium channels, inhibition of the neuronal monoamine uptake processes and inhibition of MAO (Jossain et al., 1992). The efficacy of tacrine in alleviating some of the behavioral symptoms of AD might be related to these other pharmacological actions. The purported cognitive-enhancing effects of tacrine and the CI may be difficult to disentangle from their non-specific arousal/ behavioral effects that can be expected from all classes of cholinergic stimulants.

Tacrine undergoes extensive hepatic metabolism by the cytochrome P450 1A2 isoenzyme to at least three metabolites. The major metabolite, 1-hydroxy-tacrine (velnacrine) is active. Metabolites are excreted in the urine. Tacrine has mean bioavailability of 17%, with interindividual variability ranging from 2–36% (Table 12.2) (Parnetti, 1995). This low bioavailability is thought to be secondary to large presystemic clearance. Food appears to decrease the rate, but not the extent, of absorption. Tacrine is rapidly metabolized, with mean half-lives after single doses of 25 and 50 mg of 1.6 and 2.1 h, respectively, which requires a four times a day schedule. Clinical dosages of 80–160 mg/day usually achieve AChE inhibition of approximately 30%.

Favorable effects were observed using the PDS to assess the efficacy of the drug in a 6-week study involving patients with mild-to-moderate AD (Knapp et al., 1994). Tacrine is known to produce significant elevations in liver enzymes, notably the transaminases. These elevations are generally reversible and tend to be higher in women. Serious adverse effects of tacrine include liver toxicity that has weakened its position as a drug of choice (Knapp et al., 1994).

12.2.2 Donepezil

Donepezil hydrochloride is a reversible inhibitor of AChE approved as a symptomatic therapy for mild and moderate AD. Its bioavailability is approximately 100% with peak plasma concentrations occurring between 2 and 4 h after an oral dose. Food appears to have no significant effect on absorption. Donepezil has a mean elimination half-life of 70 h with significant interindividual variation; a once-daily dosage is recommended. Donepezil is highly protein bound (93–96%) to plasma proteins although it was not found to have any significant effects on the binding of other highly protein-bound drugs, including warfarin, digoxin and furosemide. The drug is metabolized in the liver by CYP2D6 and CYP3A3/4, as well as by glucuronidation (Mihara et al., 1993; Ohnishi et al., 1993; Tiseo et al., 1998). The following metabolites were recovered in significant amounts: 6-O-desmethyl donepezil, which has similar activity to that of the parent drug; donepezil-cis-N-

Table 12.2: Pharmacokinetic and -dynamic profiles of CI used for the treatment of AD

Drugs	**Mechanism of action**	$t_{1/2}$ **(hours)**	**Starting oral dose regimen**	**F (%)**	**Protein binding (%)**	**Common adverse effects leading to treatment discontinuation**	**Comments**
Tacrine	Cholinesterase inhibition	1.6–2.14	10 mg qid, titrated up to 40 mg qid	2.4–36	75	Nausea, vomiting, increased salivation, sweating, lacrimation and elevated transaminase levels	Metabolized by CYP1A2
Donepezil	Cholinesterase inhibition	50–70	5 mg qd, titrated to 10 mg qd	100	93–96	Nausea, diarrhea, vomiting; insomnia; muscle cramps	Metabolized by CYP2D6, 3A4
Rivastigmine	Cholinesterase inhibition	–	1.5 mg bid, monthly titration up to 6 mg bid	~ 35	43	Nausea vomiting, diarrhea; dizziness and headaches	Sulfate conjugation, renal excretion
Galantamine	Cholinesterase inhibition	5–7	4 mg bid, monthly titration up to 12mg bid	90–100	Negligible	Nausea, vomiting, agitation and sleep disturbances	Partial metabolism by CYP2D6

$t_{1/2}$ = half life, F = bioavailability

oxide; 5-O-desmethyl donepezil; and the glucuronide conjugate of 5-O-desmethyl donepezil. While various unidentified metabolites were present, more of the dose was recovered in the urine (23–57%) than in the feces (9–15%) (Tiseo et al., 1998). Donepezil and its metabolites are eliminated largely by the kidney (Tiseo et al., 1998). A dosage of 5 mg/day yields steady-state AChE inhibition as determined by red blood cell cholinesterase inhibition of approximately 64% (Parnetti, 1995).

In a 30-week phase III clinical trial of donepezil, both 5 and 10 mg treatment groups had ADAS-cog scores superior to placebo throughout the 6-month trial. Moreover, more than 80% of treated patients showed either improvement or no decline during the 6-month trial (Rogers et al., 1995). Efficacy for up to 2 years was examined in patients who completed a 30-week double-blind study and who continued in a long-term, open-label phase with donepezil. For an average of 40 weeks, patients were maintained at performance levels better than their original baseline scores. ADAS-cog scores collected for more than 2 years suggest that patients treated with donepezil maintained the same magnitude of benefit as in the beginning of the study, indicating that long-term use of donepezil may be beneficial (Rogers et al., 1995). Winblad et al. (2001) evaluated the long-term clinical efficacy and safety of donepezil *versus* placebo over 1 year in patients with mild to moderate AD. The benefit of donepezil over placebo was shown by the Gottfries-Brane-Steen total score at weeks 24, 36 and 52 and at the study end point (week 52). Advantages of donepezil over placebo were also noted in cognition and ADL assessed by the MMSE at weeks 24, 36 and 52 and the end point.

The long-term efficacy of donepezil (10 mg/day) was assessed in a multi-center non-randomized open-label study of 133 patients with mild to moderate AD (Rogers & Friedhoff, 1998). Assessments were made every 3 weeks for the first 12 weeks and subsequently every 12 weeks for up to 192 weeks. The interim analysis at 98 weeks showed that donepezil improved the ADAS-cog scores that remained above baseline for 38 weeks but scores began to increase from week 50. The average increase in the score was 6.6 points/year (Rogers & Friedhoff, 1998). If patients had not been treated with donepezil, the expected impairment rate might have been 11.6 points/year in patients with a baseline score of 27.4 points, equivalent in the severity to patients recruited for this study (Stern et al., 1994; Kramer-Ginsberg et al., 1988). At the final treatment, a mean annual rate of change from baseline was 6.07 points (Rogers et al., 2000).

A study was designed to evaluate the long-term efficacy, safety and tolerability of donepezil in the treatment of patients with mild to moderate AD with donepezil 5 mg/day for 1 month, after which an increase to 10 mg/day was encouraged (Rocca et al., 2002). Cognitive assessments were carried out prior to the start of the treatment and every 3 months for a period of 1 year. Donepezil improved cognition and global functioning and was well tolerated especially considering the long duration of the observation period. A 6-month study concluded that donepezil produced functional benefit (Burns et al., 1999). Significant improvement in ADL, as measured by a modified (not independently validated) IDDD was reported in the 10-mg donepezil group compared with placebo. Scores for the 15-mg group did not differ significantly from placebo (Burns et al., 1999). In patients with moderate-to-severe

AD, donepezil improved cognition, global functioning and behavior (Feldman et al., 2002).

In a 24-week, double-blind, placebo-controlled study that assessed the efficacy and safety of donepezil with AD in the nursing home setting, the primary outcome measure was the Neuropsychiatric Inventory – Nursing Home Version (NPI-NH). Secondary efficacy measures were the Clinical Dementia Rating (Nursing Home Version) – Sum of the Boxes (CDR-SB), MMSE and the Physical Self-Maintenance Scale (PSMS) (Tariot et al., 2001). Mean NPI-NH 12-item total scores improved relative to baseline for both placebo and donepezil, with no significant differences observed between the groups at any assessment. The mean changes from baseline CDR-SB total score improved significantly with donepezil compared with placebo at week 24. The differences in mean change from baseline on the MMSE favored donepezil over placebo at weeks 8, 16 and 20. No significant differences were observed between the groups on the PSMS.

In a phase III trial in Japan, a total of 268 patients were randomized to receive placebo or 5 mg/day of donepezil treatment for 24 weeks (Homma et al., 2000). For better tolerance, patients with donepezil took a 3-mg tablet for the first week. Donepezil showed significant improvement in the Japanese version of the Alzheimer's disease assessment Scale Cognitive Subscale (ADAS-J-cog) (Homma et al., 1992) and the Japanese version of the Clinical Global impression of Change (J-CGIC) (Homma et al., 2000) as primary efficacy measures. ADAS-J-cog scores improved by 2.54 points for the 5 mg/day donepezil group compared with the placebo group. The results of the J-CGIC showed that the rates of improvement in the donepezil and placebo groups were 52 and 22%, respectively, and the aggravation rates were 17 and 43%, respectively. For the secondary measures, significant improvements were observed in the CDR-SB and Mental Function Impairment Scale scores (Homma et al., 1991).

12.2.3 Rivastigmine

Rivastigmine belongs to a generation of potent brain-selective CI, acting both on AChE and BuChE. Rivastigmine is not metabolized by the hepatic cytochrome P450 oxidative enzyme system and exhibits appropriately 40% protein binding. It is rapidly and extensively metabolized primarily at CNS receptor sites *via* cholinesterase. Rivastigmine and metabolites are eliminated primarily through the renal system. Hepatic microsomal enzymes are not involved to any significant extent (Polinsky, 1998). The plasma half-lives of rivastigmine and the major metabolite are 1 h and 2 h, respectively. However, the cholinesterase inhibition in the CNS lasts much longer (average 10 h) than the short plasma half-life would predict. It is classified as an intermediate-acting or pseudo-irreversible agent based on the fact that while its plasma half-life is ultra-short, the half-life of ChE inhibition in the CNS is about 10 h (Jann, 2000). For example, through direct CSF sampling, Cutler et al. (1998) have demonstrated rapid and sustained dose-dependent inhibition of CSF AChE following rivastigmine doses of 1–6 mg twice daily. This is due to the fact that when rivastigmine's phenolic ZNN-666 metabolite is formed, it leaves behind a carbamate moiety that stays attached to the AChE receptor for up to 10 h. This effect prevents the hydrolysis of ACh. The G1 enzymatic form of AChE is found in greater proportion than the G4 form in the hippocampus as compared to the rest of the cortex. Rivastigmine has higher affinity for the G1 than G4 and therefore has greater selectivity in those regions more vulnerable to the pathological changes of AD. Renal excretion of the metabolite is the major route of elimination. The absolute bioavailability of rivastigmine is approximately 35% after a 3-mg oral dose. Because rivastigmine exhibits nonlinear pharmacokinetics, i.e. maximum plasma drug concentration (C_{max}) and area under the curve (AUC) increase more than proportionally with an increase in dose, it is necessary to determine its absolute bioavailability at the highest approved single dose of 6 mg.

Results of phase II studies in AD patients showed good toleration up to 12 mg/day. Side effects did not include hepatotoxicity. In a study that assessed 114 patients with mild-moderate AD that were randomly assigned to either rivastigmine twice daily or three times daily or placebo in a double-blind fashion, the treatment was titrated to their maximum tolerated dose over 10 weeks followed by an 8-week maintenance phase (Forette et al., 1999). Significantly more patients taking rivastigmine b.i.d were considered improved according to CIBIC-Plus *versus* placebo (57% *versus* 16%, respectively, $p = 0.027$). The Nurses' Observation Scale for Geriatric Patients and the ADAS-cog also improved in the rivastigmine b.i.d. group compared to the placebo group. Therefore, this study suggests that rivastigmine at doses of up to 12 mg/day has useful efficacy in patients with mild-moderate AD. Adverse events were generally gastrointestinal and mild to moderate following doses of up to 12 mg/day. Results of a meta-analysis of phase III studies demonstrated significant beneficial effects on measures of cognition, global functioning and activities of daily living. The Swiss regulatory authority approved rivastigmine in August 1997 for the treatment of mild to moderate AD. Two 6-month studies have been published showing favorable effects of rivastigmine on ADL (Corey-Bloom et al., 1998; Rösler et al., 1999). In both studies, high-dose rivastigmine had a significantly better outcome than placebo on the PDS.

12.2.4 Metrifonate

Metrifonate is a CI that acts as a prodrug for the direct, long-acting inhibitor 2,2-dichlorovinyl dimethyl phosphate (DDVP). Although withdrawn from active clinical development, it is worth reviewing some of its pharmacologic characteristics. Metrifonate has a mean half-life in blood of 2.3 h; the half-life of DDVP is 3.8 h. Thus, prolonged elevation of ACh levels can be achieved (Itoh et al., 1997). Estimates of the half-life for cholinesterase recovery vary depending on study methodology with an average of 26.6 ± 15.2 h. Metrifonate undergoes little protein binding ($< 15\%$), a property that serves to minimize potential drug-drug interactions. Although both metrifonate and DDVP undergo extensive biotransformation, they are metabolized independently of the hepatic cytochrome P450 enzyme system, a property that favors the absence of drug-drug interactions involving this system.

In a prospective, multicenter, 26-week, double-blind, parallel group study, the efficacy and safely of metrifonate were assessed in AD patients (Raskind et al., 1999). Cognitive performance was analyzed using the ADAS-cog and the MMSE. Psychiatric and behavioral disturbances were analyzed using the NPI and the ADAS-Noncognitive subscale. The DAD was used to assess the ability to perform instrumental and basic activities of daily living. Additionally, global state was assessed using the CIBIC-Plus scale. A statistically significant benefit of metrifonate was observed in the cognitive performance of AD patients. Metrifonate also significantly attenuated the deterioration in activities of daily living of the patients and relieved patients' psychiatric and behavioral disturbances. In addition, metrifonate significantly improved the scores for the global state of the patients. Unfortunately, metrifonate has been withdrawn from marketing consideration due to safety concerns regarding muscle weakness.

12.2.5 Galantamine

Galantamine (also called galanthamine and marketed as Reminyl), a naturally occurring amarylidaceae alkaloid, is a selective, reversible and competitive AChE inhibitor. Galantamine potentiates cholinergic nicotinic neurotransmission by modulating the response of the nAChRs to ACh (Albuquerque et al., 1997). Because galantamine acts at a site on the nAChR that is different from the ACh binding site, it is referred to as an allosteric potentiating ligand (Maelicke, 2000). Galantamine appears to enhance presynaptic and postsynaptic nAChR function by making nAChT more sensitive to available Ach. It is observed to improve the performance of AD patients on memory tests and to be well tolerated (Thomsen et al., 1990).

It is observed to improve the performance of AD patients on memory tests and to be well tolerated (Thomsen et al., 1990). Galantamine 16 and 24 mg/day significantly benefits the cognitive, functional and behavioral symptoms of AD as compared with placebo (Tariot et al., 2000). After 5 months, the galantamine-placebo differences on ADAS-cog were 3.3 points for the 16 mg/day group and 3.6 points for the 24 mg/day group ($p < 0.001$ *versus* placebo, both doses). Slow dose escalation appears to enhance the tolerability of galantamine, minimizing the incidence and severity of adverse events. In a 5-month pivotal study, galantamine significantly slowed the progression of behavioral symptoms in patients with mild-to-moderate AD (Winblad, 2001). These behavioral benefits were associated with reduced caregiver distress and translated into reduced caregiver time. The galantamine studies used two measures of functional abilities. These are the DAD scale and the ADCS-ADL scale (Gélinas & Auer, 1996; Gélinas et al., 1999; Galasko et al., 1997). Galantamine showed evidence of functional benefit in three pivotal phase III studies of up to 6 months duration (Winblad, 2001). Furthermore, galantamine stabilized instrumental and basic ADL in an open-label 12-month study. These benefits would be expected to make an important difference to the quality of life of patients and caregivers. Overall galantamine showed significant treatment effects at daily doses of 16-32 mg for trials of 3–6 months duration. Galantamine's adverse effects appear similar to those of other CI in that it tends to produce gastrointestinal symptoms acutely and with dosage increases.

12.2.6 Controlled release physostigmine

Physostigmine reversibly inhibits the catabolic enzyme AChE, thereby augmenting central cholinergic neurotransmission. In animal studies, physostigmine improved memory in aged primates and in primates with scopolamine-induced amnesia (Bartus & Dean, 1988; Bartus, 1978). In numerous small clinical trials in AD patients, physostigmine demonstrated improved cognitive function (Thal et al., 1989; Mohs et al, 1985; Stern et al., 1987).

The development of physostigmine as a potential treatment of AD has been hindered by its extensive first-pass metabolism and its short half-life (approximately 30 min). A new extended-release formulation of physostigmine salicylate (Synapton) yields sustained blood levels, permitting twice-daily dosing (Thal et al., 1989). A recent double-blind, 12-week study which investigated the efficacy of this extended-release formulation of physostigmine salicylate in 850 patients with mild to moderate AD showed that physostigmine demonstrated a statistically significant benefit compared with placebo on an objective test of cognitive function (van Dyck et al., 2000). In the intent-to-treat analysis of the double-blind, physostigmine-treated patients scored −2.02 points better than placebo-treated patients on the ADAS-cog ($p < 0.01$) and 0.33 points higher on the CIBIC-plus ($p > 0.02$). Physostigmine is associated with high frequency of gastrointestinal side effects. Nausea and vomiting were experienced by 47% of all physostigmine-treated patients during the double-blind phase. Given the rather high rate of gastrointestinal cholinergic side effects compared with other CI, the role of physostigmine remains to be determined.

12.3 Nicotinic cholinergic strategies

AD has been shown in a variety of studies to be accompanied by a dramatic reduction in nicotinic receptors in the cortex and hippocampus (Rinne et al., 1991; Schroder et al., 1995). Several groups have found that nicotine injections or nicotine skin patches significantly improve attention, memory and learning in AD patients (Wilson et al., 1995; Newhouse et al., 1997). Acute nicotine has been found to improve memory performance on a variety of tasks (Levin, 1992). Interestingly, unlike many cholinergic agonists, nicotine retains its effectiveness with chronic administration (Levin et al., 1990). This may be due to the fact that, unlike most other agonists, nicotine causes an up-regulation of its receptor (Wonnacott, 1990). Preliminary studies have found encouraging results for the use of nicotine as a therapeutic treatment in AD (Newhouse & Hughes, 1991; Newhouse et al., 1986;

1988). Nicotine has been found to be effective in counteracting memory deficits seen in aged monkeys and rats (Buccafusco & Jackson, 1991; Cregan et al., 1989).

Nicotinic agonists can protect cerebral cortical neurons against β-amyloid-induced cell death and glutamate excitotoxicity (Kihara et al., 1998; Akaike et al., 1994). Thus, a multitude of mechanisms might account for the beneficial effects of enhancing nicotinic cholinergic function in the CNS of patients with AD even to the extent of modifying disease progression (Maelicke & Albuquerque, 2000). Recent work with more selective nicotinic receptor agonists and antagonists in animal models is providing important information concerning the neural mechanisms for nicotinic involvement in cognitive function and opening avenues for development of safe and effective nicotinic treatments for clinical use (Levin & Rezvani, 2000).

Some studies have suggested that there is an inverse relationship between smoking and AD (Ulrich et al., 1997). A dose dependency of nicotine has not yet been shown for this protective action. Subcutaneous administration of nicotine has been claimed to improve attention and information processing in AD patients. Interestingly, these effects were more evident than memory improvement. A study reported that nicotine skin patches significantly reduced errors of omission on the Continuous Performance Test (CPT) task in nonsmoking patients with AD patients (White & Levin, 1999). However, nicotine applied in the form of dermal plasters did not induce any significant effect in the treatment for memory deficits in AD patients (Snaedal et al., 1996). Further clinical investigations are clearly required.

12.4 Modulation of other neurotransmitter systems

There is a growing body of evidence that disturbances of glutamatergic neurotransmission may underlie a mechanism of neurotoxic excitatory amino acids contributing cognitive deficits of AD (Müller et al., 1995). Age-related changes of NMDA receptors have been found in cortical areas and in the hippocampus of many species. Based on these findings, several strategies have been developed to improve cognition by the use of NMDA antagonists as neuroprotective agents to slow down the progression of dementia. These antagonists include dextromethorphan, memantine (a congener of amantadine) and nitroglycerine. Evidence from randomized clinical trials has shown that memantine is able to improve symptoms of dementia and slow decline: in two pivotal studies (Winblad & Poritis, 1999; Reisberg et al, 2003) it was observed to produce a functional improvement and reduces care dependence in severely demented patients. In patients with mild to moderate vascular dementia, 20 mg/day memantine improved cognition consistently across different cognitive scales (Orgogozo et al., 2001). After 28 weeks, the mean ADAS-cog scores were significantly improved relative to placebo. In the intention-to-treat population, the memantine group mean score had gained an average of 0.4 points whereas the placebo group mean score had declined by 1.6 points. The response rate for the CIBIC-plus defined as improved or stable was 60% with memantine compared with 52% with placebo. Current drugs in preclinical development include L-701252, LY-235959 and WIN-63480-2 (Table 12.3).

The actions of neuroactive steroids at the molecular level provide the basis for their modulation of a broad spectrum of physiological and pathological conditions. For example, the NMDA-receptor antagonism produced by estrogens and 3α-reduced antagonism produced by estrogens and 3α-reduced pregnane steroids might, in part, account for their neuroprotective properties (Weaver et al., 1997). Moreover, in view of the reported memory-enhancing effects of pregnenolone and dehydroepiandrosterone (DHEA) in animal studies, steroids that are negative allosteric modulators of $GABA_A$ receptors might become novel therapeutic agents for improving cognitive deficits (Flood et al., 1988). However, although DHEA concentrations decrease with age, no consistent relationship between DHEA concentrations and cognitive function or dementia in humans has been established. Furthermore a randomized study over 6 months showed no benefit of DHEA over placebo (Wolkowitz et al, 2003). Future studies need to clarify whether steroids produce real clinical benefits in the treatment of dementias.

12.4.1 Neurotrophic growth factors

Nerve growth factor (NGF) as the prototypic neurotrophic growth factor is intimately related to the maintenance of function of the cholinergic basal forebrain system. Forebrain cholinergic neurons are the only cells in the adult brain that express high amounts of the low affinity p75 receptors for NGF. NGF increases hippocampal acetylcholine and prevents cholinergic cell loss and atrophy after fornix lesions, indicating the potential utility of NGF as a neuroprotective treatment for basal forebrain cells in AD (Holtzman et al., 1993). An innovation in AD therapy may come from NGF-mimetic drugs. Protocols for therapeutic trials in patients have been considered. A first clinical trial with NGF in AD did not bring up clear results (Seiger, 1993).

Cerebrolysin (Cere) is a compound with neurotrophic activity (Satou, 1994). It is a brain-derived peptide preparation produced by biotechnological methods, using a standardized enzymatic breakdown of purified porcine brain proteins. In a model of fimbria fornix transection, Cere was able to prevent the degeneration of medial septal cholinergic neurons (Akai, 1992). Studies with Cere in patients suffering from dementia of different etiologies have shown significant clinical benefit (Kofler, 1990). In a randomized, placebo-controlled, double-blind study investigating the effects of a 4-week treatments with 30 ml of Cere in 120 patients suffering from mild to moderate AD, improvements of cognition, global rating and activities of daily living were noted (Ruether, 1994). In a follow up evaluation of these patients 6 months after cessation of treatment, improvements were still present (Ruether, 2000). These results were confirmed in other studies (Bae, 2000; Ruether, 2001). A recent double-blind, placebo-controlled study in which AD patients were injected intravenously with placebo or 30 ml Cere 5 days per week for 4 weeks showed that Cere treatment was well tolerated and resulted in significant improvements in the global score 2 months after the end of active treatment (Panisset et al., 2002).

Table 12.3: Some drugs in early clinical development for the treatment of AD.

Classes	Drug name	Mechanism of action	Comments
Anti-excitotoxins	L-701252	NMDA antagonist	Being developed for treatment of AD, epilepsy and cerebrovascular ischemia
	LY-235959	NMDA antagonist	Competitive antagonist; potential for AD and other CNS diseases
	WIN-63480-2	NMDA antagonist	Uncompetitive antagonist; does not produce phencyclidine-like effects in animals
Neurotropic	AIT-082	NGF agonist	Phase II clinical
	AK-30-NGF	NGF agonist	Monoclonal antibody, NGF delivery system
			Potent immune stimulation and memory enhancing properties
	NBI-106	NGF agonist	Recombinant protein; also in phase III for peripheral neuropathy therapy
	rhNGF	NGF agonist	
Hormonal	ABPI-124	Estrogen	Specific for CNS, not interacting with other tissues
			An estrogen agonist developed for treatment of AD in women
	Neurestrol	Estrogen	
Anti-inflammatory	SC-110	Anti-inflammatory agent	Phase I clinical
	GR-253035	Cyclooxygenase-2 inhibitor	IC_{50} to COX-1 of over 100 μM and a COX-2 IC_{50} of 0.14 μM. Entering phase I trials for treatment of AD
	NBI-117	Cytokine modulator	Reported to bind and activate newly discovered receptors of cytokine activin

Table 12.3 (continued)

Classes	Drug name	Mechanism of action	Comments
Anti-oxidants	ARL-16556	Anti-oxidant	Spin trapping effects that scavenge free radicals as well as ability to modulate the effects of nitric oxide indicates that it may have advantage over existing compounds. Currently in phase I clinical trials
	MDL-74180DA	Anti-oxidant	Preclinical. Analogue of α-tocopherol that inhibits *in vitro* and *ex vivo* lipid oxidation and protects mice against CNS trauma
Anti-amyloid	MDL-28170	γ-secretase inhibitor?	Dipeptide aldehyde (Z-Val-Phe-H), cysteine protease inhibitor
	Calpeptin	γ-secretase inhibitor?	Dipeptide aldehyde (Z-Leu-Nle-H), cysteine protease inhibitor
	MG 132	γ-secretase inhibitor?	This tripeptide aldehyde (Z-Leu-Leu-Leu- H) and others, at low concentrations inhibit cysteine proteases and the degradatation of the cytoplasmic domains of APP; at higher concentrations inhibit γ-secretase and the proteosome
	MW 167	γ-secretase inhibitor?	Substrate-based difluoroketone, aspartyl protease inhibitor
	SKF-74652	Anti-fibrillogenic	In a model of Aβ aggregation, IC_{50} of 28 μM

12.4.2 Decreasing the cellular reaction to neurodegeneration

Microglial cells, closely related to the macrophage series of cells in the periphery, increase in size and number in the AD brain. From this, and the presence of complement in amyloid plaques, the concept of AD as an "inflammatory" disease has emerged. It has been reported that individuals taking anti-inflammatory drugs such as NSAIDs have fewer cerebral microglia, and are less likely to develop AD, with a fairly consistent risk reduction of about 50% (Mackenzie & Munoz, 1998; Breitner et al., 1995). The greatest protection is observed in those with late-onset AD (> 70 years of age) and with a strong family history of AD. In one randomized, placebo-controlled study with indomethacin in 44 patients with mild AD, after 6 months of treatment, a slight improvement in cognitive function from baseline was noted in the treated group (Rogers et al., 1993).

New products with anti-inflammatory properties in clinical development for AD include SC-110 and GR-253035, a COX-2 inhibitor. NF-AB, a DNA-binding protein, has an important role in driving transcription from inflammation-related genes, such as COX-2, that operate in stressed tissues such as the AD brain. Blocking intervention at this stage may be a potential approach for treatment.

12.4.3 Propentofylline

Propentofylline (3-methyl-1-(5-oxohexyl)-7-propyl-xanthine), a xanthine derivative and inhibitor of adenosine re-uptake, has been reported to inhibit glutamate release and may increase cerebral blood flow, and is thought to act mainly on astrocytes and microglia. It appears to delay the progression of mild-to-moderate dementia in addition to providing symptomatic relief. In a 12-month randomized, placebo-controlled study of propentofylline in AD and vascular dementia, the total patient population showed statistically significant treatment differences in favor of propentofylline for the global measures of dementia as well as for cognitive evaluations (Marcusson et al., 1997). Despite these encouraging results, propentofylline has been withdrawn from active clinical research.

12.4.4 Drugs that reduce oxidative stress

The role of oxidative stress in the etiology of AD has been hypothesized and supported by a variety of experimental and clinical studies (Markesbery, 1997; Christen, 2000). There is increasing evidence that the AD brain is under severe oxidative stress, either as a direct result of Aβ-mediated generated oxyradicals or as the result of perturbed ionic calcium balances within neurons and their mitochondria. This research has promoted interest in assessing antioxidants for their possible benefits in modifying the course, reducing the risk or delaying the onset of AD (Grundman, 2000; Launer, 2000).

Drugs such as Vitamin E, idebenone, or estrogens that have strong antioxidant properties are showing variable degrees of efficacy (Sano et al., 1997; Gutzmann et al., 1998). Well-designed intervention trials as well as observational investigations based on large cohorts studied over long periods of time with several methods for assessing antioxidant exposure are ideally suited to test the hypothesis. Nonetheless, the idea that vitamin E and vitamin C might have beneficial effects on the underlying AD process makes sense and it seems unlikely that antioxidant-rich foods would negatively affect brain aging.

Newer antioxidants in clinical development include the free radical scavenger ARL-16556, an alpha-phenyl-t-butylnitrone derivative with "spin trapping" effects.

12.4.5 Estrogen replacement therapy

Destpite the lack of consensus, many studies indicate that hormone replacement therapy may decrease the risk for or delay the onset of AD in postmenopausal women. Estrogen replacement therapy (ERT) in healthy postmenopausal women favorably affects mood and may have modest effects on cognitive function. In addition to its antioxidant properties, estrogen may promote growth of cholinergic neurons, reduce plasma levels of Apo E, modify inflammatory responses or even directly reduce Aβ generation (Xu et al., 1998). The distribution of estrogen receptors (ERs) in neurons overlap that of the brain neurons known to develop AD (Isoe et al., 1997). However, a recent study designed to determine whether ERα gene polymorphisms are associated with transcriptional activity and AD reported no association between ERα gene polymorphisms and AD (Maruyama et al., 2000). Although there were racial differences in these polymorphisms, the previously reported association between ERa gene polymorphisms and AD was not confirmed.

A randomized, double-blind placebo-controlled clinical trial which involved 120 women with mild to moderate AD was designed to determine the long-term efficacy of ERT in a large population (Mulnard et al., 2000). The results of the study showed that ERT for 1 year did not slow disease progression nor did it improve global, cognitive or functional outcome in women with mild to moderate AD. The CGIC 7-point scale difference that was the primary outcome measure for estrogen *versus* placebo was not significant (5.1 *versus* 5.0; $p = 0.043$). Of participants taking estrogen, 80 *versus* 74% of participants taking placebo worsened. This study did not support the role of estrogen for the treatment of AD. The role of estrogen therapy may be confined to primary rather than secondary prevention of AD. However, the potential role of estrogen in the prevention of AD requires further research. Ongoing clinical studies may help to determine the role of estrogen in the cognitive function of postmenopausal women and in the prevention of AD.

New products in preclinical development include Neurestrol (a sustained-release formulation) and ABPI-124, a compound without feminizing side effects. The effects of raloxifene and tamoxifene (nonsteroidal, selective estrogen receptor modulators) in AD are yet to be determined.

12.4.6 Therapeutic strategies in the APP/Aβ amyloidogenic pathway

The gene dosage effect of trisomy 21 has shown that down-regulation of APP expression is a theoretical therapeutic strategy, but our understanding of the factors which regulate APP transcription are still too rudimentary. The normal function of APP also remains uncertain, placing some restriction on our ability to predict the unintended side-effects of APP down-regulation.

Most therapeutic research has been devoted to developing inhibitors of the β- and γ secretases which are responsible for the proteolytic cleavage events which generate Aβ. While the identity of these enzymes remains unknown, several companies have developed compounds that are efficient inhibitors of γ-secretase (Table 12.3). Most are still in preclinical development. Studies that implicate the presenilins in the γ-secretase pathway are also opening up new therapeutic strategies, although the involvement of presenilin in Notch signaling has caused some concern over the potential adverse effects of γ-secretase inhibitors (Hardy & Israël, 1999).

Compounds directed at the toxic effects of Aβ or stabilizing the aggregated forms of Ab to promote its clearance from the brain are now under active development. Further insight is required in understanding the roles of other proteins or lipids (cholesterol, for example) that interact with Aβ (such as Apo E and α2-macroglobulin) or with APP as it travels through the cell towards its biogenesis of Aβ. Inhibitors of Aβ production, that is, small compounds that cross the blood-brain barrier and decrease but do not eliminate either β- or γ-secretase activity, could be therapeutic in the early clinical phases of the disease, particularly in patients with MCI. In the case of γ-secretase inhibitors, these could be designed to decrease Aβ production by some 30–40%. Hopefully, such approach would not interfere with the Notch processing.

A remarkable approach has recently been described in which transgenic mice, immunized with human Aβ, showed attenuation of amyloid plaque formation (Schenk et al., 1999). This may represent a novel mechanism for promoting the clearance of Aβ from the brain, as the rates of Aβ production were not altered. The mechanism of action with which Aβ immunization determines the inhibition of the formation of Aβ plaques and a reduction of already formed amyloid lesions is not fully understood. There are two major hypotheses explaining the increase of Aβ clearance caused by vaccination. Centrally, the vaccine activates Aβ phagocytosis by microglial monocytes. Peripherally, serum anti-Aβ antibodies appear to bind and sequester Aβ, thus altering its equilibrium between CNS and plasma.

Data from patients with AD who received a primary injection of pre-aggregated $A\beta_{42}$ followed by a booster injection in a placebo-controlled study showed that that antibodies in immune sera recognized β-amyloid plaques, diffuse Aβ deposits and vascular β-amyloid in brain vessels (Hock et al., 2002). Thus, vaccination of AD patients with $A\beta_{42}$ induces antibodies that have a high degree of selectivity for the pathogenic target structures. Long-term clinical assessment will be required to determine whether this approach will halt or improve cognition in the patients treated. The finding that vaccinated patients with AD generated antibodies against beta-amyloid encourages the further development of immunization therapy. However, this study was suspended when clinical signs of post-vaccination meningoencephalitis were reported in several patients. Although it was pointed out that the meningoencephalitis suffered by some of the trial participants is unlikely to stem from an antibody-mediated response against cellular precursor protein, this study did not clarify the cause of the inflammation in the patients. Thus, additional research and efforts are clearly warranted. While the Aβ immunization approach has opened unprecedented hopes for an effective treatment of AD, a number of issues both in terms of efficacy and safety may preclude its successful application in AD patients. The future of this approach depends on a clearer understanding of the mechanism of Aβ clearance and additional insight into the role of inflammation in the AD brain.

12.4.7 Anti-inflammatory drugs

It has been reported that the risk of AD diminishes with increased duration of NSAID use and that the onset of dementia is delayed in NSAID users (McGeer et al., 1996; Stewart et al., 1997). The pharmacological activity of NSAIDs is generally attributed to the inhibition of COX, a rate-limiting enzyme in the production of prostaglandins. COX was demonstrated to exist as two distinct isoforms (Fu et al., 1990; Xie et al., 1991). COX-1 is constitutively expressed as a housekeeping enzyme in nearly all tissues and mediates physiological responses (e.g., cytoprotection of the stomach, platelet aggregation). On the other hand, COX-2 expressed by cells that are involved in inflammation (e.g., macrophages, monocytes, synoviocytes) has emerged as the isoform that is primarily responsible for synthesis of the prostanoids involved in pathological processes such as acute and chronic inflammatory states. Accordingly, many of the side effects of NSAIDS can be ascribed to a suppression of COX-1-derived prostanoids whereas inhibition of COX-2-dependent prostaglandin synthesis accounts for the anti-inflammatory, analgesic and antipyretic effects of NSAIDS.

In AD, COX-2 appears to be up-regulated in brain areas related to memory (hippocampus, cortex) with the amount of COX-2 correlating with the deposition of β-amyloid protein in the neuritic plaques (Pasinetti, 2001). Elevation of COX-2 expression in hippocampal neurons during the early phase (mild dementia) of AD is considered to favor the later inflammatory neurodegenerative process. Emerging evidence in animal studies indicates that one of many possible therapeutic actions of NSAID in the AD brain may be the suppression of microglial activation associated with b-amyloid deposits. It has been shown that indomethacin administration significantly reduces the number of activated microglial cells surrounding the intraparenchymal β-amyloid deposits developing after infusion in the lateral ventricle of rats (Netland et al., 1998).

In a prospective study, 6889 people were followed for about seven years (in't Veld et al., 2001). All subjects were aged 55 or older and free of dementia at the beginning of the study. The mental and neurological health of the participants were assessed at the beginning, middle and end of the study. Throughout the study, the

NSAIDs use of the subjects was monitored. The results showed that the subjects who used NSAIDs for two or more years were 80% less likely to develop AD compared to individuals who did not use the drugs. During an average follow-up period of 6.8 years, dementia developed in 394 subjects, of whom 293 had AD, 56 vascular dementia and 45 other types of dementia. The relative risk of AD was 0.95 in subjects with short-term use of NSAIDS, 0.83 in those with intermediate-term use and 0.20 in those with long-term use. The risk did not vary according to age. Ibuprofen, naproxen and diclofenac were the most popular NSAIDs used. NSAIDs might decrease AD risk. Inhibition of inflammation may be one mechanism. Blockage of an enzyme involved in the production of Aβ may be another possibility. However, this study did not include mortality data. It is entirely possible that patients who received long-term treatment with NSAIDs were dying sooner from the complications. These data need to be confirmed by prospective, double-blind randomized studies to asses the ideal dose and safety involved.

12.4.8 Modulation of cholesterol homeostasis

Chronic use of cholesterol-lowering drugs such as the statins has been associated with a lower incidence of AD (Wolozin et al., 2000; Jick et al., 2000). Concurrently, high-cholesterol diets have been shown to increase Aβ pathology in animals and cholesterol-lowering drugs have been shown to reduce pathology in APP-transgenic mice (Sparks et al., 2000; Refolo et al., 2001). These effects seem to be caused by a direct effect of cholesterol on APP processing (Wahrle et al., 2002). A particular advantage of this approach is that statin drugs are generally well tolerated and have already been widely prescribed. Clinical trials are under way using the add-on design, where Lipitor or a placebo is added to stable doses of donepezil.

12.4.9 Metal ion chelators

The metal ion chelators strategy is based on the observation that Aβ aggregation is, in part, dependent on the metal ions Cu^{2+} and Zn^{2+} (Bush et al., 1994). This strategy reasons that chelation of these ions *in vivo* may prevent Aβ deposition. Aβ deposition was impeded in APP transgenic mice treated with the antibiotic clioquinol, a known Cu^{2+}/Zn^{2+} chelator (Cherny et al., 2001). This strategy has now reached the clinical trial stage.

12.5 The emerging field of pharmacogenetics

As in all complex diseases, many genetic elements are responsible for the clinical phenotype. Predicting who, in a mixed population, will respond best to any given therapeutic compound is a challenge for pharmacogeneticists. There are already some indicators that the Apo E allotype may affect responses to CI therapy (Richard et al., 1997), although this remains to be confirmed. Pharmacogenomics in AD may be even more significant in deciding which class of stabilization drug (statin *versus* amyloid-modyfing, for instance), individual patients should be getting early in the course of their disease.

12.6 Discussion

Much has been learned from the first few years of specifically targeted AD therapy. All the CI have demonstrated a statistically significant, although modest, effect *versus* placebo on the cognitive and the global performance of patients with AD. However, not all AD patients benefit equally from treatment with CI. At least, one third of treated individuals experience no benefit above baseline (Burns et al., 1999; Rösler et al., 1999). To ascertain who will benefit from treatment with CI must be an important task for researchers. Some studies have indicated that the inheritance of Apo E4 alleles is a negative predictor for treatment effect, at least after up to 30 weeks of treatment (Poirier et al., 1995; Schneider & Farlow, 1997). Apo E2-3 AD patients benefit more than Apo E4 AD patients from treatment with CI. A study designed to assess the effect of tacrine on the rate of progression in AD with relation to Apo E allele genotype showed a faster rate of decline in Apo E4 AD compared to the Apo E2-3 (Sjögren et al., 2001). These data may also suggest that Apo E4 genotype inheritance is a negative predictor of treatment effect of tacrine in AD patients.

There are no major differences yet demonstrated in terms of efficacy between the different CI over one year. The mean difference between drug and placebo effects on standardized psychometric scales is about 2 to 4 points on the ADAS-cog and 0.2 to 0.5 points on the CIBIC-plus or 5–14% of the average value of the scales (Imbimbo, 2001). The most common adverse effects observed after administration of CI are nausea, vomiting, diarrhea, asthenia and anorexia, all symptoms linked to cholinergic overstimulation. These effects are dose related and largely depend on the degree of cholinesterase inhibition. Also important is the rate of onset of cholinesterase inhibition, which depends on the kinetics of enzyme inhibition, the presence and rate of titration and the pharmacodynamic peak-to-trough fluctuations. A comparison of tacrine with other second generation CI shows that despite these drugs having modest clinical efficacy, their main differences reside in their frequency of side effects, number of drop-outs and percentage of improved patients. Although efficacy may be similar between the CI at effective doses, peripheral cholinergic adverse effects, tolerability and hepatotoxicity limit their use in some patients. However, patients not responding to a drug may be switched to another medication. Two studies showed that switching from donepezil to rivastigmine is worthwhile when donepezil has produced intolerable side effects or has failed to prove effective (Auriacombe et al, 2002; Bullock & Connolly, 2002). A beneficial effect, over and above any previous effect was demonstrated in over half the patients exposed to a second CI. The types of controlled study carried out for the CI have generally been short-term, ranging from 12 weeks to 6 months using similar kinds of cognitive outcome measures. Therefore, long-term controlled studies beyond a year need to be evaluated. Furthermore, reliable controlled data on meaningful outcomes such as dependency and institutionalization or other aspects of long-term efficacy are urgently needed.

In contrast to the CI, the beneficial effect of estrogen therapy may delay the progression of AD (Paganini-Hill & Henderson, 1996). Since combination therapies may be crucial, it will be interesting to trial combinations of drugs that possess different mechanisms of action, as for example, CI and estrogen. Clinical studies are necessary to assess the efficacy and interactive effects of these approaches. With regard to the use of NSAID in AD, it is important to note that the elderly are more susceptible to the adverse effects of NSAID and therefore, they should be used with caution. The development of COX-2 and leucotriene inhibitors might be very important. Regarding antioxidants such as Vitamin E, it is worth considering in patients with AD, since it can be obtained over the counter, and is relatively non-toxic and inexpensive.

In the last two decades, the clinical efficacy of the special extract of *Ginkgo biloba*, EGB 761® (EGb) has been extensively documented for a range of cognitive disorders (Le Bars et al., 1997; Maurer et al., 1997). Most of the studies were undertaken in Europe and reported positive results after 12-week as well as 24-week EGb treatments in patient populations suffering from different types of dementia of mild to moderate severity (Kanowski et al., 1996; Maurer et al., 1997). Recently, the outcomes of North American studies further substantiated the efficacy of treatment with EGb in AD (Le Bars et al., 1997; 2002). If these positive results are further supported by future studies, EGb may be considered a valuable agent for the treatment of AD.

AD is a complex and dynamic disease having numerous effects on both patients and their caregivers. Assessment of antidementia medication outcome is therefore difficult and requires a wide variety of evaluations to obtain a total representation of a drug's impact upon the disease. However, the knowledge gained to date has served to set the standards by which all future therapies will be measured. Progress has been remarkable, with every prospect that really effective strategies will emerge in the near future. A consideration in drug development is the definition of therapeutic benefit for AD drugs. The FDA requires that a drug show superiority to placebo on a performance-based test of cognition and a measure of global clinical function. The most common accepted instrument used to test cognitive performance is the ADAS-cog. Although other symptoms of AD (e.g. ADL impairment or behavioral symptoms) have been studied as secondary efficacy measures, the FDA has not formally accepted these symptoms as acceptable primary outcome measures. This restrictive definition of the disease is biased toward drugs that enhance performance on memory-based tests. This may pose a problem for drugs designed to slow disease progression.

There is a clear requirement in the design of clinical trials for the inclusion of scales to measure the behavioral and cognitive symptoms of AD. AD scales should be as short as possible and should cover cognitive, non-cognitive and functional items. The CERAD neuropsychological battery that usually takes 20–30 min to administer has been found to be a valuable instrument for clinical as well as research investigations and has greatly contributed to the identification and clinical understanding of AD (Welsh-Bohmer & Mohs, 1997). CERAD emerges as an interesting tool to assess the cognitive status of patients with AD at entry and during long-term follow-up in large clinical trials. Apart from the inclusion of behavioral and cognitive scales in clinical trials for the assessment of the benefits of treatment of AD, it also appears that improvements in everyday functioning and quality of life are crucial parameters to assess efficacy of a new drug (Kelly et al., 1997). Recent criticisms of clinical studies and marketing of AD drugs point out that the evaluation of new healthcare technologies must be improved (Melzer, 1998). It was suggested that it is crucial to conduct trials on patients who are representative of those for whom the drug will be licensed and marketed; to publish full trial results before the marketing of the new drug so that prescribing doctors can judge the size of clinical benefits and risks. The SCAG (and the SASG), the BDS, the GBS, the GDS and the performance, verbal, intellectual and memory quotients measured with the WAIS and the WMS were more successful than the MMSE and the ADAS in measuring some improvement in various drug trials. The general observational evaluations of cognition and behaviors that are most frequently used in drug trials, the CGIC and the CIBIC are, like the ADAS, an FDA requirement for drug approval. However, the results reported in the trials using these evaluations are not impressive. The use of the CIBIC or CGIC evaluations raises some difficulties. The nonpsychometric nature of the instruments themselves is a problem (no sensitivity, validity and interrater reliability data). In general, the scales with the best results in trials of cognition-enhancing drugs are those with a psychiatric assessment (emotional and behavioral) together with a cognitive evaluation that taps into concentration capacity and autobiographical and semantic memory (e.g., the BDS, GBS and the SCAG). However, all the scales and the clinical observational evaluations share the same problem; they do not provide information about which aspects of memory are preserved, improved or altered.

Chapter 13
Pharmacological treatment of neuropsychiatric symptoms

13.1 Introduction

Behavioral and psychological symptoms of dementia cause premature institutionalization and significant loss of quality of life for the patient and his/her family and caregivers. An imbalance of different neurotransmitters (ACh, dopamine, noradrenaline, serotonin) has been proposed as the neurochemical correlate of these neuropsychiatric symptoms. The significant decrease in cholinergic activity may result in a relative increase in monoaminergic activities, leading to hypomanic or manic symptoms and behavior including delusions, hallucinations and physical aggression (Folstein, 1997). Conversely, AChE inhibitors have shown to be able to reduce neuropsychiatric severity (see Table 13.1).

Psychoactive drugs commonly used to treat the behavioral symptoms of AD are the same classes of drugs used in the general population, including antidepressants, antipsychotics, and antianxiety medication. Because patients with AD tend to be very sensitive to the CNS effects of these drugs, they must be used especially carefully.

13.2 Depression

Any patient with dementia with significant depressive symptoms such as sleep, appetite, or energy disturbance, depressed mood or irritability, anhedonia, social withdrawal, excessive guilt, a passive death wish or suicidal ideation, or agitation should be considered for treatment of depression even if failing to meet criteria for a depressive disorder (Small et al., 1997). Some potential antidepressants used for treating depression in AD patients include some selective serotonin reuptake inhibitors (SSRI), Venlafaxine and some atypical antidepressants (see Table 13.2). In an 8-week, prospective parallel group trial, the efficacy of Paroxetine and imipramine was studied (Katona et al., 1998). It was found that both drugs were effective and there were no significant differences in the outcome between the groups, though there were trends suggesting that Paroxetine was better tolerated than imipramine in terms of adverse effects. Fluoxetine and especially trazodone have been shown to be helpful in the management of agitation (Lebert et al., 1994; Aisen et al., 1993). A large-scale clinical trial in which 694 patients with symptoms of depression and cognitive decline were treated with the reversible, type A specific MAO inhibitor moclobemide (400 mg/day) or placebo for a period of 6 weeks was performed (Roth et al.,1996). Significant benefits of moclobemide relative to place-

Table 13.1: Non-cognitive effects of AChE inhibitors in AD patients

Drugs	Effects on neuropsychiatric symptoms	References
Tacrine	↓ Delusions and apathy	Cummings et al., 1993
	↓ Disinhibition	Raskind et al., 1997
Physostigmine	↓ Agitation	Gorman et al., 1993
Donepezil	↓ Improvement of ADAS non-cog scores	Rogers & Friedhoff, 1996; Rogers et al., 1998; Schachter & Davis, 1999
Metrifonate	↓ Hallucinations	Becker et al., 1996; Morris et al., 1998
Rivastigmine	↓ Psychosis	Jan & McKeith, 1998

bo were reported in both groups of patients. The performance on the MMSE was found to improve especially among patients with depression and co-existing dementia. In summary, SSRIs and tricyclic antidepressants are both effective in treating depression in patients with dementia. SSRIs are probably safer but the choice of antidepressants used largely depends on the profile of the side-effects. In many patients, depression may remit within 12 months and the need for antidepressants should be kept under review.

Donepezil was associated with the largest number of positive observations, supporting a beneficial effect on mood and behavior in AD patients. Weiner et al. (2000) found that donepezil administration was associated with prevention of worsening depression and behavioral impairment over a 12-month period in AD patients. Cummings et al. (2000) reported that AD patients taking donepezil had lower level of behavioral disturbances, were less threatening and needed fewer sedatives than those not on donepezil as reported as caregivers after a 6-month treatment period. Hecker et al. (2000) reported improvement with donepezil in depression, anxiety and apathy areas of the NPI when compared with placebo over 4 and 24 weeks in 191 moderate-severe AD patients.

13.3 Psychosis

Neuroleptics have been the mainstay of treatment of delusions and hallucinations in dementia. Classical neuroleptics have been shown to worsen cognitive and functional ability in dementia and therefore should not be considered as first choice of drugs (Chui et al., 1994; McShane et al., 1997).

The newer atypical agents (risperidone, olanzapine, quetiapine, and clozapine) or midpotency agents are preferred because of fewer adverse affects. In the short term, they are less likely to cause extrapyramidal side effects and postural hypotension. Risperidone has been shown to be effective and is less likely to cause extra-

Table 13.2: Some antidepressants used for treating depression in AD patients

Drugs	Comments	References
SSRI		
Sertraline	Few interactions with the cytochrome P450 system and little anticholinergic activity. Good choice for patient with AD.	
Fluoxetine	Long half-life and active metabolite. Half-life allows adequate blood levels of drug to be maintained although doses are missed. Good substitute if compliance is a problem.	Lebert et al., 1994
Paroxetine	Most anticholinergic of the SSRIs and theoretically may produce more adverse effects on cognition.	Katona et al. 1998
Fluvoxamine	Has a relatively short half-life and the twice-daily dosing may impair compliance.	
SNRI		
Venlafaxine	Inhibits both serotonin and norepinephrine reuptake without having anticholinergic adverse effects.	Rudolph & Derivan, 1996
Tricyclics		
Imipramine	Should not be used in patients with AD because of the anticholinergic effects.	Katona et al. 1998
Amitriptyline	Should not be used in patients with AD because of the anticholinergic effects.	
Nortriptyline	Tricyclic antidepressant of choice because of fewer anticholinergic effects. Serum levels and electrocardiogram should be monitored at steady state before each dose increase with target levels of 50–150 ng/ml	
MAOI		
Selegiline	Two of 5 controlled trials evaluating selegiline's effect on behavior (e.g., anxiety, tension, excitement, depression) showed a positive effect	Tolbert & Fuller, 1996
Other classes		
Nefazodone	A serotonin reuptake inhibitor and $5HT_{2A}$ receptor antagonist. Administered in twice daily dosing and the most common adverse effects are lethargy, dizziness and dry mouth.	Goldberg, 1997
Bupropion	A weak norepinephrine reuptake inhibitor but is a stronger inhibitor of dopamine uptake. The dopaminergic effect may be beneficial in some AD patients and may be stimulating and effective for apathy.	Ascher et al., 1995
Mirtazepine	An $\alpha 2$-antagonist and $5HT_2$ and $5HT_3$ receptor antagonist which may be effective in treating refractory patients.	Montgomery, 1995 Raji & Brady, 2001

pyramidal side effects or impaired cognitive performance (De Deyn et al., 1999; Stoppe et al., 1999). However, risperidone can cause postural hypotension and extrapyramidal side effects at high doses. Care should be used when increasing the dose beyond 2.5 mg. Olanzapine may be a useful alternative, especially because to its sedative properties. Long-term safety of atypical neuroleptics still remains to be established.

A study showed that the typical antipsychotics are still the most commonly used despite the better safety profile of the atypicals (Condren & Cooney, 2001). More research is required into the use of psychotropic medication for dementia and there is need for consensus in this area.

13.4 Anxiety

The use of medications such as benzodiazepines, antidepressants and mood stabilizers is common with many doctors using these drugs as alternative medicines or adjunctives in the treatment of behavioral symptoms and anxiety (Condren & Cooney, 2001). There was a wide variation in the dosages used by clinicians, with some prescribing surprisingly high dosages of some drugs.

Benzodiazepines are useful in treating anxiety, tension, irritability, agitation and insomnia. Excessive sedation, ataxia, confusion and falls are the main problems associated with benzodiazepines use (Grad, 1995). If benzodiazepines are used, it may be appropriate to use shorter-acting benzodiazepines like oxazepam or lorazepam. Buspirone, a partial agonist of $5HT_{1A}$ receptor, was suggested to be useful in case-reports and open studies but failed to show effectiveness over a placebo (Lawlor et al., 1994).

13.5 Discussion

Psychotropic drugs are commonly used as part of the management plan to treat the psychological symptoms and decrease the behavioral symptoms. During the course of illness almost every dementia patient will receive at least one psychoactive agent (Terry & Katzman, 1983).

Elderly patients have increased sensitivity to the effects and side effects of medications and a tendency to accumulate drugs due to various physiological factors such as increased body fat, decreased hepatic metabolism and reduced renal clearance. Therefore, a plan should be made to assess the effects and side effects of medications. Regular review should be arranged to decide the duration of the treatment.

There is some evidence to suggest use of neuroleptics or benzodiazepines in the management of agitation in dementia but their efficacy needs to be weighed against the potentially fatal side effects such as falls. Neuroleptics may be particularly useful when agitation is associated with delusions or hallucinations. Antidepressants could be used if there are predominant depressive symptoms with agitation. Compounds that are effective for the BPSD syndrome can be identified. The identification of agents with specific efficacy in treating BPSD requires demonstration of statistically significant efficacy, independent of effects on cognition and functioning. It is better if a medication improves BPSD and cognition or if it improves BPSD and functioning. A broader indication would appear appropriate for such an agent.

More research is needed into the use of psychotropic drugs in the treatment of patients with AD including studies to determine more specific drug-responsive behaviors to maximize the benefits of these medications and the range of side effects observed with the unusually high doses of psychotropics in these patients.

Chapter 14
Psychological support and cognitive rehabilitation

14.1 Introduction

Because functional impairment is one of the main consequences of the dementing process, patients are no longer able to engage in meaningful activities on their own. Demented patients cannot pursue hobbies because of apraxia; they cannot read because of comprehension difficulties and they cannot engage in social activities because of aphasia. Meaningful activities may improve depression, diminish agitation, apathy, insomnia and repetitive vocalization and should be attempted. Meaningful activities are important at all stages of dementing illnesses. Individuals with advanced dementia need stimulation with appropriate activities because few, if any, patients with AD progress to the persistent vegetative state (Volicer et al., 1997).

Accumulating evidence suggests that certain training techniques could be beneficial to many amnesic patients. These may include practices or exercises that train direct memory, mnemonic or internal strategies that are visual imagery or verbal strategies that train information retention and recall. Other approaches include external memory aids or strategies that aim at compensating for the memory disorder. Approaches to helping people in the early stages of dementia should incorporate a range of elements that could be selected and adapted according to individual needs (see Table 14.1). One of the simplest methods to help avoid problems caused by memory impairment is to structure or rearrange the environment to enable patients to cope without adequate memory functioning. Labeling cupboards, drawers and doors; colored lines from one place to another; sign-posting and positioning material in places where it is most likely to be seen are all examples of environmental structuring. Memory rehabilitation can be classified into teaching and acquisition of domain-specific knowledge, motor coding, reality orientation and metacognition improvement.

One of the challenges for rehabilitation has been to find ways to tap into the preserved mechanisms to compensate for those that are damaged or lost. Early attempts at rehabilitation that focused on teaching mnemonic strategies were effective only for mildly impaired patients who retained some residual memory function but they provided few benefits for patients with moderate to severe memory deficits. Seeking

Table 14.1: Items that may be used to help people in the early stages of dementia

- Enhancing social contacts and support, and limiting isolation
- Giving more occasions to discuss about the experience and emotional impact of AD
- Helping patients to identify and engage in activities that they can still enjoy
- Helping patients with dementia to access information that is appropriate to their needs
- Assisting in the development of realistic and effective compensatory strategies
- Helping patients to have a sense of self-worth

to improve memory function in neurological patients in a global manner by means of mental imagery and other mnemonic devices imposes further demands upon their already impaired memory for specific episodes and events. Rehabilitation should exploit their residual learning capacities and prior knowledge to facilitate the acquisition and retention of knowledge within particular domains of practical significance to patients themselves (Glisky & Schacter, 1989; Schacter & Glisky, 1986). Social factors and individual personality need to be taken into account when designing rehabilitation. Pre- and postmorbid lifestyles may be influential in determining the nature of treatment.

14.2 Cognitive rehabilitation

Several main approaches to rehabilitation such as restoration of function, environmental control and functional adaptation have been proposed (Wilson & Patterson, 1990). Games and exercises activities can provide pleasure and a positive social atmosphere for people who may be having a difficult time because of their cognitive failures. These activities can be engineered so that even the most severely impaired person can succeed. This is important for people who frequently fail at tasks requiring a recall or new learning. They have a good face validity and may increase a feeling of satisfaction that something is being done.

Environmental control is a sensible strategy when used with severely intellectually handicapped patients. For example, reality orientation, a strategy used widely with confused elderly people employs labels, signposts and noticeboards so that the environment is structured to reduce or avoid cognitive demands for those who would otherwise fail (Holden & Woods, 1982). Positioning material or information so that it cannot be missed is another way of achieving environmental control.

14.2.1 Domain-specific knowledge

Domain-specific knowledge is considered an important method of memory rehabilitation. This approach emphasizes that rehabilitation aiming at the acquisition and maintenance of knowledge of specific domains that have practical importance for the patients in daily life is crucial (Glisky & Schacter, 1986; Schacter & Glisky, 1986). Domain-specific knowledge training can be applied to acquire various skills. It was developed utilizing various existing behavioral psychological approaches such as forward chaining (to divide a behavior into a series of steps and teach each step in turn, starting from the first step) and backward chaining (start from the last step and work towards the first step and shaping and fading the cues). The method of vanishing cues was developed. This method was based on the observation that technical learning (retention of procedural memory) and priming response such as fragment completion are conserved in amnesic patients.

Although the acquisition of the domain-specific knowledge has practical significance, some limitations are also noted. The knowledge acquired was extremely simple and small in quantity. The acquired knowledge cannot be put into practice and the knowledge is not generalized into daily life activities.

14.2.2 Errorless learning

Errorless learning has for many years been used to teach new skills to people with learning disabilities (Arkin, 1992). As the name implies errorless learning involves learning without errors or mistakes. Most of us can learn from or benefit from our errors because we remember our mistakes and therefore avoid making the same mistake repeatedly. The principle behind errorless learning is that when an individual is allowed to make errors during training those errors will interfere with the learning of the correct information because on future trials the errors may be recalled rather than the correct information. By eliminating errors and rehearsing only correct information the individual will have access solely to the correct information that will facilitate enhanced learning or re-learning of the target information.

Errorless learning principles were used to train a 72-year-old man in the early stages of dementia of the Alzheimer type to remember names of 11 members of his social club (Clare et al., 1999). Training consisted of verbal elaboration, vanishing cues and expanding rehearsal. The proportion of faces correctly named increased significantly, rising from 22% at baseline to 98% after training, generalized well from photographs to real faces in the natural environment. These cognitive improvements were maintained 9 months later. In the majority of cases errorless learning proved to be superior to trial-and-error learning.

14.2.3 Action-based memory

In people with AD, retrieval of action-based memories, for individual actions, for ADL and for procedural skills is generally superior to retrieval of verbal descriptors of the same task (Baum et al., 1996; Hirono et al., 1997). Action-based encoding and retrieval strategies are more effective than verbal strategies for people with dementia and can be used to improve episodic memory (Hutton et al., 1996). Moreover, action-based training programs can lead to some performance gains on everyday tasks (Zanetti et al., 1997). Preservation of action-based memory clearly has major practical implications. The most essential remembering required by people with AD in everyday life is activity-based and successful participation in tasks of daily living which can promote both greater functional independence and significant psychological benefits for the patient. A longitudinal study was designed to examine memory for a single routine activity (tea making) of daily living in people with AD (Rusted & Sheppard, 2002). Memory assessment in a natural setting, visiting AD patients in their homes, video-taping performance on selected tasks and analyzing the record for the presence or absence of each of its component actions over a period of 6 years was performed. Large differences in the rate of decline of the patients with substantial preservation of performed recall of the everyday task even in the more severe phases of the disease were noted. The pattern of decline shows a benign degradation of the memory trace with omissions comprising the most common category of errors and the data contrasted with the more dramatic action disorganization syndrome associated with frontal injury. Memory for action-based events or "scripts" is hierarchically structured. As memory for the event degrades over time, it is the less important details that are lost while the key elements of the event are retained. This study also showed that removing the AD patient from a familiar tea-making environment and into a novel one increased the overall likelihood of making omission and repetition errors and the disruption of the routine but did not induce qualitatively different types of error. Though a new environment made retrieval harder, the availability of motor and item cues was more important than the availability of spatial and location cues.

14.2.4 Reality orientation

Reality orientation is the teaching of specific information related to the orientation and environment in which the patient resides. It involves reminding patients with dementia about the various aspects of their environment including time, place, person in structured daily groups or in more informal ways (e.g., signposts). In practice, patients are asked questions such as "Who are you?", "Who is talking to you?" and "What is going on now?" and then their false responses are corrected by therapists. Information regarding time and place is also given to the patient. The therapists must present the information to the patient in a very clear and simple fashion. The patient is recommended to rehearse and talk to his won family and other patients. It is important that the therapist knows the details of the patient's family and their past history so that he can assess the patient's remarks and progress. Improvement in orientation is observed in well-structured formal reality orientation irrespective of whether it is conducted in groups or individually (Hanley, 1986; Greene, 1984). Combination of reality orientation with other memory training strategies may provide better results. For example, a combination of reality orientation with instruction using an alarm clock and a timetable was used successfully in training patients with early Alzheimer type dementia (Kurlychek, 1983). Reality orientation combined with attention process training in the rehabilitation for elderly dementia patients was also successfully applied (Honda & Kashima, 1992).

14.2.5 Metacognition Improvement

Many amnesic patients cannot assess their memory impairment correctly or are not aware of the insults at all. Amnesic patients' failure to use memory strategies and lack of generalization of the rehabilitation effects are often attributed to the lack of metacognition. Recently, approaches to improve metacognition or self-assessment of the memory disorder have been attempted.

14.2.6 Spaced retrieval

The spaced retrieval technique was initially introduced to assist normal subjects with long-term retention of information (Landauer & Bjork, 1978). It is believed that spaced retrieval involves little cognitive effort and therefore is also beneficial when used with memory-impaired individuals. For example, spaced retrieval was deemed successful in training motor and verbal tasks with Parkinson's dementia patients (Hayden & Camp, 1995).

Spaced retrieval has also been found to be successful in training internal memory strategies with dementia patients (Schacter et al., 1985). Spaced retrieval appears to be effective for dementia patients because it relies mostly on implicit memory, which remains relatively spared in dementia of the Alzheimer type. A group of patients described by Camp & Stevens (1990) were able to learn face-name associations when the spaced retrieval technique was implemented. Furthermore, half of the patients were able to retain the newly learned information over a one-week interval as well as at a 5-week follow-up session. McKitrick & Camp (1993) successfully used spaced retrieval to re-teach a patient with AD names of objects encountered in her everyday life. This study involved training the patient's caregiver to implement the spaced retrieval technique to maximize functionality. Researchers have also used spaced retrieval to teach dementia patients to select and redeem a coupon for money (McKitrick et al., 1992). It has also been demonstrated that spaced retrieval is successful in teaching dementia patients associations between cues and specific tasks or behaviors (Bird & Kinsella, 1996). Taken together, these results indicate spaced retrieval to be successful for dementia patients of varying severity levels (MMSE score of 11-25) (Brush & Camp, 1998).

14.3 Some other forms of treatments

Researchers have demonstrated a connection between a relaxed state and anxiety reduction in various subject populations. Scogin et al. (1992) found that both progressive and imagined relaxation alleviated subjective anxiety in an elderly population. In a controlled clinical trial of psychological therapies for depression in AD, patients with AD who met DSM-III-R criteria for major or minor depressive disorder were randomly assigned to one of two active behavioral treatments (one of them emphasizing pleasurable events and the other emphasizing carer problem solving), or to comparison treatments that included usual clinical care and a waiting list control condition (Teri et al., 1997). Patients in both active treatments showed significantly greater improvement in depression compared with those in the control groups. These benefits were maintained at 6-month follow up.

Exercise program, usually consisting of walking for 20–60 min, is valuable for as long as the patient is able to do this. There are likely benefits, but most importantly, while physical decline, frailty, and falls are to be expected, these are not inevitable, and can be attenuated. Strategies to minimize behavioral disturbance are also valuable, and, if effective, do not carry the risk of adverse effects common with psychotropic and sedating drugs, especially during the treatment of agitation.

Reminiscence group therapy involves the use of past memories to generate interest, social interaction and pleasure in patients with dementia. It may have a role in reducing aggression and demanding behavior.

14.4 Imagery mnemonics

Mnemonics, rehearsal strategies and study techniques are all ways of helping people use their damaged skills more effectively. They probably work because they encourage a deeper level of processing that results in better recall (Bower, 1972). Previously isolated items are integrated with one another and they provide inbuilt retrieval cues in the form of initial letters, locations or pegs. The real value of mnemonics is that they are useful for teaching new information and are usually better than rote rehearsal. Experimental research has demonstrated that instructions and training in the use of mental imagery lead to consistent, reliable and substantial improvements in memory performance (Richardson, 1980). This applies both to the use of simple interactive images in verbal-learning tasks and to more complicated mnemonic systems such as the pegword method or the method of loci (Yates, 1966).

Cognitive demands of everyday situations may best be handled using other strategies or devices such as diaries or notepads. In particular, imagery mnemonics may well be of very little value in learning the names of other people, in recalling messages and in remembering to do things. However, although imagery is less effective for remembering people's names than for learning paired associates, Wilson (1987) found that it could be used to remember a small set of simple, familiar names, provided that the task was highly structured. The latter might be an important and legitimate goal in the context of memory rehabilitation. When applying mnemonics, therapists should teach only one thing at a time and not expect memory-impaired people to learn several pieces of information at once. Individual styles and preferences should be taken into account. Mnemonic strategies likely enable a damaged memory system to function more efficiently, albeit in the same general way as it did premorbidly.

14.5 External aids

Diaries, notebooks, lists, alarms, wall charts, calendars, tape recorders and computers are some of the means employed to compensate for memory impairment. Most people will have used such external memory aids at one time or another. Some patients resist using such aids because they feel that it is cheating or that natural recovery might be slowed down. Such resistance should not be encouraged because it is important to use any approach that may be helpful to memory. Patients may need to be reminded that people with good memories frequently rely upon external aids. Their use will not slow down any natural recovery.

One new external memory aid that requires little in the way of learning and that could prove useful for all elderly people is NeuroPage, a system based on radiopaging technology (Hersh & Treadgold, 1994). The scheduling of reminders for each individual is entered into a computer and from then on no further human interfacing is necessary. At the appropriate day, the message is delivered to the person from the paging company. An audible or vibratory alarm sounds alerts the subject to look at the screen on which the relevant message appears. NeuroPage avoids many of the problems inherent in other external aids. It is portable and does not require large investment of time or effort to learn how to handle. Preliminary results from the USA and the UK are encouraging.

14.6 Discussion

These patients definitely need a lot of support, encouragement and understanding. Behavioral and compensatory approaches to rehabilitation clearly have great value, although they often leave theoretical understanding of the processes of recovery and the mechanism of action of rehabilitation unresolved. The ultimate goal of any cognitive rehabilitation is independence in functional living activities, thereby improving the patient's work or study skills. There remain several unsolved problems such as the extent to which improvement on trained items generalizes to untrained items and difficulty in initiation and use of the newly learned strategies without prompting. In a study that assessed the efficacy of cognitive intervention consisting of training in face-name associations, spaced retrieval and cognitive stimulation in a sample of patients with probable AD, it was shown that during the intervention, AD patients significantly improved (Davis et al., 2001). However, cognitive improvement was not observed in additional neuropsychologic measures of dementia severity, verbal memory, visual memory, word generation or motor speed or to caregiver-assessed patient quality of life. For the majority of memory-impaired people, it is unrealistic to expect complete recovery. This does not mean that nothing can be done to help. Probably, the most useful way of compensating for memory improvement is to use external memory aids.

The methodological issues in undertaking clinical studies of cognitive rehabilitation are complex. The need for a randomized design, including a group that receives no cognitive rehabilitation, raises ethical concerns despite the very limited support for this treatment approach. Comparisons of different strategies of cognitive rehabilitation or schedules of treatment (e.g., high *versus* low frequency of sessions with suitable controls for nonspecific effects of treatment) could provide important efficacy data.

Educating health professionals about AD, its consequences and care is essential at every phase of its progression and in every setting. Providing information to memory-impaired people and their relatives is crucial in memory rehabilitation. They should be informed of the nature of memory impairment and why certain problematic behaviors such as constantly repeating a question occur. Emotional support and educational information can help caregivers cope with feelings of anger and frustration in response to the patient's behavior. Caregiver support can also delay the placement of patients into nursing homes. Families can be encouraged to modulate the patient's environment in various ways to provide moderate stimulation, make use of familiar surroundings and daily routines, maintain links with the outside world and enhance communication.

Chapter 15
Discussion and conclusion

Progress in cognitive psychology, neuropsychology and more recently, brain imaging research has provided experimental tools for the objective investigation of remembering and provided a means to link cognitive-level description with underlying neural processes. Recent findings indicate that acts of remembering may be separated into component processes that are subserved by dissociable brain regions. Methods that combine techniques such as methods based on electrical activity (MEG and EEG) and hemodynamic methods (fMRI and PET) provide considerable potential for widespread characterization of the dynamic processes associated with remembering. AD can be predicted on the basis of structural changes within the medial temporal lobe and many believe that functional changes precede the gross structural changes by several years (see Buckner & Wheeler, 2001). One strategy to detect the earliest stages of dementia has been to measure medial temporal activity during memory processes using brain-imaging methods. However, structures within the medial temporal lobe are relatively small and have been a difficult challenge for imaging.

The "aging" of the brain and pathology resulting from age-associated insults to the brain are assumed to underlie the deficits in speed and memory performance. Cognitive aging is an exciting area of investigation. There is evidence that cognitive diversity increases with increasing age and that the extent of cognitive change can be predicted to some degree by a number of risk factors, including health, Apo E4 and disability (Christensen, 2001). Much is known about predictors of cognitive change but the new challenge is to identify causal relationships. A next step in this process is to investigate factors that change with cognitive change over time in large longitudinal studies. Proposed brain or cell processes capable of mediating such diverse but related changes include CBF and decreased metabolic rate, mitochondrial dysfunction, age-related myelin degeneration and various neurotransmitter depletions.

A large body of literature in the field suggests that gene mutations cause genetically inherited, early-onset AD (Tanzi,1999). Although research has grown exponentially during past 5 years and cell biological functions of APP, PSs and Apo E are beginning to be understood, we are still lacking the molecular basis and order of events involved in the disease process. The currently held view (amyloid hypothesis) is that AD begins with the deposition of large numbers of SPs. This then sets up a series of events, including alteration in tau that in turn lead to neuronal dysfunction and dementia.

Clinical trials of treatments for AD present a considerable problem both in ethical terms with regard to the notion of informed consent and in practical terms with regard to assessing behavior and cognitive function in demented patients. However, ways of dealing with these issues have been found. An objective assessment of the effects of drugs in AD patients has been developed. One area that has yet to be adequately addressed is the design of outcome measures in primary prevention trials. It is now accepted that AD pathology in the brain begins many years before clinical symptoms emerge and that delaying onset of symptoms by about 5 years would have a major impact on disease prevalence (Brookmeyer et al., 1998). A major barrier to conducting such trials is the enormous effort and cost of conducting periodic clinical evaluations to determine if subjects have declined or developed dementia. Enrichment strategies can be used to select subjects for prevention trials who are at increased risk of developing AD (e.g. high age, family history of AD, Apo E 4 genotype). Enrichment may facilitate "proof-of-concept" prevention trials by enabling designs with smaller sample sizes and shorter observation periods but with the potential risk of reduced ability to generalize the results. Even modestly effective interventions that delay the onset of AD by 1 to 3 years will substantially alleviate the growing economic and societal burden with this disease.

Early diagnosis of dementia has become a major focus of research in neuropsychiatry and related fields not least because of anticipated high individual and societal benefits in early detection of elders at risk as well as of reassurance to those normally aging. Patients with MCI should be monitored every 6 to 12 months for conversion to dementia. It is crucial that dementia be recognized and evaluated at the earliest stage so as to begin appropriate therapy and allow the patient to have a role in management decisions. Neuropsychologic assessment of AD can provide reliable and objective measures of cognitive function and be an important adjunct to the clinician who is assessing a patient for dementia. The ability of psychometric tests to identify AD, however, is greatly influenced by variables such as the patient's educational attainment, age, and social background. Cross-cultural research in cognitive assessment has been encouraged in these areas (Ardilla, 1995). Neuropsychologists can help clarify uncertainties in diagnosis and ascertain cognitive and functional impairment. Rating scales are needed for research and clinical practice. Clinicians benefit from these tools in being more able to appreciate the presentation, severity, frequency and clinical course of AD and treatment responses. However, it is important that the use of scales does not negate the appreciation of clinical judgment. It is still necessary for clinicians to make clinical judgments concerning their patient on the simple basis of what they see.

Abnormally low metabolic activity in the posterior cingulate cortex has been consistently observed in AD and may be the earliest metabolic abnormality detectable by functional imaging in that disorder. Future studies should assess the possible clinical value of autobiographical memory retrieval tasks as fMRI probes of the functional status of this region in individual patients being assessed for early AD.

Probably, the most widely used scale for measuring change in cognitive function is the ADAS-cog and for overall function, the CIBIC-plus. Assessments of the ADL are also included in some trials. Many scales are not sensitive enough to measure clinically apparent changes in behavior such as reduction in screaming from every minute to once or twice daily. In addition, few scales are available to measure the impact of the environment on a patient's behavior. Given these limitations, it may be important to concentrate on how the information is obtained rather than on the particular rating scale used. For example, having several different caregivers complete the scale may affect a study's outcome.

Assessment methods that rely on neuropsychological batteries may not detect subtle impairments because test performance may not reflect the patient's impaired ability to carry out customary activities of daily living. A history of declining cognitive and functional performance in relation to that individual's previous abilities, however, can be sensitive to early-stage dementia, even when cognitive test performance is within a normal range (Tierney et al., 1987; Crum et al., 1993; Morris et al., 1996). This history usually is obtained from someone who knows the individual well, such as a family member. Evaluation procedures that are based on neuropsychological tests rather than collateral source information may not detect subtle cognitive and functional decline in MCI or in putatively nondemented individuals. Although sensitive clinical and neuropsychological assessment methods can be helpful in this regard, the development of a biological marker that relates directly to the clinical and pathological changes of AD ultimately may be required to identify prodromal and preclinical states. These states may be the optimal targets of eventual disease-modifying treatments.

For now, clinically validated treatments for AD remain confined to symptomatic interventions such as treatment with AChE inhibitors and drugs that ameliorate behavioral disturbances. CI produce modest improvement in cognition in patients with mild to moderate AD. Recent research suggests that earlier treatment provides greater benefit in the long run and that patients continue to benefit from cholinesterase inhibitor therapy well into the severe stages of the disease. The clinical, genetic or biological variables that regulate long-term efficacy of cholinesterase inhibitors in AD are still unknown and it is not possible to predict who will benefit from the treatment. A recent study showed that high cholesterol levels correlated with faster decline at 1-year follow up in AD patients on cholinesterase inhibitors (Borroni et al., 2003). These findings suggest that serum cholesterol is a modulating factor of treatment response and additional therapies aimed at reducing treatable high cholesterol levels may represent an alternative strategy to improve cholinesterase inhibitor efficacy and slow down disease progression over time. Current research is also focused on therapies that will target the underlying pathologic mechanisms of AD. Attention should now be directed to strategies to detect preclinical AD during life to improve understanding of the pathologic role of Aβ deposition and to maximize the opportunity for prevention prior to the occurrence of dementia. Preclinical research has shown that active immunization with the human amyloid peptide (Aβ42) or passive immunization with anti-Aβ42 antibodies protects mice that express a mutant human APP transgene from cerebral amyloid deposits. Perhaps, a high titer of anti-Aβ42 antibodies may protect humans from AD. A recent study showed that patients with AD have lower levels of serum anti-amyloid peptide antibodies than healthy elderly individuals (Weksler et al., 2002).

An important issue that needs to be addressed is whether the response to pharmacological interventions of patients with probable AD but few other disorders that are being treated with few concomitant drugs, as measured in the structured setting of a clinical trial by teams of well-trained raters can be observed for individual patients in the office setting of a lone clinician. Much has been achieved but much more remains to be done to prevent the losses of cognitive function, emotional integrity, enjoyment of life and personal dignity associated with AD. Much remains to be learned about when to start treatment, how to measure response in clinical practice and when to stop treatment for lack or loss of efficacy. Doctors should provide patients and/or caregivers with realistic expectations regarding pharmacological treatments, explain the problems associated with non-adherence to therapy and recognize the importance of adherence during the first to six months of treatment to better assess and maximize potential benefit.

Neuropsychological management of memory problems consists of internal strategies and external compensations. Although the internal strategies permit the patient to learn new information, their range is limited. It has not been accepted that such training will provide a significant improvement in the general level of memory. Richardson (1992), though appreciating the utility of the visual imagery approach, recommended that the patient's efforts should be directed toward improving metacognitive skills and gaining knowledge in a field of practical value instead of seeking proficiency at strategies that are often not used in actual everyday settings. Therefore, the realistic goal of memory rehabilitation appears to assist the patient to build up knowledge in a specific field required in daily life by using external memory devices.

References

Abel T, Martin KC, Bartsch D, Kandel ER. Memory suppressor genes: inhibitory constraints on the storage of long-term memory. Science 279, 338–341, 1998

AD Collaborative Group. The structure of the presenilin 1 (S182) gene and identification of six novel mutations in early onset AD families. Alzheimer disease Collaborative Group. Nat Genet 11, 219–22, 1995

Aggleton JP, Shaw C. Amnesia and recognition memory: a reanalysis of psychometric data. Neuropsychologia 34, 51–62, 1996

Agnoli A, Martucci V, Manna V, Conti L, Fioravanti M. Effects of cholinergic and anticholinergic drugs on short-term memory in Alzheimer's dementia: a neuropsychological and computerized electroencephalographic study. Clin Neuropharmacol 14, 311–323, 1983

Aisen PS, Johannsen DJ, Marin DB. Trazodone for behavioral disturbance in dementia. Am J Geriatr Psychiatry 1, 349–350, 1993

Aisen PS, Davis KL, Berg JD, Schafer K, Campbell K, Thomas RG, Weiner MF, Farlow MR, Sano M, Grundman M, Thal LJ. A randomized controlled trial of prednisone in Alzheimer's disease. Neurology 54(3), 588–593, 2000

Akai F, Hiruma S, Sato T, Iwamoto N, Fujimoto M, Ioku M, Hashimoto S. Neurotrophic factor-like effect of FPF1070 on septal cholinergic neurons after transections of fimbria-fornix in the rat brain. Histol Histopathol 7(2), 213–221, 1992

Akaike A, Tamura Y, Yokota T, Shimohama S, Kimura J. Nicotine-induced protection of cultured cortical neurons against N-methyl-D-aspartate receptor-mediated glutamate cytotoxicity. Brain Res 644, 181–187, 1994

Albert MS. Cognitive and neurobiologic markers of early Alzheimer disease. Proc Natl Acad Sci USA 93, 13547–13551, 1996

Albert SM, Del Castillo-Castaneda C, Sano M, Jacobs DM, Marder K, Bell K, Bylsma F, Lafleche G, Brandt J, Albert M, Stern Y. Quality of life in patients with Alzheimer's disease as reported by patient proxies. J Am Geriatrics Soc 44(11), 1342–1347, 1996

Albert MS, Logsdon R. Assessing Quality of Life in Alzheimer's disease. J Ment Health Aging 5, 3–6, 1999

Albuquerque EX, Alkondon M, Pereira EF. Properties of neuronal nicotinic acetylcholine receptors: pharmacological characterization and modulation of synaptic function. J Pharmacol Exp Ther 280, 1117–1136, 1997

Albuquerque EX, Pereira EFR, Mike A, Eisenberg HM, Maelicke A, Alkondon M. Neuronal nicotinic receptors in synaptic functions in humans and rats: physiological and clinical relevance. Behav Brain Res 113, 131–141, 2000

Alexopolous GS, Abrams RC, Young RC, Shamoian CA. Cornell scale for depression in dementia. Biol Psychiatry 23, 271–284, 1988

Allen SJ, MacGowan SH, Tyler S, Wilcock GK, Robertson AG, Holden PH, Smith SK, Dawbarn D. Reduced cholinergic function in normal and Alzheimer's disease brain is associated with apolipoprotein E4 genotype. Neurosci Lett 239, 33–36, 1997

Allen NH, Gordon S, Hope T, Burns A. Manchester and Oxford Universities Scale for the Psychopathological Assessment of Dementia (MOUSEPAD). Br J Psychiatry 169(3), 293–307, 1996

Allen NH, Burns A. The non-cognitive features of dementia. Rev Clin Gerontol 5, 57–75, 1995

Almkvist O, Basun H, Backman L, Herlitz A, Lannfelt L, Small B, Viitanen M, Wahlund LO, Winblad B. Mild cognitive impairment – an early stage of Alzheimer's disease? J Neural Transm (Suppl 54), 21–29, 1998

Alvarez P, Squire LR. Memory consolidation and the medial temporal lobe: a simple network model. Proc Natl Acad Sci USA 91, 7041–7045, 1994

Alvarez V, Mata IF, Gonzalez P, Lahoz CH, Martinez C, Pena J, Guisasola LM, Coto E. Association between the TNFa-308 A/G polymorphism and the onset–age of Alzheimer disease. Am J Med Gen (Neuropsychiatric Genetics) 114, 574–577, 2002

Alzheimer A. Allgemeine Zeitschrift für Psychiatrie und Psychisch-Gerichtliche Medizin 64, 146–148, 1907

Alzheimer Disease and Related Dementias Guidelines Panel: Recognition and initial assessment of Alzheimer disease and initial assessment of Alzheimer disease and related dementias. Rockville, MD, US. Department of Health and Human Services, Public Health Service, Agency for Health Care Policy and Research, AHCPR Publication No. 97–0702, November 1996

American Psychiatric Association. Diagnostic and Statistical Manual of Mental Disorders, Fourth Edition. Washington, DC. American Psychiatric Association, 1994

American Psychiatric Association: Diagnostic and Statistical Manual of Mental Disorders, 4th edition, Text Revision. Washington, DC, American Psychiatric Association Press, 2000

Anderson JP, Esch FS, Keim PS, Sambamurti K, Lieberburg I, Robakis NK. Exact cleavage site of Alzheimer amyloid precursor in neuronal PC-12 cells. Neurosci Lett 128, 126–128, 1991

Anderson R, Higgins GA. Absence of central cholinergic deficits in ApoE knockout mice. Psychopharmacology (Berl) 132, 135–144, 1997

Anderson ND, Craik FIM. Memory in the aging brain. In: E Tulving, FIM Craik (eds), The Oxford Handbook of Memory. Oxford University Press, 2000

Andreasen N, Minthon L, Vanmechelen E, Vanderstichele H, Davidsson P, Winblad B, Blennow K. Cerebrospinal fluid tau and Abeta42 as predictors of development of Alzheimer's disease in patients with mild cognitive impairment. Neurosci Lett 273(1), 5–8, 1999

Appell J, Kertesz A, Filsman N. A study of language functioning in Alzheimer patients. Brain & Language, 17, 73–91, 1982

Arai H, Nakagawa T, Kosaka Y, Higuchi S, Matsui T, Okamura N et al. Elevated cere-

brospinal fluid tau protein level as a predictor of dementia in memory-impaired individuals. Alzheimer Dis Assoc Disord 12, 211–214, 1998
Ardilla A. Directions of research in cross-cultural neuropsychology. J Clin Exp Neuropsychol 17, 143–150, 1995
Arkin S. Audio-assisted memory training with early Alzheimer's patients: Two single subject experiments. Clinical Gerontologist 12 (2) 77–96, 1992
Ascher JA, Cole JO, Colin J-N, Feighner JP, Ferris RM, Fibiger HC, Golden RN, Martin P, Potter WZ, Richelson E, Sulser F. Bupropion: A review of its mechanism of antidepressant activity. J Clin Psychiatry 56 (7 Suppl), 395–401, 1995
Astell AJ, Harley TA. Tip-of-the-tongue states and lexical access in dementia. Brain Lang 54, 196–215, 1996
Atkinson RC a Shiffrin RM. Human memory: a proposed system and its control process. In: KW Spence (ed): Psychology of Learning and Motivation: Advances in Research and Theory. Academic Press, 89–195, 1986
Axelman K, Basun H, Winblad B, Lannfelt L. A large Swedish family with Alzheimer disease with a codon 670/671 amyloid precursor protein mutation. A clinical and genealogical investigation. Arch Neurol 51, 1193–1197, 1994
Auer SR, Monteiro IM, Reisberg B. The Empirical Behavioral Pathology in Alzheimer's Disease (E-BEHAVE-AD) Rating Scale. Int Psychogeriatr 8, 247–266, 1996
Auriacombe S, Pere JJ, Loria-Kanza Y, Vellas B. efficacy and safety of rivastigmine in patients with Alzheimer's disease who faiuled to benefit from treatment with donepezil. Curr Med Res Opin 18, 129–138, 2002
Azmitia EC, Segal M. An autoradiographic analysis of the differential ascending projections of the dorsal and median raphe nuclei in the rat. J Comp Neurol 179, 641–668, 1978
Bachman JG, O'Malley PM. When four months equal a year: Inconsistencies in student reports of drug use. Public Opinion Quaterly 45, 536–548, 1981
Baddeley AD, Hitch GJ. Working memory. In: GH Bower, The Psychology of Learning and Motivation: Advances in Research and Theory. New York, Academic Press, 1974
Baddeley AD, Logie RH, Bressi S, Della Sala S, Spinnler H. Dementia and working memory. Q J Exp Psychol 38 (A), 603–618, 1986
Baddeley AD, Lieberman K. Spatial working memory. In: RS Nickerson (ed). Attention and Performance, Vol. 8, Hillsdale NJ, Erlbaum, 521–539, 1980
Baddeley A. The episodic buffer: a new component of working memory? Trends in Cognitive Sciences 4 (11), 417–422, 2000
Baddeley A, Della Sala S, Papagno C, Spinnler H. Dual-task performance in dysexecutive and nondysexecutive patients with a frontal lesion. Neuropsychology. 11(2), 187–194, 1997
Baddeley AD, Bressi S, Della Sala S, Logie R, Spinnler H. The decline of working memory in Alzheimer disease. Brain 114, 2521–2542, 1991
Baddeley AD. Working memory. 2nd ed. Oxford University Press, 1991
Baddeley AD. Working memory. Science 255, 556–559, 1992
Bae CY, Cho CY, Cho K, Oh BH, Choi KG, Lee HS, Jung SP, Kim DH, Lee S, Choi GD, Cho H, Lee H. A double-blind, placebo-controlled, multicenter study of cerebrolysin for Alzheimer's disease. JAGS 48, 1566–1571, 2000
Bales KR, Verina T, Dodel RC, Du Y, Altstiel L, Bender M, Hyslop P, Johnstone EM. Little

SP, Cummins DJ et al. Lack of apolipoprotein E dramatically reduces amyloid beta-peptide deposition. Nat Genet 17(3), 263–264, 1997

Ball MJ. Limbic predilection in Alzheimer dementia: is reactivated herpes virus involved? Can J Neurol Sci 9, 303–306, 1982

Ballard C, Gray A, Ayre G. Psychotic symptoms, aggression and restlessness in dementia. Rev Neurol 155, 44–52, 1999

Ballard C, Bannister C, Solis M et al. The prevalence, association and symptoms of depression among dementia sufferers. J Affect Disord 36, 136–144, 1996

Bard F, Cannon C, Barbour R, Burke RL, Games D, Grajeda H, Guido T. Peripherally administered antibodies against amyloid beta-peptide enter the central nervous system and reduce pathology in a mouse model of Alzheimer disease. Natl Med 6, 916–919, 2000

Barger SW, Horster D, Furukawa K, Goodman Y, Krieglstein J, Mattson MP. TNFs alpha and beta protect neurons against amyloid beta-peptide toxicity: evidence for the involvement of a kappa B-binding factor and attenuation of peroxide and Ca^{2+} accumulation. Proc Natl Acad Sci USA 92, 9328–9332, 1995

Baringer JR, Pisani P. Herpes simplex virus genomes in human nervous system tissue analyzed by polymerase chain reaction. Ann Neurol 36 (6), 823–829, 1994

Barnett EM. Jacobsen G. Evans G. Cassell M. Perlman S. Herpes simplex encephalitis in the temporal cortex and limbic system after trigeminal nerve inoculation. J Infect Dis 169(4), 782–786, 1994

Barsalou LW. The content and organization of autobiographical memories. In: U Neisser, E. Winograd (eds), Remembering reconsidered: Ecological and traditional approaches to the study of memory. New York: Cambridge University Press, 193–243, 1988

Bartus RT, Dean RL. Tetrahydroaminoacridine, 3,4-diaminopyridine and physostigmine: direct comparison of effects on memory in aged primates. Neurobiol Aging 9, 351–356, 1988

Bartus RT. Evidence for a direct cholinergic involvement in the scopolamine-induced amnesia in monkeys effects of concurrent administration of physostigmine and methylphenidate with scopolamine. Pharmacol Biochem Behav 9, 833–836, 1978

Baum C, Edwards D, Yonan C, Storandt M. The relation of neuropsychological test performance to performance of functional tasks in dementia of the Alzheimer type. Arch of Clin Neuropsychol 14, 219–254, 1997

Baumgarten M, Becker R, Gauthier S. Validity and reliability of the Dementia Behavior Disturbance scale. J Am Geriatrics Soc 38, 221–226, 1990

Baxter MG, Lanthorn TH, Frick KM, Golski S, Wan RQ, Olton DS. D-cycloserine, a novel cognitive enhancer, improves spatial memory in aged rats. Neuroniol Aging 15, 207–213, 1994

Bayles KA, Kaszniak AW. Communication and cognition in normal ageing and dementia. Boston: College Hill Press, 1987

Bayles KA, Boone DR, Tomoeda CK, Slauson TJ, Kaszniak AW. Differentiating Alzheimer's patients from the normal elderly and stroke patients with aphasia. Journal of Speech and Hearing Disorders 54 (1), 74–87, 1989

Beach TG, otter PE, Kuo Y-M, Emmerling MR, Durham RA, Webster SD, Walker DG, Sue LI, Scott S, Layne KJ, Roher AE. Cholinergic deafferentation of the rabbit cortex: a new animal model of Aβ deposition. Neurosci Lett 283, 9–12, 2000

Beard CM, Kokmen E, Sigler C, Smith GE, Petterson T, O'Brien PC. Cause of death in Alzheimer's disease. Ann Epidemiol 6(3), 195–200, 1996

Becker RE, Colliver JA, Markwell SJ, Moriearty PL, Unni LK, Vicari S. Double-blind, placebo-controlled study of metrifonate, an acetylcholinesterase inhibitor for Alzheimer disease. Alzheimer Dis Assoc Disord 10, 124–131, 1996

Behl C. Alzheimer's disease and oxidative stress: implications for novel therapeutic approaches. Prog Neurobiol 57(3), 301–323, 1999

Behrmann M, Moscovitch M, Winocur G. Intact visual imagery and impaired visual perception in a patient with visual agnosia. J Exp Psychol Hum Percept Perform 20, 1068–1087, 1994

Belleville S, Peretz I, Malefant D. Examination of the working memory components in normal aging and in dementia of the Alzheimer type. Neuropsychologia 3, 195–207, 1996

Belli RF. The structure of autobiographical memory and the event history calendar: Potential improvements in the quality of retrospective reports in surveys. Memory 6 (4), 383–406, 1998

Bennett DA, Wilson RS, Schhneider JA, Evans DA, Beckett LA, Aggarwal NT, Barnes LL, Fox JH, Bach J. Natural shistory of mild cognitive impairment in older persons. Neurology 59, 198–205, 2002

Bentham P, Gray R, Sellwood E, Raftery J. Effectiveness of rivastigmine in Alzheimer disease. Improvements in functional ability remain unestablished [letter]. BMJ 319, 640–641, 1999

Berent S, Giordani B, Foster N, Minoshima S, Lajiness-O'Neill R, Koeppe R, Kuhl DE, Neuropsychological function and cerebral glucose utilization in isolated memory impairment and Alzheimer's disease. J Psychiatr Res 33(1), 7–16, 1999

Berg L. Clinical dementia rating (Letter). Br J Psychiatry 145, 339, 1984

Berg L, Hughes CP, Coben LA, Danziger WL, Martin RL, Knesevich J. Mild senile dementia of the Alzheimer type. Research diagnostic criteria, recruitment and description of a study population. J Neurol Neurosurg Psychiatry 45, 962–968, 1982

Berg L, Miller JP, Storandt M, Ducheck J, Morris JC, Rubin EH, Burke WJ, Coben LA. Mild senile dementia of the Alzheimer's type. 2. Longitudinal assessment. Ann Neurol 23, 477–484, 1988

Bertrand P, Poirier J, Oda T, Finch CE, Pasinetti GM. Association of apolipoprotein E genotype with brain levels of apolipoprotein E and apolipoprotein J (clusterin) in Alzheimer disease. Brain Res Mol Brain Res 33, 174–178, 1995

Beyer K, Lao JI, Fernandez-Novoa L, et al. Identification of a novel mutation (V148I) in the TM2 domain of the presenilin 2 gene in a patient with late-onset AD. Neurobiol Aging 19 (Suppl 2), 587, 1998

Bierer LM, Hof PR, Purohit DP, Carlin L, Schmeidler J, Davis KL, Perl DP. Neocortical neurofibrillary tangles correlate with dementia severity in Alzheimer disease. Arch Neurol 52, 81–88, 1995

Bigl V, Arendt T, Fischer S, Werner M, Arendt A. The cholinergic system in aging. Gerontology 33, 172–180, 1987

Binetti G, Locascio JJ, Corkin S, Vonsattel JP, Growden JH. Differences between Pick disease and Alzheimer's disease in clinical appearance and rate of cognitive decline. Arch Neurol 57, 225–232, 2000

Bird M, Kinsella G. Long-term cues recall of tasks in senile dementia. Psychol Aging 11, 45–56, 1996

Bird TD, Lampe TH, Nemens EJ, Miner GW, Sumi SM, Schellenberg GD. Familial Alzheimer disease in American descendants of the Volga Germans: probable genetic founder effect. Ann Neurol 23, 25–31, 1988

Blacker D, Wilcox MA, Laird NM, Rodes L, Horvath SM, Go RC, Perry R, Watson B Jr, Bassett SS, McInnis MG et al. Alpha-2 macroglobulin is genetically associated with Alzheimer disease. Nat Genet 19(4), 357–360, 1998

Blacker D, Haines JL, Rodes L, Terwedow H, Go RCP, Harrell LE, Perry RT, Bassett SS, Chase G, Meyers D, Albert MS, Tanzi AR. ApoE-4 and age at onset of Alzheimer's disease: The NIMH genetics initiative. Neurology 48(1), 139–147, 1997

Blacker D, Wilcox MA, Laird NM, Rodes L, Horvath SM, Go RC, Perry R, Watson B Jr, Bassett SS, McInnis MG, Albert MS, Hyman BT, Tanzi RE. Alpha-2 macroglobulin is genetically associated with Alzheimer disease. Nat Genet 19(4), 357–360, 1998

Blacker D, Crystal AS, Wilcox MA, Laird NM, Tanzi RE. An alpha-2-macroglobulin insertion-deletion polymorphism in Alzheimer disease – Reply. [Letter] Nat Genet 22(1), 21–22, 1999

Blackford RC, La Rue A. Criteria for diagnosing age-associated memory impairment: proposed improvement from the field. Developmental Neuropsychology 5, 295–306, 1989

Blasko I, Marx F, Steiner E, Hartmann T, Grubeck-Loebenstein B. TNFα plus IFNγ induce the production of Alzheimer-amyloid peptide and decrease the secretion of APPs. FASEB J 12, 63–68, 1999

Blesa R, Adroer R, Santacruz P, Ascaso C, Tolosa E, Oliva R. High apolipoprotein E epsilon 4 allele frequency in age-related memory decline. Ann Neurol 39(4), 548–551, 1996

Blessed G, Tomlinson BE, Roth M. Blessed-Roth Dementia Scale (DS). Psychopharmacol Bull 24, 705–708, 1988

Blessed G, Tomlinson BE, Roth M. The association between quantitative measures of dementia and of senile change in the cerebral grey matter of elderly subjects. Br J Psychiatry 114, 797–811, 1968

Blier P, Seletti B, Young S, Benkelfat C, de Montigny C. Serotonin1A receptor activation and hypothermia: evidence for a postsynaptic mechanism in humans (abstract). Neuropsychopharmacology 10, 92S, 1994

Bliss TV, Lomo T. Long-lasting potentiation of synaptic transmission in the dentate area of the anaesthetized rabbit following stimulation of the performant path. J Physiol (London) 232, 331–356, 1973

Bliss TV, Collingridge GL. A synaptic model of memory: long-term potentiation in thhe hippocampus. Nature 361, 31– 39, 1993

Boast C, Bartolomeo AC, Morris H, Moyer JA. 5HT anatagonists attenuate MK801-impaired radial arm maze performance in rats. Neurobiol Learn Mem 71, 259–271, 1999

Boller F, Massioui FE, Devouche E, Traykov L, Pomati S, Starkstein S. Processing emotional information in Alzheimer's disease: Effects on memory performance and neurophysiological correlates. Dement Geriatr Cogn Disord 14, 104–112, 2002

Boothby H, Mann AH, Barker A. Factors determining interrater agreement with rating global change in dementia: The CIBIC-Plus. Int J Geriatr Psychiatry 10, 1037–1045, 1995

Borchelt DR, Ratovitski T, van Lare J, Lee MK, Gonzales V, Jenkins NA, Copeland NG. Price DL, Sisodia SS. Accelerated amyloid deposition in the brains of transgenic mice coexpressing mutant presenilin 1 and amyloid precursor proteins. Neuron 19(4), 939–45, 1997

Borchelt DR, Thinakaran G, Eckman CB, Lee MK, Davenport F, Ratovitsky T, Prada CM, Kim G, Seekins S, Yager D et al. Neuron 17, 1005–1013, 1996

Borod JC, Goodglass H, Kaplan E. Normative data on the Boston diagnostic aphasia examination, parietal lobe battery and the Boston naming test. J Clin Neuropsychol 2 (3), 209–215, 1980

Borroni B, Pettenati C, Bordonali T, Akkawi N, Di Luca M, Padovani A. Serum cholesterol levels modulate long-term efficacy of cholinesterase inhibitors in Alzheimer disease. Neurosci Lett 343, 213–215, 2003

Bondi M, Monsch A, Galasko D et al. Preclinical cognitive markers of dementia of the Alzheimer's type. Neuropsychology 8, 374–384, 1994

Bourgeois MS, Burgio LD, Schulz R, Beach S, Palmer B. Modifying repetitive verbalization of community-dwelling patients with AD. The Gerontologist 37, 30–39, 1997

Bowen J, Teri L, Kukull W, McCormick W, McCurry SM, Larson EB. Progression to dementia in patients with isolated memory loss. Lancet 349(9054), 763–765, 1997

Bowen DM. Cellular aging: selective vulnerability of cholinergic neurons in human brain. Monogr Dev Biol 17, 42–59, 1984

Bower GH. Mental imagery and associative learning. In: Gregg LW (ed), Cognition in Learning and Memory. New York: John Wiley, 51–88, 1972

Bozeat S, Lambon Ralph MA, Garrard P, Patterson K, Hodges JR. Non-verbal semantic impairment in semantic dementia. Neuropsychologia 38, 1207–1215, 2000

Bozeat S, Gregory CA, Lambon Ralph MA, Hodges JR. Which neuropsychiatric and behavioural features distinguish frontal and temporal variants of frontotemporal dementia from Alzheimer's disease? J Neurol Neurosurg Psychiatry 69, 178–166, 2000

Bozoki A, Giordani B, Heidebrink JL, Berent S, Foster NL. Mild cognitive impairments predict dementia in nondemented elderly patients with memory loss. Arch Neurol 58(3), 411–416, 2001

Braak H, Del Tredici K, Schhultz C, Braak E. Vulnerability of select neuronal types to Alzheimer disease. Ann NY Acad Sci 924, 53–61, 2000

Bracco L, Gallato R, Grigoletto F, Lippi A, Lepore V, Bino G, Lazzaro MP, Carella F, Piccolo T, Pozzilli C, Giometto B, Amaducci L. Factors affecting course and survival in Alzheimer's disease: A 9-year longitudinal study. Arch Neurol 51(12), 1213–1219, 1994

Brady DR. Mufson EJ. Alz-50 immunoreactive neuropil differentiates hippocampal complex subfields in Alzheimer's disease. J Comp Neurol 305(3), 489–507, 1991

Brandt J, Rich JB. Memory disorders in the dementias. In: Baddeley AD, Wilson BA, Watts FN (eds), Handbook of Memory Disorders. Chichester, UK: Wiley, 243–270, 1995

Brandt J, Spencer M, McSorley P, Folstein MF. Semantic activation and implicit memory in Alzheimer disease. Alzheimer Dis Assoc Disord 2 (2), 112–119, 1988

Braunewell K-H, Riederer P, Spilker C, Gundelfinger ED, Bogerts B, Bernstein HG. Abnormal localization of two neuronal calcium sensor proteins, visinin-like proteins (VILIPS)-1 and -3, in neocortical brain areas of Alzheimer disease patients. Dement Geriatr Cogn Disord 12, 110–116, 2001

Breitner JCS, Welsh KA, Helms MJ, Gaskell PC, Gau BA, Roses AD, Pericak-Vance MA, Saunders AM. Delayed onset of Alzheimer's disease with nonsteroidal anti-inflammatory and histamine H2 blocking drugs. Neurobiol Aging 16(4), 523–530, 1995

Breitner JCS, Gau BA, Welsh KA, Plassman BL, McDonald WM, Helms MJ, Anthony JC. Inverse association of anti-inflammatory treatments and Alzheimer's disease: Initial results of a co-twin control study. Neurology 44(2), 227–232, 1994

Breitner JC, Welsh KA. Genes and recent development in the epidemiology of Alzheimer disease and related dementia. Epidemiol Rev 17, 39–47, 1995

Brod M, Stewart AL, Sands L, Walton P. Conceptualization and measurement of quality of life in dementia: the dementia quality of life instrument (DQoL). Gerontologist 39, 25–35, 1999

Broadbent DE. Perception and Communication. Pergamon Press, 1958

Brookmeyer R, Gray S, Kawas C. Projections of Alzheimer's disease in the United States and the public health impact of delaying disease onset. Am J Public Health 88, 1337–1342, 1998

Brown NR. Estimation strategies and the judgment of event frequency. J Exp Psychol Learn Mem Cogngnition 21, 1539–1553, 1995

Brown J. Some tests of the decay theory of immediate memory. The Q J Exp Psychol 10, 12–21, 1958

Brown DA, Goldensohn ES. The electroencephalogram in normal pressure hydrocephalus. Arch Neurol 29, 70–71, 1973

Brun A, Gustafson L. The Lund longitudinal dementia study: a 25–year perspective on neuropathology, differential diagnoses and treatment. In: Corain B, Nicolini M et al (eds), Alzheimer's Disease: Advances in Clinical and Basic Research. New York: John Wiley & Sons, 4–18, 1993

Brush JA, Camp CJ. Using spaced retrieval as an intervention during speech-language therapy. Clinical Gerontologist 19, 51–64, 1998

Bschor T, Kühl K-P, Reischies FM. Spontaneous speec of patients with dementia of the Alzheimer type and mild cognitive impairment. Int Psychogeriatr 13 (3), 289–298, 2001

Buccafusco JJ, Jackson WJ. Beneficial effects of nicotine administered prior to a delayed matching-to-sample task in young and aged monkeys. Neurobiol Aging 12, 233–238, 1991

Büchel C, Dolan RJ, Armony JL, Friston KJ. Amygdala-hippocampal involvement in human aversive trace conditioning revealed through event-related functional magnetic resonance imaging. J Neurosci 19, 10869–10876, 1999.

Buckner RL, Wheeler ME. The cognitive neuroscience of remembering. Nat Rev Neurosci 2, 624–634, 2001

Budson AE, Desikan R, Daffner KR, Schacter DL. Perceptual false recognition in Alzheimer's disease. Neuropsychology 15, 230–243, 2001

Budson AE, Michalska KJ, Rentz DM, Joubert CC, Daffner KR, Schacter DL, Sperling RA. Use of a false recognition pradigm in an Alzheimer's disease clinical trial: A pilot study. American Journal of Alzheimer's Disease and Other Dementias 17 (2), 93–100, 2002

Bulens C, Meerwaldt JD, Van der Wildt GJ, Keemink CJ. Spatial contrast sensitivity in unilateral cerebral ischhaemic lesions involving the posterior visual pathway. Brain 112 (pt 2), 507–520, 1989

Bullock R, Connolly C. Switching cholinesterase inhibitor therapy in Alzheimer's disease – donepezil to rivastigmine, is it worth it? Int J Geriatr Psychiatry 17, 288–289, 2002

Burke WJ, Miller JP, Rubin EH, Morris JC, Coben LA, Duchek J, Wittels IG, Berg L. Reliability of the Washington University Clinical Dementia Rating. Arch Neurol 45, 31–32, 1988

Burke WJ, Rubin EH, Morris JC, Berg L. Symptoms of "depression" in dementia of the Alzheimer type. Alzheimer Dis Assoc Disord 2, 356–362, 1988

Burns A, Jacoby R, Levy R. Psychiatric phenomena in Alzheimer's disease. I. Disorders of thought content. Br J Psychiatry 157, 72–76, 1990

Burns A, Jacoby R, Levy R. Psychiatric phenomena in Alzheimer's disease. IV. Disorders of behaviour. Br J Psychiatry 157, 86–94, 1990

Burns A, Rossor M, Hecker J, Gauthier S, Petit H, Moller HJ, Rogers SL, Friedhoff LT. The effects of donepezil in Alzheimer disease – results from a multinational trial. Dement Geriatr Cogn Disord 10, 237–244, 1999

Buschke H. Selective reminding for analysis of memory and learning. J Verb Learn Verb Behav 12, 543–550, 1973

Buschke H, Fuld PA. Evaluating storage, retention, and retrieval in disordered memory and learning. Neurology 24, 1019–1025, 1974

Bush AI, Pettingell WH, Multhaup G, Paradis DM, Vonsattel J-P, Gusella JF, Beyreuther K, Masters CL, Tanzi RE. Rapid induction of Alzheimer Abeta amyloid formation by zinc. Science 265(5177), 1464–1467, 1994

Butters N, Delis DC, Lucas JA. Clinical assessment of memory disorders in amnesia and dementia. Annu Rev Psychol 46, 493–523, 1995

Butters N. The clinical aspects of memory disorders: Contributions from experimental studies of amnesia and dementia. J Clin Neuropsychol 6, 17–36, 1984

Butterworth B, Cappelletti M, Kopelman M. Category specificity in reading and writing: the case of number words. Nat Neurosci 4, 784–786, 2001

Cabeza R, Grady C, Nyberg L, McIntosh A, Tulving E, Kapur S, Jennings J, Houle S, Craik FIM. Age-related differences in neural activity during memory encoding and retrieval: a positron emission tomography study. J Neurosci 17, 391–400, 1997

Cabeza R, Nyberg L. Neural bases of learning and memory: functional neuroimaging evidence. Curr Opin Neurol 13, 415–421, 2000

Cabeza R, Nyberg L. Imaging cognition II: an empirical review of 275 PET and fMRI studies. J Cogn Neurosci 12, 1–47, 2000

Cacabelos R, Hofman A, Mullan M et al. Alzheimer's disease: scientific progress for future trends. Drug News Perspect 6, 242–244, 1993

Cacabelos R. Alzheimer's disease: news and prospects. Drug News Perspect 5, 501–506, 1992

Camp CJ, Foss JW, O'Hanlon AM, Stevens AB. Mermory interventions for persons with dementia. Applied Cognitive Psychology 10, 193–210, 1996

Camp CJ, Stevens AB. Spaced-retrieval: A memory intervention for dementia of the Alzheimer's type (DAT). Clinical Gerontologist 10, 58–60, 1990

Campion D, Flaman JM, Brice A, Hannequin D, Dubois B, Martin C, Moreau V, Charbonnier F, Didierjean O, Tardieu S, et al. Mutations of the presenilin I gene in families with early-onset Alzheimer disease. Hum Mol Genet 4, 2373–2377, 1995

Campion D, Dumanchin C, Hannequin D, Dubois B, Belliard S, Puel M, Thomas-Anterion

C, Michon A, Martin C, Charbonnier F, Raux G, Camuzat A, Penet C, Mesnage V, Martinez M, Clerget-Darpoux F, Brice A, Frebourg T. Early-onset autosomal dominant Alzheimer disease: prevalence, genetic heterogeneity, and mutation spectrum. Am J Hum Genet 65, 664–70, 1999

Camus V, Schmitt L, Ousset PJ, Micas M. Depression and dementia: A French validation of Cornell Scale for Depression in Dementia (CSDD), and Dementia Mood Assessment Scale (DMAS). Encephale 21(3), 201–208, 1995

Caplan D, Waters G. Verbal working memory and sentence comprehension. Behav Brain Sci 22, 77–126, 1999

Cappelletti M, Butterworth B, Kopelman MD. Spared numerical abilities in a case of semantic dementia. Neuropsychologia 39, 1224–1239, 2001

Carlesimo GA, Fadda L, Marfia GA, Caltagirone C. Explicit memory and repetition priming in dementia: Evidence for a common basic mechanism underlying conscious and unconscious retrieval deficits. J Clin Exp Neuropsychology 17, 44–57, 1995

Cave CB, Squire LR. Equivalent impairment of spatial and nonspatial memory following damage to the human hippocampus. Hippocampus 1, 329–340, 1991

Chan D, Fox NC, Scahill RI, Crum WR, Whitewell JL, Leschziner G, Rossor AM, Stevens JM, Cipolotti K, Rossor MN. Patterns of temporal lobe atrophy in semantic dementia and Alzheimer's disease. Ann Neurology 49, 433–442, 2001

Chartier-Harlin MC, Parfitt M, Legrain S, Perez-Tur J, Brousseau T, Evans A, Berr C, Vidal O, Roques P, Gourlet V et al. Apolipoprotein E, epsilon 4 allele as a major risk factor for sporadic early and late-onset forms of Alzheimer's disease: analysis of the 19q13.2 chromosomal region. Human Mol Genet 3(4), 569–574, 1994

Checler F. Processing of the β-amyloid precursor protein and its regulation in Alzheimer disease. J Neurochem 65, 1431–1444, 1995

Chen SY, Wright JW, Barnes CD. The neurochemical and behavioral effects of β-amyloid peptide (25–35). Brain Res 720, 54–60, 1996

Chen KC, Blalock EM, Thibault O, Karninker P, Landfield PW. Expression of alpha 1D subunit mRNA is correlated with L-type Ca^{2+} channel in single neurons of hippocampal "zipper" slices. Proc Natl Acad Sci USA 97, 4357–4362, 2000

Chen P, Ratcliff G, Belle S, Cauley J, DeKosky S, Ganguli M. Patterns of Cognitive decline in presymptomatic Alzheimer Disease: A prospective Community Study. Arch Gen Psychiatry 58 (9), 853–858, 2001

Cherny RA, Atwood CS, Xilinas ME, Gray DN, Jones WD, McLean CA, Barnham KJ, Volitakis I, Fraser FW, Kim Y et al. Treatment with a copper-zinc chelator markedly and rapidly inhibits beta-amyloid accumulation in Alzheimer's disease transgenic mice. Neuron 30(3), 665–676, 2001

Chertkow H. Mild memory loss in the elderly – Can we predict development in dementia? Can J Neurol Sci 25, 1–S4, 1998

Chodosh J, Reuben DB, Albert MS, Seeman TE. Predicting cognitive impairment in high-functioning Community-dwelling older persons: MacArthur studies of successful aging. JAGS 50, 1051–1060, 2002

Christen Y. Oxidative stress and Alzheimer disease. Am J Clin Nutr 71 (Suppl), 621S-629S, 2000

Christensen H, Henderson AS, Jorm AF, Mackinnon AJ, Scott R, Korten AE. ICD-10 mild

cognitive disorder: epidemiological evidence on its validity. Psychol Med 25(1), 105–120, 1995
Christensen H. What cognitive changes can be expected with normal ageing? Aust N Z J Psychiatry 35, 768–775, 2001
Christensen H, Henderson AS, Korten AE, Jorm AF, Jacomb PA, Mackinnon AJ. ICD-10 mild cognitive disorder: its outcome three years later. Int J Geriatr Psychiatry 12(5), 581–586, 1997
Chui HC, Lyness SA, Sobel E, Schneider LS. Extrapyramidal signs and psychiatric symptoms predict faster cognitive decline in Alzheimer's disease. Arch Neurol 51(7), 676–681, 1994
Citron M, Diehl TS, Gordon G, Biere AL, Seubert P, Selkoe DJ. Evidence that the 42- and 40-amino acid forms of amyloid beta protein are generated from the beta-amyloid precursor protein by different protease activities. Proc Natl Acad Sci USA 93(23), 13170–13175, 1996
Clare L, Wilson BA, Breen EK, Hodges JR. Errorless learning of face-name associations in early Alzheimer's disease. Neurocase 5, 37–46, 1999
Clarke R, Smith AD, Jobst KA, Refsum H, Sutton L, Ueland PM. Folate, vitamin B12, and serum total homocysteine levels in confirmed Alzheimer disease. Arch Neurol 55(11), 1449–1455, 1998
Cleary J, Hittner JM, Semotuk M, Mantyh P, O'Hare E. Beta-amyloid (1–40) effects on behavior and memory. Brain Res 682, 69–74, 1995
Cobb JL, Wolf PA, Au R, White R, D'Agostino RB. The effect of education on the incidence of dementia and Alzheimer's disease in the Framingham Study. Neurology 45(9), 1707–1712, 1995
Coben LA, Danziger W, Stotandt M. A longitudinal EEG study of mild and senile dementia of Alzheimer type: changes at 1 year and 2.5 years. Electroencephalogr Clin Neurophysiol 61, 101–112, 1985
Coen RF, O'Boyle CA, Coakley D, Lawlor BA. Individual Quality of Life factors distinguishing low-burden and high-burden caregivers of dementia patients. Dement Geriatr Cogn Disord 13, 164–170, 2001
Cohen NJ, Squire LR. Preserved learning and retention of pattern-analyzing skill in amnesia: dissociation of knowing how and knowing that. Science 210, 207–210, 1980
Cogan DG. Visual disturbances with focal progressive dementing disease. Am J Opthalmol 100 (1), 68–72, 1985
Cohen D, Eisdorfer C, Gorelick P, Paveza G, Luchins DJ, Freels S, Ashford JW, Semla T, Levy P, Hirschman R. Psychopathology associated with Alzheimer's disease and related disorders. Journal of Gerontology 48(6), M255–260, 1993
Cohen-Mansfield J. Assessment of disruptive behavior/ agitation in the elderly: Function, methods and difficulties. J Geriatr Psychiatry Neurol 8, 52–60, 1995
Cohen-Mansfield J. Agitated behaviors in the elderly. II. Preliminary results in the cognitively deteriorated. J Am Geriatr Soc 34, 722–727, 1986
Cohen-Mansfield J, Marx M, Rosenthal A. A description of agitation in a nursing home. Journal of Gerontology 44, M77–M84, 1989
Cohen-Mansfield J. Conceptualization of agitation: results based on the Cohen-Mansfield

Agitation Inventory and the Agitation Behavior Mapping Instrument. Int Psychogeriatr, 8 309–315, 1996

Collette F, Van der Linden M, Bechet S, Salmon E. Phonological loop and central executive functioning in Alzheimer disease. Neuropsychologia 37, 905–918, 1999

Collie A, Maruff P. The neuropsychology of preclinical Alzheimer disease and mild cognitive impairment. Neurosci Biobehav Rev 24, 365–374, 2000

Combarros O, Leno C, Oterino A, Berciano J, Fernandez-Luna JL, Fernandez-Viadero C, Pena N, Miro J, Delgado M. Gender effect on apolipoprotein E epsilon4 allele-associated risk for sporadic Alzheimer's disease. Acta Neurologica Scandinavica 97(1), 68–71, 1998

Committee for proprietaty Medicinal Products. Note for guidance on medicinal products for the treatment for Alzeimer's disease. London, September 1977

Condren RM, Cooney C. Use of drugs by old age psychiatrists in the treatment of psychotic and behavioural symptoms in patients with dementia. Aging & Mental Health 235–241, 2001

Connor D, Salmon D, Sandy T, Galasko D, Hansen L, Thal LJ. Cognitive profiles of autopsy-confirmed Lewy Body varian *versus* pure Alzheimer disease. Arch Neurol 55 (7), 994–1000, 1998

Consensus Conference on the Behavioural Disturbances of Dementia, International Psychogeriatric Association, Spring, 1996

Convit A, De Leon MJ, Tarshish C, De Santi S, Tsui W, Rusinek H, George A. Specific hippocampal volume reductions in individuals at risk for Alzheimer's disease. Neurobiol Aging 18(2), 131–138, 1997

Conway MA. Autobiographical knowledge and autobiographical memories. In: DC Rubin (ed), Remembering our past: Studies in autobiographical memory. New York: Cambridge University Press, 1996

Cooper AJL, Sheu KFR, Burke JR, Strittmatter WJ, Gentile V, Peluso G, Blass JP. Pathogenesis of inclusion bodies in (CAG)n/Qn-expansion diseases with special reference to the role of tissue transglutaminase and to selective vulnerability. J Neurochem 72, 889–899, 1999

Copeland JR, Kelleher MJ, Kellett JM, Gourlay AJ, Gurland BJ, Fleiss JL, Sharpe L. A semi-structured clinical interview for the assessment of diagnosis and mental state in the elderly: the Geriatric Mental State Schedule. I. Development and reliability. Psychol Med 6(3), 439–449, 1976

Corder EH, Saunders AM, Strittmatter WJ, Schmechel DE, Gaskell PC, Small GW, Roses AD, Haines JL, Pericak-Vance MA. Gene dose of apolipoprotein E type 4 allele and the risk of Alzheimer's disease in late onset families. Science 261(5123), 921–923, 1993

Corder EH, Saunders AM, Risch NJ, Strittmatter WJ, Schmechel DE, Gaskell PC Jr, Rimmler JB, Locke PA, Conneally PM, Schmader KE et al. Protective effect of apolipoprotein E type 2 allele for late onset Alzheimer disease. Nat Genet 7(2), 180–184, 1994

Corey-Bloom J, Anand R, Veach J for the ENA 713 B352 Study Group. A randomized trial evaluating the efficacy and safety of ENA 713 (rivastigmine tartrate), a new acetylcholinesterase inhibitor, in patients with mild to moderately severe Alzheimer disease. Int J Geriatr Psychopharmacol 1, 55–65, 1998

Corey-Bloom J, Tiraboschi P Hansen LA, Alford M, Schoos B, Sabbagh MN, Masliah E, Thal

LJ. E4 allele dosage does not predict cholinergic activity or synapse loss in Alzheimer's disease. Neurology 54, 403–406, 2000

Correa DD, Graves RE, Costa L. Awareness of memory deficit in Alzheimer's disease patients and memory impaired older adults. Aging Neuropsychology and Cognition 3, 215–328, 1996

Costa PTJ, Williams TF, Somerfield M et al. Recognition and initial assessment of Alzheimer's disease and related disorders. AHCPR Publication 97–0702 (US Dept of Health and Human Services, Public Health, Agency for Health Care Policy and Research, Rockville, MD), 1996

Couratier P, Lesort M, Sindou P, Esclaire F, Yardin C and Hugon J. Modifications of neuronal phosphorylated tau immunoreactivity induced by NMDA toxicity. Mol Chem Neuropathol 27, 259–273, 1996

Coyle JT, Price DL, DeLong MR. Alzheimer disease: A disorder of cortical cholinergic innervation. Science 219, 1184–1190, 1983

Craik FIM. Age differences in human memory. In: JE Birren, KW Schaie (eds), Handbook of the Psychology of Aging. New York: Van Nostrand-Reinhold, 1977

Craik FIM, Lockhart R. Levels of processing: a framework for memory research. J Verbal Learn Verbal Behav 11, 671–684, 1972

Craik FIM. A functional account of age differences in memory. In F Klix, H Hangendorf (eds), Human Memory and Cognitive Capabilities: Mechanisms and Performances. Amsterdam: Elsevier, 409–422, 1986

Craik FIM. Memory changes in normal aging. Current Directions in Psychological Science 3, 155–158, 1994

Craik FIM. A functional account of age differences in memory. In: F Klix, H Hagendorf (eds), Human Memory and Cognitive Capabilities, Mechanisms and Performance, Amsterdam: Elsevier, 409–422, 1986

Craik FIM, Jennings JM. Human memory. In: FIM Craik, TA Salthouse (eds), The Handbook of Aging and Cognition. Hillsdale, NJ, Erlbaum, 1992

Cregan E, Ordy JM, Palmer E, Blosser J, Wengenack T and Thomas G. Spatial memory enhancement by nicotine of aged Long Evans rats in the T-Maze. Soc Neurosci Abstr 15, 731, 1989

Crook TH. Diagnosis and treatment of memory loss in older patients who are not demented. In: Levy R, Howard R, Burns A (eds), Treatment and Care in Old Age Psychiatry. London: Wrightson Biomedical, 95–111, 1993

Crook T, Bartus RT, Ferris SH et al. Age-associated memory impairment: proposed diagnostic criteria and measures of clinical change – Report of a National Institute of Mental Health Work Group. Dev Neuropsychology 2, 261–276, 1986

Crum RM, Anthony JC, Bassett SS, Folstein MF. Population-based norms for the Mini-Mental State Examination by age and educational level. JAMA 269, 2386–2391, 1993

Cruts M, van Duijn CM, Backhovens H, Van den Broeck M, Wehnert A, Serneels S, Sherrington R, Hutton M, Hardy J, St George-Hyslop PH, Hofman A, Van Broeckhoven C. Estimation of the genetic contribution of presenilin-1 and -2 mutations in a population-based study of presenile Alzheimer disease. Hum Mol Genet 7, 43–51, 1998

Crystal H, Dickson D, Fuld P, Masur D, Scott R, Mehler M, Masdeu J, Kawas C, Aronson

M, Wolfson L. Clinico-pathologic studies in dementia: nondemented subjects with pathologically confirmed Alzheimer's disease. Neurology 38(11), 1682–1687, 1988

Cuénod CA, Denys A, Michot JL, Jehenson P, Forette F, Kaplan D, Syrota A, Boller F. Amygdala atrophy in Alzheimer's disease: An *in vivo* magnetic resonance imaging study. Arch Neurol 50, 941–945, 1993

Cullen WK, Wu JQ, Anwyl R and Rowan MJ. Beta-amyloid produces a delayed NMDA receptor-dependent reduction in synaptic transmission in rat hippocampus. Neuroreport 8, 87–92, 1996

Cullum S, Huppert FA, McGee M, Dening T, Ahmed A, Paykel ES, Brayne C. Decline across different domains of cognitive function in normal ageing: results of a longitudinal population-based study using CAMCOG. Int J Geriatr Psychiatry 15, 853–862, 2000

Cummings JL, Benson DF. Dementia: A Clinical Approach. London: Butterworths, 1983

Cummings JL, Donohue JA, Brooks RL. The relatioship between donepezil and behavioral disturbances in patients with Alzheimer's disease. Am J Geriatr Psychiatry 8, 134–140, 2000

Cummings JL, Benson DF. Dementia: definition, prevalence, classification and approach to diagnosis. In: Cummings JL, DF Benson (eds), Dementia: A Clinical Approach. Butterworth-Heineman: Boston, MA, 1992

Cummings JL, Mega M, Gray K, Rosenberg-Thompson S, Carusi DA, Gornbein J. The Neuropsychiatric Inventory: comprehensive assessment of psychopathology in dementia. Neurology 44(12), 2308–2314, 1994

Cummings JL, Gorman DG, Shapira J. Physostigmine ameliorates the delusions of Alzheimer's disease. Biol Psychiatry 33, 536–541, 1993

Cummings JL. The Neuropsychiatric Inventory: assessing psychopathology in dementia patients. Neurology 48 (5 Suppl 6), S10–16, 1997

Cutler NR, Polinsky RJ, Sramek JJ, Enz A, Jhee SS, Mancione L, Hourani J, Zolnouni P. Dose-dependent CSF acetylcholinesterase inhibition by SDZ ENA 713 in Alzheimer's disease. Acta Neurologica Scandinavica 97(4), 244–250, 1998

Daigle I, Li C. apl-1, a *Caenorhabditis elegans* gene encoding a protein related to the human β-amyloid protein precursor. Proc Natl Acad Sci USA 90, 12045–12049, 1993

Dalla Barba G, Goldblum MC. The influence of semantic encoding on recognition memory in Alzheimer disease. Neuropsychologia 34 (12), 1181–1186, 1996

Daly E, Zaitchik D, Copeland M, Schmahmann J, Gunther J, Albert M. Predicting conversion to Alzheimer disease using standardized clinical information. Arch Neurol 57, 675–680, 2000

Damasio AR. Descartes' Error. GP Putnam, New York, 1994

Dannenbaum SE, Parkinson SR, Inman VW. Short-term forgetting: Comparison between patients with dementia of the Alzheimer type, depressed and normal elderly. Cognitive Neuropsychology 5, 213–233, 1988

Davies CA, Mann DMA. Is the "preamyloid" of diffuse plaques in Alzheimer disease really nonfibrillar? Am J Pathol 143, 1594–1605, 1993

Davis RN, Massman PJ, Doody RS. Cognitive intervention in Alzheimer disease: A randomized placebo-controlled study. Alzheimer Dis Assoc Disord 15, 1, 1–9, 2001

Daw EW, Payami H, Nemens EJ, Nochlin D, Bird TO, Schellenberg GD, Wijsman EM. The number of trait loci in late-onset Alzheimer disease. Am J Hum Genet 66, 196–204, 2000

DeBettignies BH, Mahurin RK, Pirozzolo FJ. Insight for impairment in independent living skills in Alzheimer disease and multi-infarct dementia. J Clin Exp Neuropsychol 12, 355–363, 1990

De Deyn PP, Rabheru K, Rasmussen A, Bocksberger JP, Dautzenberg PL, Eriksson S, Lawlor BA. A randomized trial of risperidone, placebo, and haloperidol for behavioral symptoms of dementia. Neurology 53(5), 946–955, 1999

DeJong R, Osterlund OW, Roy GW. Measurement of quality-of-life changes in patients with Alzheimer disease. Clin Ther 11, 545–554, 1989

DeKosky ST, Ikonomovic MD, Styren SD, Beckett L, Wisniewski S, Bennett DA, Cochran EJ, Kordower JH, Mufson EJ. Upregulation of choline acetyltransferase activity in hippocampus and frontal cortex of elderly subjects with mild cognitive impairment. Ann Neurol 51(2), 145–155, 2002

Del Ser T, Hachinski V, Merskey H, Munoz DG. An autopsy-verified study of the effect of education on degenerative dementia. Brain 122, 2309–2319, 1999

DeMattos RB, Bales KR, Cummings DJ, Dodart J-C, Paul SM and Holtzman DM. Peripheral antiAβ antibody alters CNS and plasma Aβ clearance and decreased brain Aβ burden in a mouse model of Alzheimer disease. Proc Natl Acad Sci USA 17, 8850–8855, 2001

Dempster FN. Memory span: Sources of individual and developmental differenes. Psychol Bull 89, 63–100, 1981

De Renzi E, Faglioni P. Normative data and screening power of a shortened version of the Token test. Cortex 14, 41–49, 1978

Dermaut B, Cruts M, Slooter AJC, Van Gestel S, De Jonghe C, Vanderstichele H, Vanmechelen E, Breteler MM, Hofman A, Van Duijn CM, Van Broeckhoven C. The Glu318Gly substitution in presenilin-1 is not causally related to Alzheimer disease. Am J Hum Genet 64, 290–292, 1999

Deslauriers S, Landreville P, Dicaire L, Verreault R. Validité et fidélité de l"inventaire d'agitation de Cohen-Mansfield. Canadian Journal on Aging / La Revue Canadienne du vieillissement 20 (3), 373–384, 2001

De Strooper B, Saftig P, Craessaerts K, Vanderstichelle H, Guhde G, Annaert W, Von Figura K, Van Leuven F. Deficiency of presenilin-1 inhibits the normal cleavage of amyloid precursor protein. Nature 391, 387–390, 1998

De Strooper B, Umans L, Van Leuven F, Van der Berghe H. Study of the synthesis and secretion of normal and artificial mutants of murine amyloid precurosr protein (APP): cleavage of APP occurs in a late compartment of the default secretion pathway. J Cell Biol 121, 295–304, 1993

Deutsch LH, Rovner BW. Agitation and other non-cognitive abnormalities in Alzheimer disease. Psychiatr Clin North Am 14, 341, 1991

Devanand DP, Jacobs DM, Tang M-X et al. The course of psychopathologic symptoms in mild-to-moderate Alzheimer's disease. Arch Gen Psychiatry 54, 257–263, 1997

Deweer B, Pillon B, Michon A, Dubois B. Mirror reading in Alzheimer's disease: normal skill learning and acquisition of item-specific information. J Clin Exp Neuropsychol 15(5), 789–804, 1993

Deweer B, Ergis AM, Fossati P, Pillon B, Boller F, Agid Y and Dubois B. Explicit memory, procedural learning and lexical priming in Alzheimer disease. Cortex 30, 113–126, 1994

Dickson DW. The pathogenesis of senile plaques. J Neuropathol Exp Neurol 56, 321–339, 1997

Diesfeldt HFA. The distinction between long-term and short-term memory in senile dementia: an analysis of free recall and delayed recognition. Neuropsychologia 16, 115–119, 1978

Di Luca M, Pastorino L, Bianchetti A, Perez J, Vignolo LA, Lenzi GL, Trabucchi M, Cattabeni F, Padovani A. Differential level of platelet amyloid b precursor protein isoforms. An early marker for Alzheimer disease. Arch Neurol 55, 1195–1200, 1998

Dodart J-C, Bales KR, Gannon KS, Greene SJ, Demattos RB, Mattis C, DeLong CA, Wu S, Wu X, Holtzman DM, Paul SM. Immunization reverse memory deficits without reducing brain Aβ burden in Alzheimer's disease model. Nature Neurosci 5 (5), 452– 457, 2002

Dolan RJ, Fletcher PC. Dissociating prefrontal and hippocampal function in episodic memory encoding. Nature 388, 582–585, 1997

Doody RS, Vacca JL, Massman PJ, Liao TY. The influence of handedness on the clinical presentation and neuropsychology of Alzheimer disease. Arch Neurol 56 (9), 1133–1137, 1999

Dooneief G, Marder K, Tang M-X, Stern Y. The Clinical Dementia Rating Scale: Community-based validation of "profound" and "terminal" stages. Neurology 46, 1746–1749, 1996

Drachman DA. Memory and cognitive function in man: does the cholinergic system have a specific role? Neurology 27, 783–786, 1977

Dritschel BH, Williams JM, Baddeley AD, Nimmo-Smith I. Autobiographical fluency: a method for the study of personal memory. Memory & Cognition 20(2), 133–140, 1992

Du Y, Dodel R, Hampel H, Buerger K, Lin S, Eastwood B, Bales K, Gao F, Moeller H-J, Oertel W, Farlow M, Paul S. Reduced levels of amyloid β-peptide antibody in Alzheimer disease. Neurology 57, 801–805, 2001

Dubinsky RM, Stein AC, Lyons K. Practice parameter: risk of driving and Alzheimer's disease. Neurology 54, 2205–2211, 2000

Duff K, Eckman C, Zehr C, Yu X, Prada CM, Perez-Tur J, Hutton M, Buee L, Harigaya Y, Yager D, Morgan D, Gordon MN, Holcomb L, Refolo L, Zenk B, Hardy J, Younkin S. Increased amyloid-beta42(43) in brains of mice expressing mutant presenilin 1. Nature 383(6602), 710–713, 1996

Dunn LM, Dunn LM. Peabody picture vocabulary test – revisited. Circle Pines, MN: American Guidance Service, 1981

Earnst KS, Marson DC, Harrell LE. Cognitive models of physician's legal standard and personal judgments of competency in patients with Alzheimer's disease. J Am Geriatr Soc 48, 949–957, 2000

Ebly E, Hogan D, Parhad I. Cognitive impairment in the non-demented elderly: Results from the Canadian study of health and aging. Arch Neurol 52, 612–619, 1995

Eggert A, Crismon ML, Ereshefsky L. Alzheimer disease. In: JT Dipiro, RL Talbert (eds), Pharmacotherapy, 3rd. New York: Elsevier: 1325–1344, 1997

Eichenbaum H. A cortical-hippocampal system for declarative memory. Nat Neurosci 1, 41–50, 2000

Einstein GO, Smithh RE, McDaniel MA, Shaw P. Aging and prospective memory: The influence of increased task demands at encoding and retrieval. Psychol Aging 12, 479–488, 1997

Eldridge LL, Knowlton BJ, Furmanski CS, Bookheimer SY, Engel SA. Remembering episodes: a selective role for the hippocampus during retrieval. Nat Neurosci 3, 1149–1152, 2000

Elias MF, Beiser A, Wolf PA, Au R, White RF, D'Agostino RB. The preclinical phase of Alzheimer disease. A 22–year prospective study of the Famingham cohort. Arch Neurol 57, 808–813, 2000

Ergis AM, Van der Linden M, Deweer B. Priming for new associations in normal aging and in mild dementia of the Alzheimer type. Cortex 34, 357–373, 1998

Ericsson KA, Kinsch W. Long-term working memory. Psychol Rev 102, 211–245, 1995

Esch FS, Keim PS, Beaatie EC, Blacher RW, Culwell AR, Oltersdorf T, McClure D, Ward PJ. Cleavage of amyloid b peptide during constitutive processing of its precurosr. Science 248, 1122–1124, 1990

Eustache F, Lambert J. Modèles neurocognitifs de l'écriture et maladie d'Alzheimer: éclairages mutuels. Rev Neurol 152, 658–668, 1996

Evans DA, Funkenstein HH, Albert MS, Scherr PA, Cook NR, Chown MJ, Herbert LE. Prevalence of Alzheimer disease in a community population higher than previously reported. JAMA 262, 2251–2256, 1989

Evans D, Scherr PA, Cook NR et al. Estimated prevalence of Alzheimer disease in the United States. Milbank Q 68, 267–289, 1990

Ezquerra M, Carnero C, Blesa R, Oliva R. A novel presenilin 1 mutation (Leu166Arg) associated with early-onset Alzheimer disease. Arch Neurol 57, 485–488, 2000

Ezquerra M, Carnero C, Blesa R, Gelpf JL, Ballesta F, Oliva R. A presenilin 1 mutation (Ser169Pro) associated with early-onset AD and myoclonic seizures. Neurology 52, 566–570, 1999

Fabrigoule C, Rouch I, Taberly A, Letenneur L, Commenges D, Mazaux J-M, Orgogozo J-M, Dartigues J-F. Cognitive process in preclinical phase of dementia. Brain 121, 135–141, 1998

Farah MJ. The neurological basis of mental imagery: a componentional analysis. Cognition 18, 242–272, 1984

Farber SA, Nitsch RM, Schulz JG, Wurtman RJ. Regulated secretion of β-amyloid precursor protein in rat brain. J Neurosci 15, 7442–7451, 1995

Farrer LA, Cupples LA, Haines JL et al. Effects of age, sex and ethnicity on the association between apolipoprotein E genotype and Alzheimer disease. JAMA 278, 1349–1356, 1997

Faustman WO, Moses JA, Csernansky JG. Limitations of the Mini-Mental State Examination in predicting neuropsychological functioning in a psychiatric sample. Acta Psychiatr Scan 81, 126–131, 1990

Feher EP, Mahurin RK, Doody RS et al. Establishing the limits of the Mini-Mental State: Examination of "subtests". Arch Neurol 49, 87–92, 1992

Feher EP, Larrabee GJ, Sudilovsky A, Crook TH. Memory self-report in Alzheimer's disease and in age_associated memory impairment. J Geriatr Psychiatry Neurology 6, 58–65, 1994

Feldman H, Gauthier S, Hecker J et al. Donepezil improved the clinical state and quality of life in moderate to severe Alzheimer's disease. Neurology 57, 613–620, 2001

Ferlazzo F, Conte S, Gentilomo A. Event-related potentials and recognition memory: the effects of word imagery value. Int J Psychophysiol 15, 115–122, 1993

Ferraro FR, Balota DA, Connor T. Implicit memory and the formation of new associations in nondemented Parkinson's disease individuals and individuals with senile dementia of the Alzheimer type: a serial reaction time (SRT) investigation. Brain Cogn 21, 163–180, 1993

Fillenbaum GG, Peterson B, Morris JC. Estimating the validity of the Clinical Dementia Rating Scale: The CERAD experience. Aging Clin Exp Res 8, 379–385, 1996

Finkel SI. The signs of the behavioural and psychological symptoms of dementia. Clinician 16, 33–42, 1998

Finkel SI. New focus on behavioral and psychological signs and symptoms of dementia. Int Psychogeriatr 8 (Suppl. 3), 215–216, 1996

Finkel SI, Costa E, Silva JC, Cohen GD, Miller S, Sartorius N. Behavioral and psychological symptoms of dementia: A consensus statement on current knowledge and implications for researc and treatment. Am J Geriatr Psychiatry 6, 97–100, 1998

Fisher SA, Fischer TM, Carew TJ. Multiple overlapping processes underlying short-term synaptic enhancement. Trends Neurosci 20, 170–177, 1997

Fleischman DA, Gabrieli JDE, Reminger S, Rinaldi J, Morrell F, Wilson R. Conceptual priming in perceptual identification for patients with Alzheimer disease and a patient with a right occipital lobectomy. Neuropsychology 9, 187–197, 1995

Fleischman DA, Gabrieli JDE, Rinaldi JA et al. Word-stem completion priming for perceptually and conceptually encoded words in patients with Alzheimer disease. Neuropsychologia 1996, 35 (1), 25–35

Fleischman DA, Gabrieli JDE. Repetition priming in normal aging and Alzheimer's disease. A review of findings and theories. Psychol Aging 13, 88–119, 1998

Fleischman DA, Gabrieli JDE, Rinaldi JA, Reminger SA, Grinnell ER, Lange KL, Shapiro R. Word-stem completion priming for perceptually and conceptually encoded words in patients with Alzheimer's disease. Neuropsychologia 35, 25–35, 1997

Fleischman DA, Gabrielli JDE, Reminger SL, Vaidya CJ and Bennett DA. Object decision priming in Alzheimer disease. J Intern Neuropsychol Soc 4, 435–446, 1998

Fletcher PC, Frith CD, Grasby PM, Shallice T, Frackowiak RSJ, Dolan RJ. Brain systems for encodin and retrieval of auditory-verbal memory: an *in vivo* study in humans. Brain 118, 401–416, 1995

Fletcher PC, Frith CD, Rugg MD. The functional neuroanatomy of episodic memory. Trends Neurosci 20, 213–218, 1997

Fletcher PC, Henson RNA. Frontal lobes and human memory. Insights from functional neuroimaging. Brain 124, 849–881, 2001

Flicker C, Ferris SH, Reisberg B. A two-year longitudinal study of cognitive function in normal aging and Alzheimer disease. J Geriatr Psychiatry Neurol 6, 84–96, 1993

Flicker C. Ferris SH. Crook T. Bartus RT. Implications of memory and language dysfunction in the naming deficit of senile dementia. Brain Lang 31(2), 187–200, 1987

Flood JF, Morley JE, Roberts E. Amnestic effects in mice of four synthetic peptides homologous to amyloid b protein from patients with Alzheimer disease. Proc Natl Acad Sci USA 88, 3363–3366, 1991

Flood JF, Smith GE, Roberts E. Dehydroepiandrosterone and its sulfate enhance memory retention in mice. Brain Res 447(2), 269–278, 1988

Flood JF,Morley JE, Roberts E. Amyloid β-protein fragment, Aβ (12–28), equipotently impairs post-training memory processing when injected into different limbic system structures. Brain Res 663, 271–276, 1994

Flynn FG, Cummings JL, Gornbein J. Delusions in dementia syndromes: investigations of behavioral and neuropsychological correlates. J Neuropsych Clin Neurosci 3, 364–370, 1991

Folstein MF, Folstein SE, McHugh PR. Mini-mental state: A practical method for grading the cognitive state of patients for the clinician. J Psychiatr Res 12, 189–198, 1975

Folstein MF, Anthony C, Perhed I, Duffy B, Gruenberg EM. The meaning of cognitive impairment in the elderly. J Am Geriatr Soc 33, 228–235, 1985

Folstein MF. Differential diagnosis of dementia. The clinical process. Psychiatr Clin North Am 20, 45–57, 1997

Forette F, Anand R, Gharabawi G. A phase II study in patients with Alzheimer's disease to assess the preliminary efficacy and maximum tolerated dose of rivastigmine (Exelon®). Eur J Neurol 6, 423–429, 1999

Forsell Y, Fratiglioni L, Grut M, Viitanen M, Winblad B. Clinical staging of dementia in a population survey: comparison of DSM-III-R and the Washington University clinical dementia rating scale. Acta Psychiatr Scand 86, 49–54, 1992

Förstl H, Burns A, Levy R, Cairne N. Neuropathological correlates of psychotic phenomena in confirmed Alzheimer's disease. Br J Psychiatry 165, 53–59, 1994

Förstl H, Besthorn C, Burns A, Geiger-Kabisch C, Levy R, Sattel A. Delusional misidentification in Alzheimer's disease: A summary of clinical and biological aspects. Psychopathology 27, 194–199, 1994

Fox N, Freeborough PA, Rossor MN. Visualization and quantification of rates of atrophy in Alzheimer disease. Lancet 348, 94–97, 1996

Fox PT, Mintun MA, Raichle ME, Herscovitch P. A noninvasive approach to quatitative functional brain mapping with H2 (15)O and positron emission tomography. J Cereb Blood Flow Metab 4, 329–333, 1984

Fox NC, Warrington EK, Seiffer AL, Agnew SK and Rossor MN. Presymptomatic cognitive deficits in individuals at risk of familial Alzheimer disease. A longitudinal prospective study. Brain 121, 1631–1339, 1998

Francis PT, Sims NR, Procter AW and Bowen DM. Cortical pyramidal neurone loss may cause glutamatergic hypoactivity and cognitive impairment in Alzheimer disease – investigative and therapeutic perspectives. J Neurochem 60, 1589–1604, 1993

Frenkel D, Kariv N, Solomon B. Generation of autoantibodies toward Alzheimer's disease vaccination. Vaccine 19, 2615–2619, 2001

Fried I, Wilson CL, Morrow JW, Cameron KA, Behnke ED, Ackerson LC, Maidment NT. Increased dopamine release in the human amygdala during performance of cognitive tasks. Nat Neurosci 4, 201–206, 2001

Friedland RP, Fritsch T, Smyth KA, Koss E, Lerner AJ, Chen CH, Petot GJ, Debanne SM.

Patients with Alzheimer's disease have reduced activities in midlife compared with healthy control-group members. Proc Natl Acad Sci USA 98(6), 3440–3445, 2001

Fritz JB, Becker D, Mishkin M, Saunders RC. A comparison of the effects of medial temporal and cortical lesions on auditory recognition memory in the rhesus monkey. Abstr Soc Neurosci 25, 789, 1999

Fromholt P, Larsen SF. Autobiographical memory in normal aging and primary degenerative dementia (dementia of Alzheimer type). J Gerontol 46 (3), P85–91, 1991

Fu JY, Masferrer JL, Seibert K, Raz A, Needleman P. The induction and suppression of prostaglandin H2 synthase (cyclooxygenase) in human monocytes. J Biol Chem 265, 16737–16740, 1990

Fujimori M, Imamura T, Yamashita H, Hirono N, Ikejiri Y, Shimomura T and Mori E. Age at onset and visuocognitive disturbances in Alzheimer disease. Alzheimer Dis Assoc Disord 12, 163–166, 1998

Fuld PA. The Fuld object memory evaluation. Chicago, IL: Stoelting Instrument Company, 1981

Fuld PA, Dickson D, Crystal HA, Aronson MK. Primitive plaques and memory dysfunction in normal and impaired elderly person (Letter). N Eng J Med 316, 756, 1987

Fuld PA, Masur DM, Blau AD, Crystal H, Aronson MK. Object memory evaluation for prospective detection of dementia in normal functioning elderly: predictive and normative data. J Clin Exp Neuropsychol 12, 520–528, 1990

Fuld PA, Katzman R, Davies P, Terry RD. Intrusions as a sign of Alzheimer's dementia: chemical and pathological verification. Ann Neurol 11, 155–159, 1982

Fuld PA. Psychological testing in the differential diagnosis of the dementias. In: R Katzman, RD Terry, KL Bick (eds), Alzheimer disease: senile dementia and related disorders. New York: Raven Press, 185–193, 1978

Funnell E. Objects and properties: a study of the breakdown of semantic memory. Memory 3, 497–518, 1995

Gabrieli JDE, Keane MM, Stanger BZ et al. Dissociations among structural-perceptual, lexical-semantic, and event-fact memory systems in Alzheimer, amnesic, and normal subjects. Cortex 30, 75–103, 1994

Gabrieli JDE, Brewer JB, Desmond JB, Glover GH. Separate neural bases of two fundamental memory processes in human medial temporal lobe. Science 276, 264–266, 1997

Galasko D, Bennett D, Sano M et al. An inventory to assess the activities of daily living for clinical trials in Alzheimer disease. The Alzheimer disease Cooperative Study. Alzheimer Dis Assoc Disord 11 (suppl 2), S33–39, 1997

Galton CJ, Patterson K, Graham KS, Lambon Ralph MA, Williams G, Antoun N, Sahakian BJ, Hodges JR. Differing patterns of temporal lobe atrophy in Alzheimer's disease and semantic dementia: diagnostic and theoretical implications. Neurology 57, 216–225, 2001

Games D, Adams D, Alessandrini R, Barbour R, Berthelette P, Blackwell C, Carr T, Clemens J, Donaldson T, Gillespie F et al. Alzheimer-type neuropathology in transgenic mice overexpressing V717F beta-amyloid precursor protein. Nature 373(6514), 523–527, 1995

Ganguli M, Seaberg EC, Ratcliff GC and Belle SH. Cognitive stability over 2 years in a rural elderly population: the MoVIES Project. Neuroepidemiology 15, 42–50, 1996

Ganguli M, Ratcliff G, Chandra V et al. A Hindi version of the MMSE: the development of a cognitive screening instrument for a largely illiterate rural elderly population in India. Int J Geriatr Psychiatry 10, 367–377, 1995

Gathercole SE. Cognitive approaches to the development of short-term memory. Trends Cognit Sci 3, 410–419, 1999

Gearing M, Mirra SS, Hedreen JC, Sumi SM, Hansen LA, Heyman A. The consortium to establish a registry for Alzheimer disease (CERAD). Part X. Neurology 45, 461–466, 1995

Gelb DJ, St. Laurent RT. Alternative calculation of the global Clinical Dementia Rating. Dis Ass Disord 7, 202–211, 1993

Gelb DJ, St. Laurent RT. Clinical dementia rating. Letter: Comment. Neurology 44, 1983–1984, 1994

Geldmacher DS, Whitehouse PJ. Evaluation of dementia. N Engl J Med 335, 330–336, 1996

Gelinas I, Auer S. Functional autonomy. In: S Gauthier (ed), Clinical Diagnosis and Management of Alzheimer Disease. London: Martin Dunitz Ltd, 1996

Gelinas I, Gauthier L, McIntyre M, Gauthier S. Development of a functional measure for persons with Alzheimer disease: the disability assessment for dementia. Am J Occup Ther 53, 471–481, 1999

Gélinas I, Auer S. Functional autonomy. In: Gauthier S (ed), Clinical Diagnosis and Management of Alzheimer Disease. London: Martin Dunitz Ltd, 1996

Gélinas I, Gauthier L, McIntyre M, Gauthier S. Development of a functional measure for persons with Alzheimer disease: the disability assessment for dementia. Am J Occup Ther 53, 471–481, 1999

Gellerstedt N. Zur Kenntnis der Hirnveränderungen bei der normalen Altersinvolution. Ups. LaKaref, Forhh, 38, 193, 1933

Genoux D, Haditsch U, Knobloch M, Michalon A, Storm D, Mansuy IM. Protein phosphatases 1 is a molecular constraint on learning on memory. Nature 418, 29, 970–975, 2002

Giacobini E. Selective inhibitors of butyrylcholinesterase. A valid alternative for therapy of Alzheimer's disease? Drugs Aging 18 (12), 891–898, 2001

Gilley DW, Whalen ME, Wilson RS, Bennett DA. Hallucinations and associated factors in Alzheimer's disease. J Neuropsych Clin Neurosci 3, 371–376, 1991

Gilley DW, Wilson RS, Beckett LA, Evans DA. Psychotic symptoms and physically aggressive behavior in Alzheimer disease. J Am Geriatr Soc 45, 1074–1079, 1997

Giovannelli L, Casamenti F, Scali C, Bartolini L, Papeu G. Differential effects of amyloid peptides β-(1–40) and β-(25–35) injections into the rat nucleus basalis. Neuroscience 66, 781–792, 1995

Glenner GG, Wong CW. Alzheimer's disease: initial report of the purification and characterization of a novel cerebrovascular amyloid protein. Biochem Biophys Res Comm 120(3), 885–890, 1984

Glisky EL, Schacter DL, Tulving E. Learning and retention of computer-related vocabulary in memory-impaired patients: method of vanishing cues. J Clin Exp Neuropsychol 8, 282–312, 1986

Glisky EL, Schacter DL. Models and methods of memory rehabilitation. In: F Boller, J Grafman (eds): Handbook of Neuropsychology 3. Elsevier, Amsterdam, 233–246, 1989

Glisky EL. Prospective memory and the frontal lobes. In: M Brandimonte, GO Einstein, MA McDaniel (eds), Prospective Memory: Theory and Applications. Hillsdale, NJ: Erlbaum, 249–266, 1996

Glosser G, Wolfe N, Albert ML, Lavine L, Steele JC, Calne DB, Schoenberg BS. Cross-Cultural Cognitive Examination: Validation of a dementia screening instrument for neuroepidemiological research. J Am Geriatr Soc 41(9), 931–939, 1993

Gnanalingham KK, Byrne EJ, Thornton A, Sambrooks MA, Bannister P. Motor and cognitive function in Lewy body dementia: comparison with Alzheimer's and parkinson's disease. J Neurol Neurossurg Psychiatry 6 (2), 243–252, 1997

Golebiowski M, Barcikowska M, Pfeffer A. Magnetic resonance imaging-based hippocampal volumetry in patients with dementia of the Alzheimer type. Dementia Geriatr Cognit Disord 10, 284–288, 1999

Goodglass H, Kaplan E. The assessment of aphasia and related disorders. Philadelphia: Lea and Febiger, 1972

Goate A, Chartier-Harlin MC, Mullan M, Brown J, Crawford F, Fidani L, Giuffra L, Haynes A, Irving N, James L et al. Segregation of a missense mutation in the amyloid precursor protein gene with familial Alzheimer's disease. Nature 349 (6311), 704–706, 1991

Goldberg RJ. Antidepressant use in the elderly: current status of Nefazodone, venlafaxine and moclobemide. Drugs Aging 11, 119–131, 1997

Goldgaber D, Lerman MI, McBride OW, Saffiotti U, Gajdusek DC. Characterization and chromosomal localization of a cDNA encoding brain amyloid of Alzheimer's disease. Science 235(4791), 877–880, 1987

Goldman WP. Winograd E. Goldstein FC. O'Jile J. Green RC. Source memory in mild to moderate Alzheimer's disease. J Clin Exp Neuropsychol 16(1), 105–116, 1994

Goldman-Rakic PS. The cortical dopamine system: role in memory and cognition. Adv Pharmacol 42, 707–711, 1998

Golebiowski M, Barcikowska M, Pfeffer A. Magnetic resonance imaging-based hippocampal volumetry in patients with dementia of the Alzheimer type. Dementia Geriatr Cognit Disord 10, 284–288, 1999

Golob EJ, Johnson JK, Starr A. Auditory event-related potentials during target detection are abnormal in mild cognitive impairment. Clinical Neurophysiology 113, 151–161, 2001

Gomez-Vargas M, Ogawa N. Clinical applications of neurotransmitter-receptor studies in geriatric neuropharmacology. Acta Med Okayama 50, 173–190, 1996

Gonnerman LM, Andersen ES, Devlin JT, Kempler D, Seidenberg MS. Double dissociation of semantic categories in Alzheimer's disease. Brain Lang 57 (2), 254–279, 1997

Gorman DG, Read S, Cummings JL. Cholinergic therapy of behavioral disturbances in Alzheimer's disease. Neuropsychiatry Neuropsychol Behav Neurol 6, 229–234, 1993

Gosche KM, Mortimer JA, Smith CD, Markesbery WR, Snowdon DA. Hippocampal volume as an index of Alzheimer neuropathology. Findings from the Nun Study. Neurology 58, 1476–1482, 2002

Gottfries CG, Brane G, Gullberg B, Steen G. A new rating scale for dementia syndromes. Archives of Gerontology & Geriatrics 1(4), 311–330, 1982

Gottfries CG, Nyth AL. Effects of citalopram, a selective 5–HT reuptake blocker, in emotionally disturbed patients with dementia. Ann NY Acad Sci 640, 276–279, 1991

Grad R. Benzodiazepines for insomnia in community dwelling elderly: a review of benefit and risk. J Fam Pract 41, 473–481, 1995

Grady C, Haxby J, Horowitz B, Sundarum M, Berg G, Shapiro M, Friedland R, Rapoport S. A longitudinal study of the early neuropsychological and cerebral metabolic changes in dementia of the Alzheimer's type. J Clin Exp Psychol 10, 576–596, 1988

Grady CL, Haxby JV, Horwitz B, Berg G, Rapoport SI. Neuropshological and cerebral metabolic function in early *versus* late onset dementia of the Alzheimer type. Neuropsychologia 25, 807–816, 1987

Graf P. Life-span changes in implicit and explicit memory. Bulletin of the Psychonomic Society 28, 353–358, 1990

Grafton ST, Mazziotta JC, Presty S, Friston KJ, Frackowiak RS, Phelps ME. Functional anatomy of human procedural learning determined with regional cerebral blood flow and PET. J Neurosci 12(7), 2542–2548, 1992

Graham DI, Gentleman SM, Lynch A, Roberts GW. Distribution of beta-amyloid protein in the brain following severe head injury. Neuropathol Appl Neurobiol 21, 27–34, 1995

Graham KS. Semantic dementia: a challenge to the multiple-trace theory? Trends in Cognitive Sciences 3, 85–87, 1999

Graham KS, Hodges JR. Differentiating the roles of the hippocampal complex and the neorortex in long-term memory storage: evidence from the study of semantic dementia and Alzheimer's disease. Neuropsychology 11, 77–89, 1997

Graves AB, Larson EB, Edland SD, Bowen JD, McCormick WC, McCurry SM, Rice MM, Wenzlow A, Uomoto JM. Prevalence of dementia and its subtypes in the Japanese American population of King County, Washington state. The Kame Project. Am J Epidemiol 144(8), 760–771, 1996

Grayson LWD, Jorm AF, Creasey H, Cullen J, Bennett H, Casey B, Broe GA. Informant-based staging of dementia using the Clinical Dementia Rating. Alzheimer Dis Assoc Disord 13 (1), 34–37, 1999

Green RC, Woodard JL, Green J. Validity of the Mattis Dementia Rating Scale for the detection of cognitive impairment in the elderly. J Neuropsychiatry Clin Neurosci 7, 357–360, 1995

Green RC, Cupples LA, Kurz A, Auerbach S, Go R, Sadovnick D, Duara R, Kukull WA, Chui H, Edeki T, Griffith PA, Friedland PR, Backman D, Farrer L. Depression as a risk factor for Alzheimer disease. The MIRAGE Study. Arch Neurol 60, 753–759, 2003

Greene JDW, Hodges JR, Baddeley AD. Autobiographical memory and executive function in early dementia of the Alzheimer type. Neuropsychologia 33 (12), 1647–1670, 1995

Greene JG. The evaluation of reality orientation. In: Hanley I, Hodge J (eds), Psychological Approaches to the Care of the Elderly, London, Croom, Helm, 192–212, 1984

Greenamyre JT, Young AB. Excitatory amino acids and Alzheimer disease. Neurobiol Aging 10, 593–602, 1989

Greenough WT. Bailey CH. The anatomy of a memory: Convergence of results across a diversity of tests. Trends Neurosci 11(4), 142–146, 1988

Greenwald B. Depression in Alzheimer disease and related dementias. In: Lawlor (ed), Behavioural Complications in Alzheimer Disease. Washington: American Psychiatric Association Press, 19–54, 1995

Griffin WST, Stanley LC, Ling C et al. Brain interleukin 1 and S-100 immunoreactivity are elevated in Down syndrome and Alzheimer disease. Proc Natl Acad Sci USA 86, 7611–7615, 1989

Griffin WST, Sheng JG, Royston MC, Gentleman SM, Mckenzie JE, Graham DI, Roberts GW, Mrak RE. Glial-neuronal interactions in Alzheimers disease – The potential role of a cytokine cycle in disease progression. Brain Pathol 8(1), 65–72, 1998

Grigsby J, Stevens D. Neurodynamics of Personality. Guildford, 2000

Grimaldi LME, Casadei VM, Ferri C, Veglia F, Licastro F, Annoni G, Biunno I, De Bellis G, Sorbi S, Mariani C, Canal N, Griffin WST, Franceschi M. Association of early-onset Alzheimer's disease with an interleukin-1alpha gene polymorphism. Ann Neurol 47(3), 361–365, 2000

Grober E, Buschke H. Genuine memory deficits in dementia. Dev Neuropsychol 3, 13–36, 1987

Grober E, Buschke H, Crystal H, Bang S, Dresner R. Screening for dementia by memory testing. Neurology 38, 900–903, 1988

Grober E, Dickson DW, Sliwinski M et al. Free and cued selective reminding is sensitive to neuropathology of early Alzheimer disease. Neurology 48, A338, 1997

Grundman M. Vitamin E and Alzheimer disease: the basis for additional clinical trials. Am J Clin Nutr 71 (Suppl), 630S-636S, 2000

Grunwald M, Busse F, Hensel A, Riedel-Heller S, Kruggel F, Arendt T, Wolf H, Gertz H-J. Theta-power differences in patients with mild cognitive impairment under rest condition and during haptic tasks. Alzheimer Dis Assoc Disord 16 (1), 40–48, 2002

Gsell W, Conrad R, Hickethier M, Sofic E, Frölich L, Wichart I, Jellinger K, Moll G, Ransmayr G, Beckmann H, Riederer P. Decreased catalase activity but unchanged superoxide dismutase activity in brains of patients with dementia of Alzheimer type. J Neurochem 64, 1216–1223, 1995

Guillozet AL, Smiley JF, Mash DC, Mesulam MM. Butyrylcholinesterase in the life cycle of amyloid plaques. Ann Neurol 42(6), 909–918, 1997

Guo Z, Cupples LA, Kurz A, Auerbach SH, Volicer L, Chui H, Green RC, Sadovnick AD, Duara R, DeCarli C et al. Head injury and the risk of AD in the MIRAGE study. Neurology 54(6), 1316–23, 2000

Gurland BJ, Copeland JR, Kuriansky J et al. The mind and mood of ageing. Mental health problems of community elderly in New York and London. Beckenham: Croom Helm, 1982

Gutman CA, Strittmatter W, Weisgraber KH, Matthew WD. Apolipoprotein E binds to and potentiates the biological activity of ciliary neurotrophic factor. J Neurosci 17, 6114–6121, 1997

Gutman CA, Jolesz F, Kikinis R, Killiany R, Moss M, Sandor T, Albert M. White matter changes with normal aging. Neurology 50, 972–978, 1998

Gutzmann H, Hadler D. Sustained efficacy and safety of idebenone in the treatment of Alzheimer disease: update on a 2-year double-blind multicentre study. J Neural Transm (Suppl) 54, 301–310, 1998

Guy W (ed). Clinical Global Impressions (CGI). ECDEU assessment manual for psychopharmacology. Rockville, MD: US Department of Health and Human Services, Public Health Service, Alcohol Drug Abuse and Mental Health Service, Alcohol Drug Abuse and Mental Health Administration. NIMH Psychopharmacology Research Branch, 218–222, 1976

Haan MN, Shemanki L, Jagust WJ, Manolio TA. The role of APOE E4 in modulating effects of other risk factors for cognitive decline in elderly persons. JAMA 282, 40–46, 1999

Haass C, Selkoe DJ. Cellular processing of β-amyloid precurosr protein and the genesis of amyloid β-peptide. Cell 75, 1039–1042, 1993

Hachinski VC, Lassen NA, Marshall J. Multi-infarct dementia: a cause of mental deterioration in the elderly. Lancet 2, 207–210, 1974

Hale S, Bronik MD, Fry AF. Verbal and spatial working memory in school-age children: Developmental differences in susceptibility to interference. Dev Psychol 33, 364–371, 1997

Hall KS, Gao S, Emsley CL et al. Community Screening Interview for Dementia (CSI "D"): performance in five disparate study sites. Int J Geriatr Psychiatry 15, 521–531, 2000

Halliday GM, McCann HL, Pamphlett R, Brooks WS, Creasey H, McCusker E, Cotton RG, Broe GA, Harper CG. Brain stem serotonin-synthesizing neurons in Alzheimer's disease: a clinicopathological correlation. Acta Neuropathologica 84(6), 638–650, 1992

Hampel H, Teipel SJ, Bayer W, Alexander GE, Schwarz R, Schapiro MB, Rapoport SI, Möller H-J. Age transformed of combined hippocampus and amygdala volume improves diagnostic accuracy in Alzheimer disease. J Neurol Sci 194, 15–19, 2002

Hanley I. Reality orientation in the care of the elderly patient with dementia – three case studies. In: Hanley I, Gilhoorly M (eds) Psychological Therapies for the Elderly, London, Cromm Helm, 65–79, 1986

Hanninen T, Hallikainen M, Koivisto K, Helkala EL, Reinikainen KJ, Soininen H, Mykkanen L, Laakso M, Pyorala K, Riekkinen PJ Sr. A follow-up study of age-associated memory impairment: neuropsychological predictors of dementia. J Am Geriatr Soc 43(9), 1007–1015, 1995

Hansen L, Salmon D, Galasko D, Masliah E, Katzman R, DeTeresa R et al. The Lewy body variant of Alzheimer's disease: a clinical and pathologic entity. Neurology 40, 1–8, 1990

Harasty JA, Halliday GM, Code C, Brooks WS. Quantification of cortical atrophy in a case of progressive fluent aphasia. Brain 119, 181–190, 1996

Hardy J, Israël A. Alzheimer disease: In search of γ-secretase. Nature 398, 466–467, 1999.

Hardy J, Duff K, Gwinn-Hardy K, Perez-Tur J, Hutton M. Genetic dissection of Alzheimer disease and related dementias: amyloid and its relationship to tau. Nat Neurosci 1, 355–358, 1998

Harkany T, O'Mahony S, Kelly JP, Soos K; Töro I, Penke B, Luiten PGM, Nyakas C, Gulya K, Leonard BF. β-Amyloid (Phe(SO3H)24)25–35 in rat nucleus basalis induces behavioral dysfunctions, impairs learning and memory and disrupts cortical cholinergic innervation. Behav Brain Res 90, 133–145, 1998

Harner RN. EEG evaluation of the patient with dementia. In: Benson DF, Blumer D (eds), Psychiatric Aspects of Neurologic Disease. New York, Grune and Stratton, 1975

Harris ME, Wang YN, Pedigo NW, Hensley K, Butterfield DA and Carney JM. Amyloid beta peptide (25–35) inhibits Na^+ dependent glutamate uptake in rat hippocampal astrocyte cultures. J Neurochem 67, 277–286, 1996

Hasegawa I, Fukushima T, Ihara T, Miyushita Y. Callosal window between prefrontal cortices: cognitive interactions to retrieve long-term memory. Science 281, 814–818, 1998

Haug H, Eggers R. Morphometry of the human cortex cerebri and corpus striatum during aging. Neurobiol Aging 12, 336–338, 1991

Haxby JV, Raffaele K, Gillette J, Schapiro MB, Rapoport SI. Individual trajectories of cognitive decline in patients with dementia of the Alzheimer type. J Clin Exp Neuropsychol 14, 575–592, 1992

Haxby JV. Cognitive deficits and local metabolic changes in dementia of the Alzheimer type. In: Rapoport SI, Petit H, Leys D, Christen Y (eds), Imaging, Cerebral Tomography and Alzheimer's Disease. Berlin, Germany: Springer-Verlag, 109–119, 1990

Haxby JV, Petit L, Ungerleider LG, Courtney SM. Distinguishing the functional roles of multiple regions in distributed neural systems for visual working memory. Neuroimage 11, 380–391, 2000

Hebert LE, Scherr PA, Beckett LA, Funkenstein HH, Albert MS, Chown MJ, Evans DA. Relation of smoking and alcohol consumption to incident Alzheimer's disease. Am J Epidemiol 135(4), 347–355, 1992

Hayden CM, Camp CJ. Spaced-retrieval: A memory inervention for dementia in Parkinson's disease. Clinical Gerontologist 16, 80–82, 1995

Hecker J, Fon D, Gauthier S et al. Benefits of donepezil in the treatment of behavioral problems in moderate to severe Alzheimer's disease. Alzheimer's Disease and Related Disorders Association World Alzheimer Congress. Washington, DC, July 9–18, 2000

Heckers S, Weiss A, Alpert MM, Schacter DL. Hippocampal and brain stem activation during word retrieval after repeated and semantic encoding. Cereb Cortex 12, 900–907, 2002

Heindel WC, Salmon DP, Shults CW, Walicke PA, Butters N. Neuropsychological evidence for multiple implicit memory systems: a comparison of Alzheimer's, Huntington's, and Parkinson's disease patients. J Neurosci 9, 582–587, 1989

Heizmann CW, Braun K. Changes in Ca(2+)-binding proteins in human neurodegenerative disorders. Trends Neurosci 15, 259–264, 1992

Helmuth L. Alzheimer's congress. Further progress on a β-amyloid vaccine. Science 289, 375, 2000

Henderson VW, Buckwalter JG. Cognitive deficits of men and women with Alzheimer's disease. Neurology 44, 90–96, 1994

Hendriks L, Cras P, Martin JJ, Van Broeckhoven C. Alzheimer disease and hemorrhagic stroke: Their relationship to beta amyloid deposition. In: KS Kosik (ed), Alzheimer Disease: Lessons from Cell Biology, 37–48, Springer-Verlag, Berlin/Heidelberg, 1995

Herlitz A, Nilsson L, Backman L. Gender differences in episodic memory. Memory & Cognition 25, 801–811, 1997

Herlitz A, Small BJ, Fratiglioni L et al. Detection of mild dementia in community surveys: is it possible to increase the accuracy of our instruments ? Arch Neurol 54, 319–324, 1997

Hersh N, Treadgold L. Neuropage: the rehabilitation of memory dysfunction by prosthetic memory and cueing. Neurorehabil 4, 187–197, 1994

Heun R, Mazanek M, Atzor KR, Tintera J, Gawehn J, Burkart M et al. Amygdala-hippocampal atrophy and memory performance in dementia of Alzheimer type. Dementia Geriatr Cognit Disord 8(6), 329–336, 1997

Heun R, Maier W, Muller H. Gender and AD (Letter). Neurology 47, 1357–1358, 1996

Hier DB, Hagenlocker K, Shindler AG. Language disintegration in dementia: Effects of etiology and severity. Brain Lang 25 (1), 117–133, 1985

Hillis AE, Wityk RJ, Tuffiash E, Beauchamp NJ, Jacobs MA, Barker PB et al. Hypoperfusion of Wernicke's area predicts severity of semantic deficit in acute stroke. Ann Neurol 50, 561–566, 2001

Hirono N, Mori E, Ikejiri Y, Imamura T, Shimomura T, Ikeda M, Yamashita H, Takatsuki Y, Tokimasa A, Yamadori A. Procedural memory in patients with mild Alzheimer's disease. Dement Geriatr Cogn Disord 8(4), 210–216, 1997

Hock C, Konietzko U, Papassotiropoulos A, Wollmer A, Streffer J, Von Rotz RC, Davey G, Moritz E, Nitsch RM. Generation of antibodies specific for β-amyloid by vaccination of patients with Alzheimer disease. Nat Med, Advance online publication, October 15, 2002

Hodges JR, Miller BL. The classification, genetics and neuropathology of frontotemporal dementia. Neurocase 7, 31–35, 2001

Hodges JR, Graham N, Patterson K. Charting the progression in semantic dementia. Memory 3, 463–495, 1995

Hodges JR, Salmon DP, Butters N. The nature of the naming deficit in Alzheimer's and Huntington's disease. Brain 114, 1547–1558, 1991

Hodges JR, Patterson K, Oxbury S, Funnell E. Semantic dementia: progressive fluent aphasia with temporal lobe atrophy. Brain 115, 1783–1806, 1992

Hodges JR, Patterson K, Ward R, Garrard P, Bak T, Perry R, Gregory C. The differentiation of semantic dementia and frontal lobe dementia (temporal and frontal variants of frontotemporal dementia) from early Alzheimer's disease: a comparative neuropsychological study. Neuropsychology 13, 31–40, 1999

Hof PR, Bouras C, Morrison JH. Cortical neuropathology in aging and dementing disorders. In: Peters A, Morrison JH (eds), Cerebral Cortex. Vol. 14, New York (NY): Kluwer Academic / Plenum Publishers, 175–311, 1999

Hogervorst E, Barnetson L, Jobst KA, Nagy Zs, Combrinck M, Smith AD. Diagnosing dementia: Interrater reliability assessment and accuracy of the NINCDS/ADRDA criteria *versus* CERAD histopathological criteria for Alzheimer disease. Dement Geriatr Cogn Disord 11, 107–113, 2000

Holcomb L, Gordon MN, McGowan E, Yu X, Benkovic S, Jantzen P, Wright K, Saad I, Mueller R, Morgan D, Sanders S et al. Accelerated Alzheimer-type phenotype in transgenic mice carrying both mutant amyloid precursor protein and presenilin 1 transgenes. Nat Med 4, 97–100, 1998

Holden VP, Woods RT. Reality Orientation: Psychological Approaches to the Confused Elderly. London: Churchill Livingstone, 1982

Holmes C, Arranz MJ, Powell JF, Collier DA, Lovestone S. 5HT2A and 5HT2C receptor polymorphisms and psychopathology in late onset Alzheimer disease. Hum Mol Genet 7, 1507–1509, 1998

Holsinger RM, McLean CA, Beyreuther K, Masters CL, Evin G. Increased expression of the amyloid precursor beta-secretase in Alzheimer's disease. Ann Neurol 51(6), 783–786, 2002

Holtzman DM, Li Y, Chen K, Gage FH, Epstein CJ, Mobley WC. Nerve growth factor reverses neuronal atrophy in a Down syndrome model of age-related neurodegeneration. Neurology 43, 2668–2673, 1993.

Holtzman DM, Bales KR, Wu S, Bhat P, Parsadanian M, Fagan AM, Chang LK, Sun Y, Paul SM. Expression of human apolipoprotein E reduces amyloid-beta deposition in a mouse model of Alzheimer's disease. J Clin Invest 103(6), R15–R21, 1999

Hom J, Turner MB, Risser R, Bonte FJ, Tintner R. Cognitive deficits in asymptomatic first-degree relatives of Alzheimer's disease patients. J Clin Exp Neuropsychol 16(4), 568–576, 1994

Homma A, Fukuzawa K, Tsukada Y, Ishii T, Hasegawa K, Mohs RC. Development of a Japanese version of Alzheimer's disease Assesment Scale (ADA). Jpn J Geriatr Psychiatry 3, 647–655, 1992

Homma A, Niina R, Ishii T, Hasegawa K. Development of a new rating scale for dementia in the elderly: Mental Function Impairment Scale (MENFIS). Jpn J Geriatr Psychiatry 2, 1217–1222, 1991

Homma A, Takeda M, Imai Y, Udaka F, Hasegawa K, Kameyama M, Nishimura T. Clinical efficacy and safety of donepezil on cognitive and global function in patients with Alzheimer's disease. A 24-week, multicenter, double-blind, placebo-controlled study in Japan. E2020 Study Group. Dement Geriatr Cogn Disord 11(6), 299–313, 2000

Honda T, Kashima H. Cognitive rehabilitation program for demented patients. Ronen Seishin Igaku Zasshi (Jpn J Geriatr Psychiatry) 3, 301–306, 1992

Honey RC, Watt A, Good M. Hippocampal lesions disrupt an associative-mismatch process. J Neurosci 18, 2226–2230, 1998

Hope T, Fairburn CG. The present behavioural examination (PBE): the development of an interview to measure current behavioural abnormalities. Psychol Med 22, 223–230, 1992

Hope RA, Fairburn CG. The nature of wandering in dementia: a community-based study. Int J Geriatr Psychiatry 5, 239–245, 1990

Hörtnagl H, Hellweg R. Pathophysiological aspects of human neurodegenerative diseases. Wien Klin Wochenschr 109, 623–635, 1997

Howard D, Patterson K. Pyramids and Palm Trees: a test of semantic access from pictures and words. Bury St Edmunds: Thames Valley Publishing Company, 1992

Howieson DB, Holm LA, Kaye JA, Oken BS. Neurologic function in the optimally healthy oldest people. Neuropsychological evaluation. Neurology 43, 1882–1886, 1993

Howieson DB, Dame A, Camicioli R, Sexton G, Payami H and Kaye JA. Cognitive markers preceding Alzheimer's dementia in the healthy oldest old. JAGS 45, 584–589, 1997

Hoyer D, Clarke DE, Fozard JR, Hartig PR, Martin GR, Mylecharane EJ, Saxena PR and Humphrey PPA. International Union of Pharmacology classification of receptors for 5-hydroxytryptamine (serotonin). Pharmacol Rev 46, 157–203, 1994

Hsiao K, Chapman P, Nilsen S, Eckman C, Harigaya Y, Younkin S, Yang F, Cole G. Correlative memory deficits, Abeta elevation, and amyloid plaques in transgenic mice. Science 274 (5284), 99–102, 1996

Huemer HP, Menzel HJ, Potratz D, Brake B, Falke D, Utermann G, Dierich MP. Herpes simplex virus binds to human serum lipoprotein. Intervirology 29(2), 68–76, 1988

Hughes CP, Berg L, Danziger WL et al. A new clinical scale for the staging of dementia. Br J Psychiatry 140, 566–572, 1982

Huppert FA, Beardsall L. Prospective memory impairments as an early indicator of dementia. J Clin Exp Neuropsychol 1994, 15 (5), 805–821

Huppert FA, Brayne C, Gill C, Paykel ES, Beardsall L. CAMCOG – a concise neuropsychological test to assist dementia diagnosis: socio-demographic determinants in an elderly population sample. Br J Clin Psychology 34, 529–541, 1995

Hutton M, Lendon CL, Rizzu P, Baker M, Froelich S, Houlden H, Pickering-Brown S, Chakraverty S, Isaacs A, Grover A et al. Association of missense and 5'-splice-site mutations in tau with the inherited dementia FTDP-17. Nature 393(6686), 702–705, 1998

Hutton S, Sheppard L, Rusted JM, Ratner HH. Structuring the acquisition and retrieval environment to facilitate learning in individuals with dementia of the Alzheimer type. Memory 4, 113–130, 1996

Hyman BT, Van Hoesen GW, Damasio AR, Barnes CL. Alzheimer disease: cell-specific pathology isolates the hippocampal formation. Science 225, 1168–1170, 1984

Hyman G, Van Hoesen G, Kromer C, Damasio A. Alzheimer disease: Cell specific pathology isolates the hippocampal formation. Science 225, 1168–1170, 1985

Hyman BT, Van Hoesen G, Kromer L, Damasio A. Perforant pathway changes and the memory impairment of Alzheimer disease. Ann Neurol 20, 472–481, 1986

Hyman BT, Smith C, Buldyrev I, Whelan C, Brown H, Tang MX, Mayeux R. Autoantibodies to amyloid b and Alzheimer's disease. Ann Neurol 49, 808–810, 2001

Hyman BT, Van Hoesen W, Damasio AR. Memory-related neural systems in Alzheimer's disease. An anatomic study. Neurology 40, 1721–1730, 1990

Hughes C, Berg L, Danziger W, Coben L, Markan R. A new clinical scale for the staging of dementia. Br J Psychiatry 140, 566–572, 1982

Hutton M, Perez-Tur J, Hardy J. Genetics of Alzheimer disease. Essays Biochem 33, 117–131, 1998

Imbimbo BP. Pharmacodynamic-tolerability relationship of cholinesterase inhibitors for Alzheimer's disease. CNS Drugs 15 (5), 375–390, 2001

In't Veld BA, Ruitenberg A, Hofman A, Launer LJL, Van Duijn CM, Stijnen T, Bretelier MMB, Stricker BHC. Nonsteroidal anti-inflammatory drugs and the risk of Alzheimer disease. N Engl J Med 345 (21), 1515–1521, 2001

Ionescu MD. Sex differences in memory estimates revisited. Psychological Reports 91, 167–172, 2002

Irwin W, Davidson RJ, Lowe MJ, Mock BJ, Sorenson JA, Turski PA. Human amygdala activation detected with echo-planar functional magnetic resonance imaging. NeuroReport 7, 1765–1769, 1996

Ishi I, Sasaki M, Yamaji S, Sakamoto S, Kitagaki H, Mori E. Demonstration of decreased posterior cingulate perfusion in mild Alzheimer's disease by means of H2O15 positron emission tomography. Eur J Nucl Med 24, 670–673, 1997

Isoe K, Ji Y, Urakami K, Adachi Y, Nakashima K. Genetic association of estrogen receptor gene polymorphisms with Alzheimer disease. Alzheimer's Res 3, 195–197, 1997

Itoh A, Nitta A, Katono Y, Usui M, Naruhashi K, Iida R, Hasegawa T, Nabeshima T. Effects of metrifonate on memory impairment and cholinergic dysfunction in rats. [erratum appears in Eur J Pharmacol 332(1), 113, 1997]. Eur J Pharmacol 322(1), 11–19, 1997

Itzhaki RF, Maitland NJ, Wilcock GK et al. Detection by polymerase chain reaction of herpes simplex virus type 1 (HSV1) DNA in brain of aged normals and Alzheimer disease

patients. In: Corain B, Iqbal K, Nicolini M (eds), Alzheimer Disease: Advances inClinical and Basic Research. Chichester (NY): John Wiley & Sons Ltd, 98–102, 1993
Itzhaki RF, Lin WR, Shang D, Wilcock GK, Faragher B, Jamieson GA. Herpes simplex virus type 1 in brain and risk of Alzheimer's disease. Lancet 349(9047), 241–244, 1997
Iwatsubo T, Odaka A, Suzuki N, Mizusawa H, Nukina N, Ihara Y. Visualization of A beta 42 (43) and A beta 40 in senile plaques with end-specific A beta monoclonals: Evidence that an initially deposited species is A beta 42 (43). Neuron 13, 45–53, 1994
Jack CR Jr, Petersen RC, Xu YC, O'Brien PC, Smith GE, Ivnik RJ, Boeve BF, Waring SC, Tangalos EG, Kokmen E. Prediction of AD with MRI-based hippocampal volume in mild cognitive impairment. Neurology 52(7), 1397–1403, 1999
Jack DB. Herpes simplex virus and Alzheimer disease. Mol Med Today 2, 270, 1996
Jackendorff RS. Semantic structures. Ambridge (MA): MIT Press, 1990
Jacobs DM, Sano M, Dooneief G, Marder K, Bell KL, Stern Y. Neuropsychological detection and characterization of preclinical Alzheimer disease. Neurology 45, 957–962, 1995
Jacobs D, Troster AI, Butters N. Intrusion errors in the figural memory of patients with Alzheimer's and Huntington's disease. Arch Clin Neuropsychology 5, 49–57, 1990
Jacoby LL. A process dissociation framework: Separating automatic from intentional uses of memory. Journal of Memory and Language 30, 513–541, 1991
Jacoby LL, Dallas M. On the relationship between autobiographical memory and perceptual learning. J Exp Psychology Gen 100, 306–340, 1981
Jan G, McKeith IG. Special workshop in dementia with Lewy bodies (abstract). Neurobiol Aging 19, 45, 1998
Jann MW. Rivastigmine, a new-generation cholinesterase inhibitor for the treatment of Alzheimer's disease. Pharmacotherapy 20, 1–12, 2000
Janus C, Pearson J, Mclaurin J, Mathews PM, Jiang Y, Schmidt SD, Azhar Christi M, Horne P, Heslin D, French J et al. Aβ peptide immunization reduces behavioural impairment and plaques in a model of Alzheimer disease. Nature 408, 979–982, 2000
Jarrett JT, Berger EP, Lansbury PT Jr. The carboxy terminus of the beta amyloid protein is critical for the seeding of amyloid formation: implications for the pathogenesis of Alzheimer's disease. Biochemistry 32, 4693–4697, 1993
Jarvik GP, Wijsman EM, Kukull WA, Schellenberg GD, Yu C, Larson EB. Interactions of apolipoprotein E genotype, total cholesterol level, age, and sex in prediction of Alzheimer disease: a case-control study. Neurology 45, 1092–1096, 1995
Jelic V, Shigeta M, Julin P et al. Quantitative electroencephalography power and coherence in Alzheimer's disease and mild cognitive impairment. Dementia 7, 314–323, 1996
Jelicic M, Bonebakker AE, Bonke B. Implicit memory in patients with Alzheimer disease: a concise review. Tijdschrift voor Gerontologie en Geriatrie 25 (1), 22–27, 1994
Jenkins R, Fox NC, Rossor AM, Harvey RJ, Rossor MN. Intracranial volume and Alzheimer disease. Evidence against the cerebral reserve hypothesis. Arch Neurol 57, 220–224, 2000
Jennings JM, McIntosh AR, Kapur S, Tulving E, Houle S. Cognitive subtractions may not add up: the interaction between semantic processing and response mode. Neuroimage 5, 229–239, 1997
Jeste DV, Wragg RE, Salmon DP, Harris MJ, Thal LJ. Cognitive deficits of patients with Alzheimer disease with and without delusions. Am J Psychiatry 149, 184–189, 1992

Jeste DV, Finkel SI. Psychosis of Alzheimer's disease and related dementias: diagnostic criteria for a distinct syndrome. Am J Geriatr Psychiatry 8, 29–34, 2000

Jick H, Zornberg GL, Jick SS, Seshadri S, Drachman DA. Statins and the risk of dementia. Lancet 356, 1627–1631, 2000

Jo Li-Wen, Sheu K-F R, Thaler HT, Markesbery WR, Blass JP. Selective loss of KGDHC-enriched neurons in Alzheimer temporal cortex. Does mitochondrial variation contribute to selective vulnerability. J Mol Neurosci 17, 361–369, 2001

Joffres C, Graham J, Rockwood K. Qualitative analysis of the Clinician Interview-Based Impression of Change (Plus): methodological issues and implications for clinical research. Int Psychogeriatr 12 (3), 403–413, 2000

Jope RS, Song L, Powers RE. Cholinergic activation of phosphoinositide signaling is impaired in Alzheimer disease brain. Neurobiol Aging 18, 111–120, 1997

Jorm AF, Jolley A. The incidence of dementia. Neurology 51, 728–733, 1998

Jorm AF. Assessment of cognitive impairment and dementia using informant report. Clin Psychol Rev 16, 51–73, 1996

Jorm AF, Broe GA, Creasey H, Sulway MR, Dent O, Fairley MJ, Kos SC, Tennant C. Further data on the validity of the Informant Questionnaire on Cognitive Decline in the Elderly (IQ-CODE). Int J Geriatric Psychiatry 11, 131–139, 1996

Jorm AF, Scott R, Cullen JS, Mackinnon AJ. Performance of the Informant Questionnaire on Cognitive Decline in the Elderly (IQCODE) as a screening test for dementia. Psychol Med 21, 785–790, 1991

Josselyn SA, Shi C, Carlezon WA Jr, Neve RL, Nestler EJ, Davis M. Long-term memory is facilitated by cAMP response element-binding protein overexpression in the amygdala. J Neurosci 21, 2404–2412, 2001

Joyce CRB, O'Boyle, McGee H. Individual Quality of life: Approaches to Conceptualisation and Assessment. Reading, Harwood Academic, 1999

Juottonen K, Laakso MP, Insausti R, Lehtovirta M, Pitkanen A, Partanen K, Soininen H. Volumes of the entorhinal and perirhinal cortices in Alzheimer's disease. Neurobiol Aging 19(1), 15–22, 1998

Jossain SS, Adem A, Winblad B, Oreland L. Characterisation of dopamine and serotonin uptake inhibitory effects of thetrahydroaminoacridine in rat brain. Pharmacol Toxicol 71, 213–215, 1992

Juva K, Sulkava R, Erkinjuntti T, Valvanne J, Tilvis R. Clinical dementia rating (CDR)-scale in screening dementia. Neurobiol Aging 13 (Suppl 1), S2–S3, 1992

Juva K, Sulkava R, Erkinjuntti T, Ylikoski R, Valvanne J, Tilvis R. Usefulness of the clinical dementia rating scale in screening for dementia. Int Psychogeriatr 7, 17–24, 1995

Kabani NJ, Sled JG, Chertkow H. Magnetization transfer ratio in mild cognitive impairment and dementia of Alzheimer's type. Neuroimage 15, 604–610, 2002

Kabir ZN, Herlitz A. The Bangla adaptation of the Mini-Mental State Examination (BAMSE): an instrument to assess cognitive function in illiterate and literate individuals. Int J Geriatr Psychiatry 15, 441–450, 2000

Kanowski S, Herrman WM, Stephan K, Wierich W, Hörr R. Proof of efficacy of the Ginkgo biloba special extract EGb 761 in outpatients suffering from mild to moderate primary degenerative dementia of the Alzheimer type or multi-infarct dementia. Pharmacopsychiatry 29, 47–56, 1996

Kapur S, Craik FI, Tulving E, Wilson AA, Houle S, Brown GM. Neuroanatomical correlates of encoding in episodic memory: levels of processing effect. Proc Natl Acad Sci USA 91, 2008–2011, 1994

Karlawish JHT, Casarett D, Klocinski J, Clark CM. The relationship between caregivers global ratings of Alzheimer's disease patients quality of life, disease, severity and the caregiving experience. J Am Geriatr Soc 49, 1066–1070, 2001

Kasa P, Rakonczay Z, Gulya K. The cholinergic system in Alzheimer disease. Prog Neurobiol52, 511–535, 1997

Katona CL, Hunter BN, Bray J. A double-blind comparison of the efficacy and safety of paroxetine and imipramine in the treatment of depression with dementia. Int J Geriatr Psychiatry 13, 100–108, 1998

Katona CL, Aldridge CR. The dexamethasone suppression test and depressive signs in dementia. J Affect Disord 8, 83–89, 1985

Katz IR, Jeste DV, Mintzer JE, Clyde C, Napolitano J, Brecher M and the Risperidone Study Group. Comparison of risperidone and placebo for psychosis and behavioral disturbances associated with dementia: a randomized, double-blind trial. J Clin Psychiatry 60, 107–115, 1999

Katzman R, Aronson M, Fuld P, Kawas C, Brown T, Morgenstern H, Frishman W, Gidez L, Eder H, Ooi WL. Development of dementing illnesses in an 80-year-old volunteer cohort. Ann Neurol 25(4), 317–324, 1989

Katzman R, Rowe JW (eds). Principles of Geriatric Neurology. Philadelphia, Pa: FA/Co Publishers Davis, 1992

Katzman R, Zhang MY, Ouang-Ya-Qu, Wang ZY, Liu WT, Yu E, Wong SC, Salmon DP, Grant I. A Chinese version of the Mini-Mental State Examination; impact of illiteracy in a Shanghai dementia survey. J Clin Epidemiol 41(10), 971–978, 1988

Kauer JA, Malenka RC, Nicoll RA. A persistent postsynaptic modification mediates long-term potentiation in the hippocampus. Neuron 1, 911–917, 1988

Kazui H, Hashimoto M, Hirono N, Mori E. Nature of personal semantic memory: Evidence from Alzheimer's disease. Neuropsychologia 41 (8), 981–988, 2003

Keane MM, Gabrieli JDE, Fennema AC, Growdon JH and Corkin S. Evidence for a dissociation between perceptual and conceptual priming in Alzheimer disease. Behav Neurosci 105, 326–342, 1991

Keane MM, Gabrieli JDE, Growdon JH, Corkin S. Priming in perceptual identification of pseudowords is normal in Alzheimer's disease. Neuropsychologia 32, 343–356, 1994

Kelley WM, Miezin FM, McDeremott KB, Buckner RL, Raichle ME, Cohen NJ, Ollinger JM, Akbudak E, Conturo TE, Snyder AZ, Petersen SE. Hemispheric specialization in human dorsal frontal cortex and medial temporal lobe for verbal and nonverbal memory encoding. Neuron 20, 927–936, 1998

Kelly CA, Harvey RJ, Cayton H. Drug treatments for Alzheimer disease. BMJ 314, 693–694, 1997

Kempler D. Language changes in dementia of the Alzheimer type. In: R Lubinski (ed), Dementia and Communication: Research and Clinical Implications. San Diego: Singular, 98–114, 1995

Kertesz A. Language deterioration in dementia: In: VOB Emery, TE Oxman (eds), Dementia:

Presentations, Differential Diagnosis and Nosology. Baltimore: Johns Hopkins University Press, 108–122, 1994

Kertesz A, Appell J, Fishman M. The dissolution of language in Alzheimer's disease. Can J Neurol Sci 13, 415–418, 1986

Khachaturian ZS. Diagnosis of Alzheimer disease. Arch Neurol 42, 1097–1105, 1985

Kia HK, Brisorgueil MJ, Daval G, Langlois X, Hamon M, Verge D. Serotonin 1A receptors are expressed by a subpopulation of cholinergic neurons in the rat medial septum and diagonal band of Broca-a double immunocytochemical study. Neuroscience 74, 143–154, 1996

Kihara T, Shimohama S, Urushitani M, Sawada H, Kimura J, Kume T, Maeda T, Akaike A. Stimulation of alpha4beta2 nicotinic acetylcholine receptors inhibits beta-amyloid toxicity. Brain Res 792(2), 331–334, 1998

Kijak HA, Teri L, Borson S. Physical and functional health assessment in normal aging and in Alzheimer disease: self-reports *versus* family reports. Gerontologist 34, 324–330, 1994

Killiany R, Moss MB, Albert MS, Sandor T, Tieman J, Jolesz F. Temporal lobe regions on magnetic resonance imaging identify patients with early Alzheimer's disease. Arch Neurol 50, 949–954, 1993

Kimberly WT, Xia W, Rahmati T, Wolfe MS, Selkoe DJ. The transmembrane aspartates in presenilin 1 and 2 are obligatory for gamma-secretase activity and amyloid beta-protein generation. J Biol Chem 275(5), 3173–3178, 2000

Kirchoff BA, Wagner AD, Maril A, Stern CE. Prefrontal temporal circuitry for episodic encoding and subsequent memory. J Neurosci 20, 6173–6180, 2000

Kirshner HS. Progressive aphasia and other focal presentations of Alzheimer disease, Pick disease, and other degenerative disorders. In: VOB Emery, TE Oxman (eds), Dementia: Presentations, Differential Diagnosis and Nosology. Baltimore: Johns Hopkins University Press, 108–122, 1994

Kitamura Y, Shimohama S, Koike H, Kakimura J, Matsuoka Y, Nomura Y, Gebicke-Haerter PJ, Taniguchi T. Increased expression of cyclooxygenases and peroxisome proliferator-activated receptor-gamma in Alzheimer's disease brains. Biochem Biophys Res Comm 254(3), 582–586, 1999

Kivipelto M, Helkala EL, Laakso MP, Hanninen T, Hallikainen M, Alhhainen K, Soininen H, Tuomilehto J, Nissinen A. Midlife vascular risk factors and Alzheimer disease in later life: longitudinal, population based study. BMJ 322, 1447–1451, 2001

Klapper PE, Cleator GM, Longson MM. Mild forms of herpes encephalitis. J Neurol Neurosurg Psychiatry 47, 1247–1250, 1984

Klein WL, Kraft GA, Finch CE. Targeting small Abeta oligomers: the solution to an Alzheimer's disease conundrum? Trends Neurosci 24, 219–224, 2001

Klein WL. Aβ toxicity in Alzheimer's disease: globular oligomers (ADDLs) as new vaccine and drug targets. Neurochem Int 41 (5), 345–352, 2002

Kleinsmith LJ, Kaplan S. Paired-associate learning as a function of arousal and interpolated interval. J Exp Psychol, 67, 124–126, 1963

Kluger A, Ferris SH. Scales for the assessment of Alzheimer disease. Psychiatr Clin North Am 14, 309–327, 1991

Kluger A, Gianutsos JG, Golomb J, Ferris SH, Reisberg B. Motor/psychomotor dysfunction in normal aging, mild cognitive decline, and early Alzheimer disease: diagnostic and differential diagnostic features. Int Psychogeriatr 1, 307–316, 1997

Kluger A, Gianutsos JG, Golomb J, Ferris SH, George AE, Franssen E, Reisberg B. Patterns of motor impairement in normal aging, mild cognitive decline, and early Alzheimer's disease. J Gerontol B Psychol Sci Social Sci 52(1), P28–39, 1997

Knapp MJ, Knopman DS, Solomon PR, Pendlebury WW, Davis CS, Gracon SI and the Tacrine Study Group. A 30–week randomized controlled trial of high-dose tacrine in patients with Alzheimer disease. JAMA 271, 985–991, 1994

Knapp MJ, Knopman DS, Solomon PR. A 30–week randomized controlled trial of high dose Tacrine in patients with Alzheimer's disease. J Am Med Association 271, 985–991, 1994

Knight RG. Controlled and automatic memory process in Alzheimer disease. Cortex 34, 427–435, 1998

Knopman D, Schneider L, Davis K, Talwalker S, Smith F, Hoover T, Gracon S, Apter JT, Barnett M, Baumel B et al. Long-term tacrine (Cognex) treatment: effects on nursing home placement and mortality, Tacrine Study Group. Neurology 47, 166–177, 1996

Knopman DS, Knapp MJ, Gracon SI, Davis CS. The Clinician Interview-Based Impression (CIBI): a clinician's global change rating scale in Alzheimer's disease. Neurology 44(12), 2315–2321, 1994

Knopman DS, DeKosky ST, Cummings JL, Chui H, Corey-Bloom J, Relkin N, Small GW, Miller B, Stevens JC. Practice parameter: diagnosis of dementia (an evidence-based review). Report of the Quality Standards Subcommittee of the American Academy of Neurology. Neurology 56(9), 1143–1153, 2001

Knopman DS. The differential diagnosis of dementia in the elderly. Mediguide Geriatric Neurol 1, 2–9, 1997

Kofler B, Erhart C, Erhart P, Harrer G. A multidimensional approach in testing nootropic drug effects (Cerebrolysin). Arch Gerontol Geriatr 10, 129–140, 1990

Kokmen E, Beard CM, O'Brien PC, Offord KP, Kurland LT. Is the incidence of dementing illness changing? A 25-year time trend study in Rochester, Minnesota (1960–1984) [erratum appears in Neurology 44(7), 1368, 1994]. Neurology 43(10), 1887–1892, 1993

Kolb B, Whishaw IQ. Fundamentals of human neuropsychology, New York, Freeman, 1990

Kolsch H, Ludwig M, Lütjohann D, Rao ML. Neurotoxicity of 24–hydroxycholesterol, an important cholesterol elimination product of the brain, may be prevented by vitamin E and estradiol-17b. J Neural Transm 108, 475–488, 2001

Kopelman MD, Wilson BA, Baddeley AD. The autobiographical memory interview: a new assessment of autobiographical and personal semantic memory in amnesic patients. J Clin Exp Neuropsychol 11 (5), 724–744, 1989

Kopelman MD. Rates of forgetting in Alzheimer type dementia and Korsakoff's syndrome. Neuropsychology 23, 623–638, 1985

Kopelman MD. Amnesia: organic and psychogenic. Br J Psychiatry 150, 428–442, 1987

Kopelman MD. Non-verbal, short-term forgetting in the alcoholic Korsakoff syndrome and Alzheimer-type dementia. Neuropsychologia 8, 737–747, 1991

Kordower JH, Chu Y, Stebbins GT, Dekosky ST, Cochran EJ, Bennett D and Mufson EJ. Loss and atrophy of layer II entorhinal cortex neurons in elderly people with mild cognitive impairment. Ann Neurol 49, 202–213, 2001

Kosaka K, Yoshimura M, Ikeda K, Budka H. Diffuse type of Lewy body disease: progressive dementia with abundant cortical Lewy bodies and senile changes of varying degree – a new disease? Clin Neuropathol 3, 185–192, 1984

Kosaka K. Diffuse Lewy body disease in Japan. J Neurol 237, 197–204, 1990

Koss E, Weiner M, Ernesto C, Cohen-Mansfield J, Ferris SH, Grundman M, Schafer K, Sano M, Thal LJ, Thomas R, Whitehouse PJ and The Alzheimer's Disease Cooperative Study. Assessing Patterns of Agitation in Alzheimer's Disease patients with the Cohen-Mansfield Agitation Inventory. Alzheime Dis Assoc Disord 11 (Suppl 2), S45–S50, 1997

Kosslyn SM, Behrmann M, Jeannerod M. The cognitive neuroscience of mental imagery. Neuropsychologia 33, 1335–1344, 1995

Kotler-Cope S, Camp CJ. Anosognosia in Alzheimer disease. Alzheimer's Dis 9, 52–56, 1995

Kral VA. Senescent forgetfulness, benign and malignant. J Can Med Asso 86, 257–260, 1962

Kramer JH, Delis DC, Blusewicz MJ, Brandt MJ, Ober BA, Strauss M. Verbal memory errors in Huntington's and Alzheimer's dementias. Dev Neuropsychol 4, 1–15, 1988

Kramer-Ginsberg E, Mohs RC, Aryan M, Lobel D, Silverman J, Davidson M, Davis KL. Clinical predictors of course for Alzheimer patients in a longitudinal study: a preliminary report. Psychopharmacol Bull 24(3), 458–462, 1988

Krause BJ, Schmidt D, Mottaghy FM, Taylor J, Halsband U, Herzog H, Tellmann L, Muller-Gartner HW. Episodic retrieval activates the precuneus irrespective of the imagery content of word pair associates. A PET study. Brain 122 (Pt 2), 255–263, 1999

Kukull WA, Larson EB, Bowen JD, McCormick WC, Teri L, Pfanschmidt ML, Thompson JD, O'Meara ES, Brenner DE, van Belle G. Solvent exposure as a risk factor for Alzheimer's disease: a case-control study. [erratum appears in Am J Epidemiol 142(4), 450, 1995]. Am J Epidemiol 141(11), 1059–1071; discussion 1072–1079, 1995

Kumar-Singh S, Dewachter I, Moechars D, Lübke U, De Jonghe C, Ceuterick C, Checler F, Naidu A, Cordell B, Cras P, Van Broeckhoven C, Van Leuven F. Behavioral disturbances without amyloid deposits in mice overexpressing human amyloid precursor protein with Flemish (A692G) or Dutch (E693Q) mutation. Neurobiol Dis 7, 9–22, 2000

Kuo Y-M, Kokjohn TA, Beach TG, Sue LI, Brune D, Lopez JC, Kalback MM, Abramowski D, Sturchler-Pierrat C, Staufenbiel M, Roher AE. Comparative analysis of amyloid b chemical structure and amyloid plaque morphology of transgenic mouse and Alzheimer disease brains. J Biol Chem 276, 12991–12998, 2001

Kuo YM, Emmerling MR, Bisgaier CL, Essenburg AD, Lampert HC, Drumm D and Roher AE. Elevated low-density lipoprotein in Alzheimer's disease correlates with brain abeta 1–42 levels. Biochem Biophys Res Commun 252, 711–715, 1998

Kurlychek RT. Use of a digital alarm chronograph as a memory aid in early dementia. Clin Gerontol 1, 93–94, 1983

Laakso MP, Lehtovirta M, Partanen K, Riekkinen PJ Sr, Soininen H. Hippocampus in Alzheimer disease: A 3–year follow-up MRI study. Biol Psychiatry 47, 557–561, 2000

LaBar KS, Gatenby JC, Gore JC, LeDoux JE, Phelps EA. Human amygdala activation during conditioned fear acquisition and extinction: a mixed-trial fMRI study. Neuron 20, 937–945, 1998

LaBar KS, LeDoux JE, Spencer DD, Phelps EA. Impaired fear conditioning following unilateral temporal lobectomy in humans. J Neurosci 15, 6846–6855, 1995

Lambert AK, Barlow BA, Chromy C, Edwards R, Freed M, Liosatos TE, Morgan I, Rozovsky B, Trommer KL, Viola P, Wals C, Zhang CE, Finch GA, Krafft GA, Klein WL. Diffusible, nonfibrillar ligands derived from Abeta 1–42 are potent central nervous system neurotoxins. Proc Natl Acad Sci USA 95, 6448–6453, 1998

Lambon Ralph MA, Howard D. Gogi aphasia or semantic dementia? Simulating and assessing poor verbal comprehension in a case of progressive fluent aphasia. Cognitive Neuropsychology 17, 437–465, 2000

Lamy PP. The role of cholinesterase inhibitors in Alzheimer disease. CNS Drugs 1, 146–165, 1994

Landauer TK, Bjork RA. Optimal rehearsal patterns and name learning. In: M Gruneberg, P Morris, R Sykes (eds), Practical Aspects of Memory. London, Academic Press, 1978

La Rue A, Matsuyama SS, Sherman J, Jarvik LF. Cognitive performance in relatives of patients with probable Alzheimer disease: An age at onset effect? J Clin Exp Neuropsychology 14, 533–538, 1997

Launer IJ. Is there epidemiologic evidence that anti-oxidants protect against disorders in cognitive function? J Nutr Health Aging 4, 197–201, 2000

Launer LJ, Andersen K, Dewey ME, Letenneur L, Ott A, Amaducci LA, Brayne C, Copeland JR, Dartigues JF, Kragh-Sorensen P et al. Rates and risk factors for dementia and Alzheimer's disease: results from EURODEM pooled analyses. EURODEM Incidence Research Group and Work Groups. European Studies of Dementia. Neurology 52(1), 78–84, 1999

Lauro-Grotto R, Piccini C, Shallice T. Modality-specific operations in semantic dementia. Cortex 33, 593–622, 1997

Lavenu I, Pasquier F, Lebert F, Pruvo JP, Petit H. Explicit memory in frontotemporal dementia: The role of medial temporal atrophy. Dementia Geriatr Cognit Disord 9, 99–102, 1998

Lawlor BA, Radcliffe J, Molechan SE, Martinez RA, Hill JL, Sunderland T. A pilot placebo-controlled study of trazodone and buspirone in Alzheimer's disease. Int J Geriatr Psychiatry 9, 55–59, 1994

Lawlor BA, Ryan TM, Schmeidler J, Mohs RC, Davis KL. Clinical symptoms associated with age at onset in Alzheimer's disease. Am J Psychiatry 151, 1646–1649, 1994

Lawton MP. Quality of life in Alzheimer disease. Alzheimer Dis Assoc Disord 8 (Suppl 3), 138–150, 1994

Le Bars PL, Katz MM, Berman N, Itil TM, Freedman AM, Schatzberg AF. A placebo-controlled, double-blind, randomized trial of an extract of Ginkgo biloba for dementia. North American EGb Study Group. JAMA 278, 1327–1332, 1997

Le Bars PL, Velasco FM, Ferguson JM, Dessain EC, Kieser M, Hoerr R. Influence of the severity of cognitive impairment on the effect of the Ginkgo biloba extract EGb 761 in Alzheimer's disease. Neuropsychobiology 45, 19–26, 2002

Leber P. Guidelines for the clinical evaluation of anti-dementia drugs. 1st draft. Rockville: US Food and Drug Administration, 1990

Leber PD. Developing safe and effective antidementia drugs. In: R Becker, E Giacobini (eds), Alzheimer Disease: From Molecular Biology to Therapy. Boston, Birkhäuser, 579–584, 1997

Lebert F, Pasquier F, Petit H. Behavioural effects of fluoexetine in dementia of Alzheimer type. Int J Geriatr Psychiatry 9, 590–591, 1994

LeDoux JE. Emotion: Clues from the brain. Ann Rev Psychology 46, 209–264, 1995

LeDoux LE. Emotion circuits in the brain. Ann Rev Neurosci 23, 155–184, 2000

Lee PN. Smoking and Alzheimer disease: a review of the epidemiological evidence. Neuroepidemiology 13, 131–144, 1995

Lehmann DJ, Johnston C, Smith AD. Synergy between the genes for butyrylcholinesterase K variant and apolipoprotein E4 in late-onset, confirmed Alzheimer disease. Hum Mol Genet 6, 1933–1936, 1997

Lekeu F, Van der Linden M, Chicherio C, Collette F, Degueldre C, Franck G, Moonen G, Salmon E. Brain correlates of performance in a free/cued recall task with semantic encoding in Alzheimer disease. Alzheimer Dis Assoc Disord 17 (1), 35–45, 2003

Lendon CL, Ashall F, Goate AM. Exploring the etiology of Alzheimer disease using molecular genetics. JAMA 277, 825–831, 1997

Lenkkeri U, Kestila M, Lamerdin J, McCready P, Adamson A, Olsen A, Tryggvason K. Structure of the human amyloid-precursor-like protein gene APLP1 at 19q13.1. Hum Genet 102, 192–196, 1998

Lepage M, Ghaffar O, Nyberg L, Tulving E. Prefrontal cortex and episodic memory retrieval mode. Proc Natl Acad Sci USA 97, 506–511, 2000

Lerner AJ, Koss E, Patterson M, Ownby RL, Chatterjee A, Friedland RP, Whitehouse PJ. Visual hallucinations in Alzheimer's Disease and the BEHAVE-AD. Neurology 42 (suppl. 3), 140, 1992

Lesch K-P, Mössner R. Genetically driven variation in serotonin uptake: Is there a link to affective spectrum, neurodevelopmental, and neurodegenerative disorders. Biol Psychiatry 44, 179–192, 1998

Letemendia F, Pampiglione G. Clinical and electroencephalographic observations in Alzheimer disease. J Neurol Neurosurg Psychiatry 21, 167–172, 1958

Leturman JD, Haroutunian V, Yemul S, Ho L, Purohit D, Aisen PS, Mohs R, Pasinetti GM. Cytokine gene expression as a function of the clinical progression of Alzheimer disease dementia. Arch Neurol 57, 1153–1160, 2000

Levin ED, Rezvani AH. Development of nicotinic drug therapy for cognitive disorders. Eur J Pharmacol 393, 141–146, 2000

Levin ED. Nicotinic systems and cognitive functions. Psychopharmacology 108, 417–431, 1992

Levin ED, Lee C, Rose JE, Reyes A, Ellison G, Jarvik M, Gritz E. Chronic nicotine and withdrawal effects on radial-arm maze performance in rats. Behav Nerol Biol 53, 269–276, 1990

Lewis RL. Accounting for the fine structure of syntactic working memory: Similarity-based interference as a unifying principle. Behav Brain Sci 22, 105–106, 1999

Levy R. Ageing-associated cognitive decline. Int Psychogeriatr 6, 63–68, 1994

Levy ML, Cummings J, Fairbanks LA, Bravi D, Calvani M, Carta A. Longitudinal assessment of symptoms of depression, agitation and psychosis in 181 patients with Alzheimer disease. Am J Psychiatry 153, 1438–1443, 1996

Levy ML, Cummings JL, Fairbanks LA et al. Apathy is not depression. J Neuropsychiatry Clin Neurosci 10, 314–319, 1998

Levy-Lahad E, Wasco W, Poorkaj P et al. Candidate gene for chromosome 1 familial Alzheimer disease locus. Science 269, 973–977, 1995

Levy-Lahad E, Wasco W, Poorkaj P, Romano DM, Oshima J, Pettingell WH, Yu CE, Jondro

PD, Schmidt SD, Wang K et al. Candidate gene for the chromosome 1 familial Alzheimer disease locus Science 269, 973–7, 1995

Levy-Lahad E, Bird TD. Genetic factors in Alzheimer disease: a review of recent advances. Ann Neurol 40, 829–840, 1996

Lewis DA, Campbell MJ, Terry RD, Morrison JH. Laminar and regional distributions of neurofibrillary tangles and neuritic plaques in Alzheimer's disease: a quantitative study of visual and auditory cortices. J Neurosci 7 (6), 1799–1808, 1987

Lewy FH. Paralysis agitans. In: Lawandowsky M (ed), Handbuch der Neurologie. Berlin: Springer-Verlag, 920–958, 1912

Li QX, Maynard C, Cappai R, McLean CA, Cherny RA, Lynch T, Culvenor JG, Trevaskis J, Tanner JE, Bailey KA et al. Intracellular accumulation of detergent-soluble amyloidogenic A beta fragment of Alzheimer's disease precursor protein in the hippocampus of aged transgenic mice. J Neurochem 72(6), 2479–2487, 1999

Li QX, Evin G, Small DH, Multhaup G, Beyreuther K, Masters CL. Proteolytic processing of Alzheimer disease beta A4 amyloid precursor protein in human platelets. J Biol Chem 270(23), 14140–7, 1995

Li T, Holmes C, Sham PC, Vallada H, Birkett J, Kirov G et al. Allelic functional variation of serotonin transporter expression is a susceptibility factor for late-onset Alzheimer disease. Neuroreport 8, 683–686, 1997

Li S, Mallory M, Alford M, Tanaka S and Masliah E. Glutamate transporter alterations in Alzheimer disease are possibly associated with abnormal APP expression. J Neuropathol Exp Neurol 56, 901–911, 1997

Liao A, Gomez-Isla T, Clatworthy A, Hyman BT. Lack of association of presenilin-1 intron-8 polymorphism with neuropathological features of Alzheimer disease. Brain Res 816, 295–298, 1999

Liao A, Nitsch RM, Greenberg SM, Finckh U, Blacker D, Albert M, Rebeck GW, Gomez-Isla T, Clatworthy A, Binetti G et al. Genetic association of an α2-macroglobulin (Val1000Ile) polymorphism and Alzheimer disease. Hum Mol Genet 7, 1953–1956, 1998

Lichtenthaler SF, Wang R, Grimm H, Uljon SN, Masters CL, Beyreuther K. Mechanism of the cleavage specificity of Alzheimer's disease gamma-secretase identified by phenylalanine-scanning mutagenesis of the transmembrane domain of the amyloid precursor protein. Proc Natl Acad Sci USA 96(6), 3053–3058, 1999

Light LL, La Voie D. Direct and indirect measures of memory in old age. In: P Graf, MEJ Masson (eds), Implicit Memory, Hillsdale, NJ: Erlbaum, 207–230, 1993

Linas J, Vilalta J, Lopez PS, Amiel J, Vidal C. The Cambridge Mental Disorder of the Elderly Examination. Validation of the Spanish adaptation. Neurologia 5, 117–120, 1990

Lindeboom J, Schmand B, Tulner L, Wastra G, Jonker C. Visual association test to detect early dementia of the Alzheimer type. J Neurol Neurosurg Psychiatry 73, 126–133, 2002

Lindsay J, Laurin D, Verreault R, Hébert R, Helliwell B, Hill GB, McDowell I. Risk factors for Alzheimer's Disease: A prospective analysis from the Canadian study of Health and Aging. Am J Epidemiol 156 (5), 445 –453, 2002

Lindstrom J. Nicotinic acetylcholine receptors in health and disease. Mol Neurobiol 15, 193–222, 1997

Lines CR, Dawson C, Preston GC, Reich S, Foster C, Traub M. Memory and attention in patients with senile dementia of the Alzheimer type and in normal elderly subjects. J Clin Exp Neuropsychol 13, 691–702, 1991

Lin X, Koelschh G, Wu S, Downs D, Dashhti A, Tang J. Human aspartic protease memapsin 2 cleaves the β-secretase site of β-amyloid precursor protein. Proc Natl Acad Sci USA 97, 1456–1460, 2000

Linn RT, Wolf PA, Bachman DL, Knoefel JE, Cobb JL, Belanger AJ, Kaplan EF, D'Agostino RB. The "preclinical phase" of probable Alzheimer disease. A 13–year prospective study of the Framingham cohort. Arch Neurol 52, 485–490, 1995

Liu H-C, Teng EL, Chuang Y-Y, Lin K-N, Fuh J-L, Wang P-N. The Alzheimer's Disease Assessment Scale: Findings from a low-education population. Dementia Geriatr Cognit Disord 13, 21–26, 2002

Locascio JJ, Growdon JH, Corkin S. Cognitive test performance in detecting, staging and tracking. Alzheimer disease. Arch Neurol 52, 1087–1099, 1995

Logie RH, Marchetti C. Visuo-spatial working memory: visual, spatial or central executive ? In: RH Logie and M. Denis (eds), Mental Images in Human Cognition. Amsterdam: North Holland Press, 105–115, 1991

Logsdon R, Gibbons LE, McCurry SM, Teri L. Quality of life in Alzeimer's disease. Patient and caregiver reports. In: SM Albert, RG Logsdon (eds), Assessing Quality of Life in Alzheimer's Disease. New York, NY. Springer Publishing Company, 17–30, 2000

Logsdon RG, Teri L, Weiner MF, Gibbons LE, Raskind M, Peskind E, Grundman M, Koss E, Thomas RG, Thal LJ and members of the Alzheimer's Disease Cooperative Study. Assessment of agitation in Alzheimer's disease: the agitated behavior in dementia scale. JAGS 47, 1354–1358, 1999

Logsdon RG, Gibbons LE, McCurry SM, Teri L. Assessing quality of life in older adults with cognitive impairment. Psychosom Med 64, 510–519, 2002

Looi JCL, Franz CP, Sachdev PS. Differentiation of vascular dementia from AD on neuropsychological tests. Neurology 53, 670–678, 1999

Lopez OL, Wisniewski SR, Becker JT, Boller F, DeKosky ST. Psychiatric medication and abnormal behavior as predictors of progression in probable Alzheimer disease. Arch Neurol 56, 1266–1272, 1999

Loske C, Gerdemann A, Schepl W, Wycislo M, Schinzel R, Palm D, Riederer P, Münch G. Transition metal mediated glycoxidation accelerates crosslinking of β-amyloid peptide. Eur J Biochem 267, 4171–4178, 2000

Lukatela K, Malloy P, Cohen R. The nature of the naming deficit in early Alzheimer's and vascular disease. Neurology A232, 1997

Lund and Manchester Groups. Clinical and neuropathological criteria for frontotemporal dementia. J Neurol Neurosurg Psychiatry 57, 416–418, 1994

Lundberg C, Johansson K. Dementia and driving: an attempt at consensus. Alzheimer Dis Assoc Disord 45, 949–953, 1997

Luo LQ, Martin-Morris LE, White K. Identification, secretion, and neural expression of APPL, a Drosophila protein similar to human amyloid protein precursor. J Neurosci 10, 3849–3861, 1990

Luria A. The Working Brain: An Introduction to Neuropsychology. New York: Basic Books, 1973

Lyketsos CG, Steele CR, Baker L, Galik E, Kopunek S, Steinberg M, Warren A. Major and minor depression in Alzheimer disease: prevalence and impact. J Neuropsychiatry 9, 556–561, 1997

Lyketsos CG, Steinberg M, Tschantz JT et al. Mental and behavioral disturbances in dementia: findings from the Cache County Study on Memory in Aging. Am J Psychiatry 157, 708–714, 2000

Maccoby EE, Jacklin CN. The Psychology of Sex Differences. Palo Alto, CA: Standard Univer Press, 1974

Mack JL, Patterson MB. The evaluation of behavioral disturbances in Alzheimer disease: the utility of three rating scales. J Geriat Psychiatry Neurol 7, 101–117, 1994

Mackenzie IR, Munoz DG. Nonsteroidal anti-inflammatory drug use and Alzheimer-type pathology in aging. Neurology 50, 986–990, 1998

Mackinnon A, Mulligan R. Combining cognitive testing and informant report to increase accuracy in screening for dementia. Am J Psychiatry 155, 1529–1535, 1998

Maelicke A. Allosteric modulation of nicotinic receptors as a treatment strategy for Alzheimer disease. Dement Geriatr Cogn Disord 11 (Suppl 1) 11–18, 2000

Maelicke A, Albuquerque EX. New approach to drug therapy in Alzheimer disease. Drug Disc Today 1, 53–59, 1996

Magnusson KR. Aging of glutamate receptors: correlations between binding and spatial memory performance in mice. Mech Ageing Dev 104, 227–248, 1998

Matsuoka N, Aigner TG. FK960 (N-(4-acetyl-1-piperazinyl)-p-fluorobenzamide monohydrate), a novel potential antidementia drug, improves visual recognition memory in rhesus monkeys: comparison with physostigmine. J Pharmacol Exp Ther 280, 1201–1209, 1997

Malleret G, Haditsch U, Genoux D, Jones MW, Bliss TV, Vanhoose AM, Weitlauf C, Kandel ER, Winder DG, Mansuy IM. Inducible and reversible enhancement of learning, memory, and long-term potentiation by genetic inhibition of calcineurin. Cell 104(5), 675–686, 2001

Mann MB, Wu S, Rostamkhani M, Tourtellotte W, MacMurray J, Comings DE. Phenylethanolamine N-methyltransferase (PNMT) gene and early-onset Alzheimer disease. Am J Med Genet 105, 312–316, 2001

Mansuy IM, Mayford M, Jacob B, Kandel ER, Bach ME. Restricted and regulated overexpression reveals calcineurin as a key component in the transition from short-term to long-term memory. Cell 92, 39–49, 1998

Marcusson J, Rother M, Kittener B, et al. A 12–month, randomized, placebo-controlled trial of propentofylline (HWA 285) in patients with dementia according to DSM IIIR. The European Propentofylline Study Group. Dement Geriatr Cogn Disord 8, 320–328, 1997

Markesbery W. Oxidative stress hypothesis in Alzheimer's disease. Free Radic Biol Med 23, 134–147, 1997

Markowitsch HJ. Organic and psychogenic retrograde amnesia: two sides of the same coin? NeuroCase 2, 357–371, 1996

Marslen-Wilson WD, Teuber HL. Memory for remote events in anterograde amnesia: recognition of public figures from news photographs. Neuropsychologia 13, 353–364, 1975

Martin-Ruiz CM, Court JA, Molnar E et al. α4 but not α3 and α7 nicotinic acetylcholine receptor subunits are lost from the temporal cortex in Alzheimer disease. J Neurochem 73, 1635–1640, 1999

Maruyama H, Toji H, Harrington CR, Sasaki K, Izumi Y, Ohnuma T, Arai H, Yasuda M, Tanaka C, Emson PC, Nakamura S, Kawakami H? Lack of an association of estrogen receptor a gene polymorphisms and transcriptional activity with Alzheimer disease. Arch Neurol 57, 236–240, 2000

Martin A, Haxby JV, Lalonde FM, Wiggs CL, Ungerleider LG. Discrete cortical regions associated with knowledge of color and knowledge of action. Science 270, 102–105, 1995

Martin A, Ungerleider L, Haxby JV. Category specificity and the brain: the sensory/motor model of semantic representations of objects. In: MS Gazzaniga (ed), The New Cognitive Neuroscience. 2nd ed, Cambridge (MA): MIT Press, 1023–1036, 2000

Martin A, Chao LL. Semantic memory and the brain: structure and processes. Curr Opin Neurobiol 11(2), 194–201, 2001

Marwick C. Promising vaccine treatment for Alzheimer disease found. JAMA 284, 1503–1505, 2000

Masur DM, Blau AD, Fuld PA, Crystal H, Dickson D, Aronson MK. Neuropsychological predictors of dementia in the old-old (abstract). Neurobiol Aging 11, 263, 1990a

Masur DM, Sliwinski M, Lipton RB, Blau AD, Crystal HA. Neuropsychological prediction of dementia and the absence of dementia in healthy elderly persons. Neurology 44, 1427–1432, 1994

Masur DM, Fuld PA, Blau AD, Crystal H, Aronson MK. Predicting development of dementia in the elderly with the selective reminding test. J Clin Exp Neuropsychol 12, 529–538, 1990

Masur DM, Sliwinsi M, Lipton R et al. Neuropsychological prediction of dementia and the absence of dementia in healthy elderly persons. Neurology 44, 1427–1432, 1994

Masur DM, Sliwinski M, Lipton RB, Blau AD, Crystal HA. Neuropsychological prediction of dementia and the absence of dementia in healthy persons. Neurology 44, 1427–1432, 1994

Mattis S. Mental status examination for organic mental syndrome in the elderly patients. In: L Bellak, TB Karasu (eds), Geriatric Psychiatry: A Handbook for Psychiatrists and Primary Care Physicians. Grune & Stratton, New York, 77–121, 1976

Mattis S. Dementia Rating Scale. Odessa, Psychological Assessment Resources, 1989

Maunoury C, Micot J-L, Caillet H, Parlato V, Leroy-Willig A, Jehenson P, Syrota A, Boller F. Specificity of temporal amygdala atrophy in Alzheimer's disease. Quantitative assessment with magnetic resonance imaging. Dementia 7, 10–14, 1996

Maurer K, Ihl R, Dierks T, Frolich L. Clinical efficacy of Ginkgo biloba special extract EGb 761 in dementia of the Alzheimer type. J Psychiatr Res 31, 645–655, 1997

Maurice T, Lockhart BP, Privat A. Amnesia induced in mice by centrally administered β-amyloid peptides involves cholinergic dysfunction. Brain Res 706, 181–193, 1996

Martin A, Fedio P. Word production and comprehension in Alzheimer disease: The breakdown of semantic knowledge. Brain Lang 19, 124–141, 1983

Mayes AR, Roberts N. Theories of episodic memory. Phil Trans R Soc Lond B, 356, 1395–1408, 2001

Mayes AR. Human Organic Memory Disorders. Cambridge University Press, 1988

Mayes AR. Aware and unaware memory: does unaware memory underlie aware memory? In: C Hoerl, T McCormack (eds), Time and Memory. Oxford: Clarendon Press, 187–212, 2001

Mayeux R, Ottman R, Maestre G, Ngai C, Tang MX, Ginsberg H, Chun M, Tycko B, Shelanski M. Synergistic effects of traumatic head injury and apolipoprotein-epsilon 4 in patients with Alzheimer's disease. Neurology 45(3 Pt 1), 555–557, 1995

Mazziotta JC, Phelps ME. Human sensory stimulation and deprivation: positron emission tomographic results and strategies. Ann Neurol 15 (Suppl), 550–560, 1984

Mazziotta JC, Phelps ME, Carson RE, Kuhl DE. Tomographic mapping of human cerebral metabolism: auditory stimulation. Neurology 32, 921–937, 1982

Mazzoni M, Ferroni L, Lombardi L, Del Torto E, Vista M, Moretti P. Mini-Mental State Examination (MMSE): sensitivity in an Italian sample of patients with dementia. Ital J Neurol Sci 13(4), 323–329, 1992

McClelland JL, McNaughton BL, O'Reilly RC. Why there are complementary learning systems in the hippocampus and neocortex: insights from the successes and failures of connectionist models of learning and memory. Psychol Rev 102(3), 419–457, 1995

McCullar MM, Coates M, Van Fleet N, Duchek J, Grant E, Morris JC. Reliability of clinical nurse specialists in the staging of dementia. Arch Neurol 46, 1210–1211, 1989

McDermott KB, Buckner RL, Petersen SE, Kelley WM, Sanders AL. Set- and code-specific activation in frontal cortex: an fMRI study of encoding and retrieval of faces and words. J Cogn Neurosci 11, 631–640, 1999

McDonald MP, Dahl EE, Overmier JB. Effects of an exogenous β-amyloid peptide on retention for spatial learning. Behav Neurol Biol 62, 60–67, 1994

McEwen BS. Clucocorticoids and hippocampus: Receptors in search of a function. Curr Top Neuroendocrin 2, 1–22, 1982

McGaugh JL. Memory: A century of consolidation. Science 287, 248–251, 2000

McGeer PL, Schulzer M, McGeer EG. Arthritis and anti-inflammatory agents as possible protective factors for Alzheimer disease: a review of 17 epidemiological studies. Neurology 47 (2), 425–432, 1996

McKee RD, Squire LR. On the development of declarative memory. J Exp Psychol Learn Mem Cogn 19, 397–404, 1993

McKeith IG, Galasko D, Kosaka K, Perry EK, Dickson DW, Hansen LA, Salmon DP, Lowe J, Mirra SS, Byrne EJ et al. Consensus guidelines for the clinical and pathologic diagnosis of dementia with Lewy bodies (DLB): report of the consortium on DLB international workshop. Neurology 47(5), 1113–1124, 1996

McKhann G, Drachman D, Folstein M, Katzman R Price D, Stadlan EM. Clinical diagnosis of Alzheimer disease: report of the NINCDS-ADRDA Work Group under the auspices of the Department of Health and Human Services Task Force on Alzheimer disease. Neurology 34, 939–944, 1984

McKitrick LA, Camp CJ, Black FW. Prospective memory intervention in Alzheimer disease. J Gerontol 47 (5), 337–343, 1992

McKitrick LA, Camp CJ. Relearning the names of things: The spaced-retrieval intervention implemented by a caregiver. Clinical Gerontologist 14, 60–62, 1993

McRae A, Dahlstrom A, Polinsky R, Ling EA. Cerebrospinal fluid microglial antibodies: potential diagnostic markers for immune mechanisms in Alzheimer disease. Behav Brain Res 57, 225–234, 1993

McShane R, Keene J, Fairburn C, Jacoby R, Hope T. Issues in drug treatment for Alzheimer's disease. Lancet 350(9081), 886–887, 1997

Meares R. Intimacy and Alienation. Memory, Trauma and Personal Being. London, Routledge, 2000

Mecocci P, Cherubini A, Bregnocchi M, Chionne F, Cecchetti R, Lowenthal DT, Senin U. Tau protein in cerebrospinal fluid: a new diagnostic and prognostic marker in Alzheimer disease? Alzheimer Dis Assoc Disord 12(3), 211–214, 1998

Mega MS, Masterman DM, O'Connor SM, Barclay TR, Cummings JL. The spectrum of behavioral responses to cholinesterase inhibitor therapy in Alzheimer disease. Arch Neurol 56(11), 1388–1393, 1999

Mehlhorn G, Hollborn M, Schliebs R. Induction of cytokines in glial cell surrounding cortical beta-amyloid plaques in transgenic Tg2576 mice with Alzheimer pathology. Int J Dev Neurosci 18, 423–431, 2000

Melzer D. New drug treatment for Alzheimer disease: Lessons for healthcare policy. BMJ 316, 762–764, 1998

Mendez MF, Underwood KL, Zander BA, Mastri AR, Sung JH and Frey WH. Risk factors in Alzheimer disease – a clinicopathologic study. Neurology 42, 770–775, 1992

Mendez MF, Mendez MA, Martin R, Smyth KA, Whitehouse PJ. Complex visual disturbances in Alzheimer's disease. Neurology 40(3 Pt 1), 439–443, 1990

Mendez MF, Perryman KM, Miller BL, Cummings JL. Behavioral differences between frontotemporal dementia and Alzheimer's disease: A comparison on the BEHAVE-AD Rating Scale. Int Psychogeriatr 10 (2), 155–162, 1998

Mendez MF, Cherrier M, Perryman KM, Pachana N, Miller BL, Cummings JL. Frontotemporal dementia *versus* Alzheimer's disease: differential cognitive features. Neurology 47(5), 1189–1194, 1996

Mendez MF, Doss RC, Cherrier MM. Use of the cognitive estimations test to discriminate frontotemporal dementia from Alzheimer disease. J Geriartr Psychiatry Neurol 11, 2–6, 1998

Meneses A. 5-HT system and cognition. Neurosci Biobehav Rev 23, 1111–1125, 2000

Mentis MJ, Alexander GE, Krasuski J, Pietrini P, Furey ML, Schapiro MB and Rapoport SI. Increasing required neural response to expose abnormal brain function in mild *versus* moderate or severe Alzheimer disease: PET study using parametric visual stimulation. Am J Psychiatry 155, 785–794, 1998

Meyer JS, Li Y, Xu G, Thornby J, Chowdhury M, Quach M. Feasibility of treating mild cognitive impairment with cholinesterase inhibitors. Int J Geriatr Psychiatry 17, 586–588, 2002

Meyer JS, Rauc GM, Rauch RA, Haque A, Crawford K. Cardiovascular and other risk factors for Alzheimer disease and vascular dementia. Ann NY Acad Sci 903, 411–423, 2000

Mielke R, Herholz K, Grond M, Kessler J, Heiss WD. Differences of regional cerebral glucose metabolism between presenile and senile dementia of Alzheimer type. Neurobiol Aging 13, 93–98, 1991

Mihara M, Ohnishi A, Tomono Y, Hasegawa J, Shimamura Y, Yamazaki K, Morishita N. Pharmacokinetics of E2020, a new compound for Alzheimer's disease, in healthy male volunteers. Int J Clin Pharmacol Ther Toxicol 31(5), 223–229, 1993

Miller BL, Cummings JL. The Human Frontal Lobes: Functions and Disorders. New York, The Guildford Press, 1999

Miller GA, Johnson-Laird PN. Language and Perception. Cambridge (MA): Cambridge University Press, 1976

Miller BL, Darby AL, Swartz JR, Yener GG, Mena I. Dietary changes, compulsions and sexual behavior in frontotemporal degeneration. Dementia 6, 195–199, 1995

Milner B. Disorders of learning and memory after temporal lobe lesions in man. Clin Neurosurg 19, 421–446, 1972

Milner B. In: C Millikan, F Darley (eds), Basic Mechanisms Underlying Speech and Language. Grune and Stratton, New York, 122–131, 1967

Milner B. Interhemispheric differences in the localization of psychological processes in men. Br Med Bull 27, 272–277, 1971

Minoshima S, Giordani B, Berent S, Frey K, Foster NL, Kuhl DE. Metabolic reduction in the posterior cingulate cortex in very early Alzheimer's disease. Ann Neurol 42, 85–94, 1997

Mintzer JE, Brawman-Mintzer O. Agitation as possible expression of generalized anxiety disorder in demented elderly patients: towards a treatment approach. J Clin Psychiatry 57 (Suppl 7), 55–63, 1996

Mirra SS, Heyman A, McKeel DW, Sumi SM, Crain BJ, Brownlee LM et al. The Consortium to Establish a Registry for Alzheimer disease (CERAD). Part II. Standardization of the neuropathologic assessment of Alzhheimer's disease. Neurology 41, 479–486, 1991

Moak GS, Fisher WH. Alzheimer disease and related disorders in state mental hospitals: Data from a nationwide survey. Gerontologist 30, 798–802, 1990

Moechars D, Lorent K, De Strooper B, Dewachter I, Van Leuven F. Expression in brain of amyloid precursor protein mutated in the alpha-secretase site causes disturbed behavior, neuronal degeneration and premature death in transgenic mice. EMBO J 15, 1265–1274, 1996

Mohr E, Feldman H, Gauthier S. Canadian guidelines for the development of antidementia therapies: a conceptual summary. Can J Neurol Sci 22, 62–71, 1995

Mohs RC, Davis BM, Johns CA, Mathé AA, Greenwald BS, Horvath TB, Davis KL. Oral physostigmine treatment of patients with Alzheimer disease. Am J Psychiatry 142, 28–33, 1985

Mohs RC, Cohen L. Alzheimer disease assessment scale (ADAS). Psychopharmacol Bull 24, 627–628, 1988

Mohs RC, Knopman D, Petersen RC, Ferris SH, Ernesto C, Grundman M, Sano M, Bieliauskas L, Geldmacher D, Clark C, Thal LJ. Development of cognitive instruments for use in clinical trials of antidementia drugs: additions to the Alzheimer's Disease Assessment Scale that broaden its scope. The Alzheimer's Disease Cooperative Study. Alzheimer Dis Assoc Disord 11 (Suppl 2), S13–21, 1997

Molchan SE, Vitiello B, Minichiello M, Sunderland T. Reciprocal changes in psychosis and mood after physostigmine in a patient with Alzheimer's disease. Arch Gen Psychiatry 48(12), 1113–1114, 1991

Monsch AU, Bondi MW, Salmon DP et al. Clinical validity of the Mattis Dementia Rating Scale in detecting dementia of the Alzheimer type: a double cross-validation and application to a community-dwelling sample. Arch Neurol 52, 899–904, 1995

Montgomery SA. Safety of Mirtazepine: a review. Int Clin Psychopharm 10 (Suppl 4), 37–45, 1995

Monti LA, Gabrieli JDE, Reminger SL, Rinaldi JA, Wilson RS, Fleischman DA. Differential effects of aging and Alzheimer disease upon conceptual implicit and explicit memory. Neuropsychology 10, 101–112, 1996

Monti LA, Gabrieli JDE, Wilson RS, Reminger SL. Intact text specific implicit memory in patients with Alzheimer disease. Psychol Aging 9, 64–71, 1994

Montine TJ, Sidell KR, Crews BC, Markesbery WR, Marnett LJ, Roberts LJ 2nd, Morrow JD. Elevated CSF prostaglandin E2 levels in patients with probable AD. Neurology 53(7), 1495–1498, 1999

Morgan D, Diamond DM, Gottschall PE, Ugen KE, Dickey C, Hardy J, Duff K, Jantzen P, Di Carlo G, Wilcock D, Connor K, Hatcher J, Hope C, Gordon M, Arendash GW. Aβ peptide vaccination prevents memory loss in an animal model of Alzheimer disease. Nature 408, 982–985, 2000

Morris CD et al. Levels of processing *versus* transfer appropriate processing. J Verbal Learn Verbal Behav 16, 519–533, 1977

Morris JC, Storandt M, McKeel DW et al. Cerebral amyloid deposition and diffuse plaques in "normal" aging: evidence for presymptomatic and very mild Alzheimer disease. Neurology 46, 707–719, 1996

Morris JC, Storandt M, Miller JP, McKeel DW, Price JL, Rubin EH, Berg L. Mild cognitive impairment represents early-stage Alzheimer disease. Arch Neurol 58(3), 397–405, 2001

Morris JC, Price JL. Pathologic correlates of nondemented aging, mild cognitive impairment and early-stage Alzheimer disease. J Mol Neurosci 17, 101–118, 2001

Morris JC, McKeel DW Jr, Storandt M, Rubin EH, Price JL, Grant EA, Ball MJ, Berg L. Very mild Alzheimer disease: Informant-based clinical, psychometric and pathologic distinction from normal aging. Neurology 41, 469–478, 1991

Morris JC, McKeel DW, Fulling K, Torack RM, Berg L. Validation of clinical diagnostic criteria for Alzheimer's disease. Ann Neurol 24, 17–22, 1988

Morris JC, Edland SD, Clark C et al. The Consortium to establish a registry for Alzheimer disease (CERAD). Part IV. Rates of cognitive change in the longitudinal assessment of probable Alzheimer disease. Neurology 43, 2457–2465, 1993

Morris JC. Relationship of plaques and tangles to Alzheimer disease phenotype. In: AM Goate, F Ashall (eds), The Pathophysiology of Alzheimer Disease. San Diego: Academic Press, 194, 1995

Morris JC, Price JL. Pathologic correlates of nondemented aging, mild cognitive impairment, and early stage Alzheimer disease. J Mol Neurosci 17, 101–118, 2001

Morris JC, Cyrus PA, Orazem J, Mas J, Bieber F, Ruzicka BB, Gulanski B. Metrifonate benefits cognitive, behavioral and global function in patients with Alzheimer's disease. Neurology 50, 1222–1230, 1998

Morris JC. The clinical dementia rating (CDR): Current version and scoring rules. Neurology 43, 2412–1314, 1993

Morris JC. Clinical assessment of Alzheimer disease. Neurology 49, S7–S10, 1997

Morris JF, Frith CD, Perrett DI, Rowland D, Young AW, Calder AJ, Dolan RJ. A differential neural response in the human amygdala to fearful and happy facial expressions. Nature 383, 812–815, 1996

Morris RG. Dementia and the functioning of the articulatory loop system. Cognitive Neuropsychology 1, 143–157, 1984

Morris RG, Frey U. Hippocampal synaptic plasticity: role in spatial learning or the automatic recording of attended experience? Phil Trans R Soc Lond B352, 1489–1503, 1997

Morris RG. Short-term forgetting in senile dementia of the Alzheimer's type. Cognitive Neuropsychology 3, 77–97, 1986

Morris RG. Articulatory rehearsal in Alzheimer-type dementia. Brain Lang 30, 351–362, 1987

Morris RG, Baddeley AD. Comment: primary and working memory functioning in Alzheimer-type dementia. J Clin Exp Neuropsychol 10 (2), 279–296, 1988

Mortimer JA, Ebbitt B, Jun SP, Finch MD. Predictors of cognitive and functional progression in patients with probable Alzheimer disease. Neurology 42, 1689–1696, 1992

Moscovitch M. In: M Gazzaniga (ed), The Cognitive Neurosciences. MIT Press, Cambridge, MA, 1341–1356, 1995

Moscovitch M, Nadel L. Consolidation and the hippocampal complex revisited: in defense of the multiple-trace model – discussion point. Curr Opin Neurobiol 8, 297–300, 1998

Moscovitch M. A neuropsychological model of memory and consciousness. In: LR Squire, N Butters (eds), Neuropsychology of Memory, 2nd ed. New York: Guildford Press, 5–22, 1992

Moss M, Albert M, Butters N, Payne M. Differential patterns of memory loss among patients with Alzheimer disease, Huntington's Disease, and alcoholic Korsakoff's syndrome. Arch Neurol 43, 239–246, 1986

Moss M, Albert M. Alzheimer disease and other dementing disorders. In: M Albert, M Moss (eds), Geriatric Neuropsychology, New York, Guildford Press, 145–178, 1988

Motter R, Vigo-Pelfrey C, Kholodenko D, Barbour R, Johnson-Wood K, Galasko D, Chang L, Miller B, Clark C, Green R et al. Reduction of beta-amyloid peptide42 in the cerebrospinal fluid of patients with Alzheimer's disease. Ann Neurol 38(4), 643–648, 1995

Moulin CJA, Perfect TJ, Conway MA, North AS, Jones RW, James N. Retrieval-induced forgetting in Alzheimer's disease. Neuropsychologia 40, 862–867, 2002

Moyer JR Jr, Disterhoft JR. Nimodipine decreases calcium action potentials in rabbit hippocampal CA1 neurons in age-dependent and concentration-dependent manner. Hippocampus 4, 11–17, 1994

Mu Q, Xie J, Wen Z, Weng Y, Shuyun Z. A quantitative MR study of the hippocampal formation, the amygdala, and the temporal horn of the lateral ventricle in healthy subjects 40 to 90 years of age. Am J Neuroradiol 20, 207–211, 1999

Müller WE, Mutschler E, Riederer P. Noncompetitive NMDA receptor antagonists with fast open-channel blocking kinetics and strong voltage-dependency as potential therapeutic agents for Alzheimer's dementia. Pharmacopsychiatry 28, 113–124., 1995

Muller-Hill B. Beyreuther K. Molecular biology of Alzheimer's disease. Ann Rev Biochem 58, 287–307, 1989

Muller G, Weisbrod S, Klinberg F. Test battery for objective assessment and differential diagnosis in the early stage of suspected development of presenile dementia of the Alzheimer type. Zeitschrift für Gerontologie 26 (2), 70–80, 1993

Mulligan R, Mackinnon A, Jorm AF, Giannakopoulos P, Michel J-P. A comparison of alternative methods of screening for dementia in clinical settings. Arch Neurol 53, 532–537, 1996

Mulnard RA, Cotman CW, Kawas C, van Dyck CH, Sano M, Doody R, Koss E, Pfeiffer E, Jin S, Gamst A, Grundman M, Thomas R, Thal LJ. Estrogen replacement therapy for treatment of mild to moderate Alzheimer disease. A randomized controlled trial. JAMA 283, 1007–1015, 2000

Mungas D, Reed BR, Ellis WG, Jagust WJ. The effects of age on rate of progression of Alzheimer disease and dementia with associated cerebrovascular disease. Arch Neurol 58, 1243–1247, 2001

Nabeshima T, Nitta A. Memory impairment and neuronal dysfunction induced by β-amyloid protein in rats. Tohoku J Exp Med 174, 241–248, 1994

Nacmias B, Tedde A, Forleo P, Piacentini S, Latorraca S, Guarnieri BM, Ortenzi L, Bartoli A, Petruzzi C. Psychosis, serotonin receptor polymorphism and Alzheimer disease. Arch Gerontol Geriatr (Suppl) 7, 279–283, 2001

Nadel L, Moscovitch M. Memory consolidation, retrograde amnesia and the hippocampal complex. Curr Opin Neurobiol 7, 217–227, 1997

Nadel L, Samsonovitch A, Ryan L, Moscovitch M. Multiple trace theory of human memory: computational neuroimaging, and neuropsychological results: Hippocampus 10, 352–369, 2000

Nakayama M, Uchimura K, Zhu RL, Nagayama T, Rose ME, Stetler RA, Isakson PC, Chen J, Graham SH. Cyclooxygenase-2 inhibition prevents delayed death of CA1 hippocampal neurons following global ischemia. Pro Natl Acad Sci USA 95(18), 10954–10959, 1998

Naslund J, Haroutunian V, Mohs R, Davis KL, Davies P, Greengard P, Buxbaum JD. Correlation between elevated levels of amyloid beta-peptide in the brain and cognitive decline. JAMA 283(12), 1571–1577, 2000

National Institute on Aging and Reagan Institute Working Group on Diagnostic Criteria for the Neuropathological Assessment of Alzheimer disease. Consensus recommendations for the postmortem diagnosis of Alzheimer disease. Neurobiol Aging 18, S1–S2, 1997

Naugle RI, Kawczak K. Limitations of the Mini-Mental State Examination. Cleve Clin J Med 56, 277–281, 1989

Nebes RD, Boller F, Holland A. Use of semantic context by patients with Alzheimer disease. Psychol Aging 1, 261–269, 1986

Nebes RD, Brady CB, Huff FJ. Automatic and attentional mechanisms of semantic priming in Alzheimer disease. J Clin Exp Neuropsychol 11, 219–230, 1989

Nebes RD. Semantic memory in Alzheimer disease. Psychol Bull 106, 377–394, 1989

Neisser U. Nested structure in autobiographical memory. In: DC Rubin (ed), Autobiographical Memory. New York: Cambridge University Press, 71–81, 1986

Neri M, Andermarcher E, Spano A, Salvioli G, Cipolli C. Validation study of the Italian version of the Cambridge Mental Disorders of the Elderly Mental Disorders of the Elderly Examination: preliminary findings. Dementia 3, 70–77, 1992

Netland EE, Newton JL, Majocha RE, Tate BA. Indomethacin reverses the microglial response to amyloid beta-protein. Neurobiol Aging 19(3), 201–204, 1998

Neufeld MY, Rabey MJ, Parmet Y, Sifris P, Treves TA, Korczyn AD. Effects of a single intravenous dose of scopolamine on quantitative EEG in Alzheimer's disease patients and age-matched controls. Electroencephalogr Clin Neurophysiol 91, 407–412, 1994

Neuroinflammation Working Group. Inflammation and Alzheimer's disease. Neurobiol Aging 21, 383–421, 2000

Neve RL, McPhie DL, Chen Y. Alzheimer's disease. A dysfunction of the amyloid precursor protein. Brain Res 886, 54–66, 2000

Newhouse PA, Potter A, Levin ED. Nicotinic systems and Alzheimer disease: Implications for therapeutics. Drugs Aging 11, 206–228, 1997

Newhouse PA, Hughes JR. The role of nicotine and nicotinic mechanisms in neuropsychiatric disease. Br J Addict 86, 521–526, 1991

Newhouse PA, Sunderland T, Tariot PN, Blumhardt CL, Weingartner H, Mellow A, Murphy DL. Intravenous nicotine in Alzheimer's disease: a pilot study. Psychopharmacology 95, 171–175, 1988

Newhouse PA, Sunderland T, Thompson K, Tariot PN, Weingartner H, Mueller ER, Cohen RM, Murphy DL. Intravenous nicotine in a patient in a patient with Alzheimer's disease. Am J Psychiat 143, 1494–1495, 1986

Newhouse PA, Tatro A, Naylor M, Quealey K, Delgado P. Alzheimer disease, serotonin systems and tryptophan depletion. Am J Geriatr Psychiatry 10 (4), 483–484, 2002

Newman S, Warrington E, Kennedy A, Rossor M. The earliest cognitive change in a person with familial Alzheimer disease: Presymptomatic neuropsychological features in a pedigree with familial Alzheimer disease confirmed at necropsy. J Neurol Neurosurg Psychiatry 57, 967–972, 1994

Nicholl CG, Lynch S, Kelly CA, White L, Simpson PM, Wesnes KA, Pitt BMN. The Cognitive Drug Research Computerized Assessment system in the evaluation of early dementia – Is speed of the essence? Int J Geriatr Psychiatry 10, 199–206, 1995

Nicoll JAR, Mrak RE, Graham DI, Stewart J, Wilcock G, MacGowan S, Esiri MM, Murray LS, Dewar D et al. Association of interleukin-1 gene polymorphisms with Alzheimer's disease. Ann Neurology 47(3), 365–368, 2000

Nishizawa S, Benkelfat C, Young SN, Leyton N, Mzengeza S, de Montigny C, Blier P, Diksic M. Differences between males and females in rates of serotonin synthesis in human brain. Proc Natl Acad Sci USA 94, 5308–5313, 1997

Nitsch RM, Slack BE, Wurtman RJ, Growdon JH. Release of Alzheimer amyloid precursor derivatives stimulated by activation of muscarinic acetylcholine receptors. Science 258: 304–307, 1992

Nitsch RM, Deng M, Growdon JH, Wurtman RJ. Serotonin 5HT2a and 5HT2c receptors stimulate amyloid precursor protein ectodomain secretion. J Biol Chem 271, 4188–4194, 1996

Nitsch RM, Deng A, Wurtman RJ, Growdon JH. Metabotropic glutamate receptor subtype mGuR1a stimulates the secretion of the amyloid β-protein precursor ectodomain. J Neurochem 69, 704–712, 1997

Nitsch RM, Kim C, Growdon JH. Vasopressin and bradykinin regulate secretory processing of the amyloid protein precursor of Alzheimer disease. Neurochem Res 23, 807–814, 1998

Nogawa S, Zhang F, Ross ME et al. Cyclo-oxygenase-2 gene expression in neurons contributes to ischemic brain damage. J Neurosci 17, 2746–2755, 1997

Nolde SF, Johnson MK, Raye CL. The role of prefrontal cortex during tests of episodic memory. Trends in Cognitive Sciences 2, 399–406, 1998

Nordberg A. PET studies and cholinergic therapy in Alzheimer disease. Rev Neurol (Paris), 155, 4S, 53–63, 1999

Notkola IL, Sulkava R, Pekkanen J et al. Serum total cholesterol, apolipoprotein E e4 allele and Alzheimer's disease. Neuroepidemiology 17, 14–20, 1998

Novella JL, Jochum C, Jolly D, Morrone I, Ankri J, Bureau F, Blanchard F. Agreement between patients' and proxies' reports of quality of life in Alzheimer's disease. Qual Life Res 10, 443–452, 2001

Nunnally JC. Psychometric Theory. New York, McGraw-Hill, 1978

Nyback H, Nyman H, Blomqvist G, Sjogren I, Stone-Elander S. Brain metabolism in Alzheimer's dementia: Studies of C-deoxyglucose accumulation, CSF monoamine metabolites and neuropsychological test performance in patients and healthy subjects. J Neurol Neurosurg Psychiatry 54, 672–678, 1991

Nyberg L, Persson J, Habib R, Tulving E, McIntosh AR, Cabeza R, Houle S. Large scale neurocognitive networks underlying episodic memory. J Cogn Neurosci 12(1), 163–173, 2000

Nyberg L, McIntosh AR, Houle S, Nilsson LG, Tulving E. Activation of medial temporal structures during episodic memory retrieval. Nature 380, 715–717, 1996

Ober BA, Shenaut GK. Semantic priming in Alzheimer's disease. Meta-analysis and theoretical evaluation. In: PA Allen, TR Bashore (eds), Age Differences in Word and Language Processing. Amsterdam, Elsevier Science 247–271, 1995

O'Connor DW, Pollit PA, Hyde JB, Fellows JL, Miller ND, Brook CP, Reiss BB, Roth M. The prevalence of dementia as measured by the Cambridge Mental Disorders of the Elderly Examination. Acta Psychiatr Scand 79, 190–198, 1989

O'Connor DW, Pollit PA, Hyde JB, Fellows JL, Miller ND, Roth M. A follow-up study of dementia diagnosed in the community using the CAMDEX. Acta Psychiatr Scan 81, 78–82, 1990

Ogeng'o JA, Cohen DL, Sayi JG, Matuja WB, Chande HM, Kitinya JN, Kimani JK, Friedland RP, Mori H, Kalaria RN. Cerebral amyloid beta protein deposits and other Alzheimer lesions in non-demented elderly east Africans. Brain Pathol 6(2), 101–107, 1996

Ohnishi A, Mihara M, Kamakura H, Tomono Y, Hasegawa J, Yamazaki K, Morishita N, Tanaka T. Comparison of the pharmacokinetics of E2020, a new compound for Alzheimer's disease, in healthy young and elderly subjects. J Clin Pharmacol 33(11), 1086–1091, 1993

Ojemann GA, Dodrill CB. Intraoperative techniques for reducing language and memory deficits with left temporal lobectomy. Adv Epileptol 16, 327–330, 1987

O'Keefe J, Nadel L. The hippocampus as a cognitive map. Oxford: University Press, 1978

Okuizumi K, Onodera O, Namba Y et al. Genetic association of the very low density lipoprotein (VLDL) receptor gene with sporadic Alzheimer disease. Nat Genet 11, 207–209, 1995

Olin JT, Schneider LS, Doody RS, Clark CM, Ferris SH, Morris JC, Reisberg B, Schmitt FA. Clinical evaluation of global change in Alzheimer's disease: identifying consensus. J Geriatr Psychiatry Neurol 9(4), 176–180, 1996

Olson JM, Goddard KAB and Dudek DM. A second locus for Very-Late Onset Alzheimer disease: A Genome scan reveals linkage to 20p and epistasis between 20p and the amyloid precursor protein region. Am J Hum Genet 71, 154–161, 2002

Orgogozo J-M, Rigaud A-S, Stöffler A, Möbius H-J, Forette F. Efficacy and safety of memantine in patients with mild to moderate vascular dementia. A randomized, placebo-controlled trial (MMM 300). Stroke 33, 1834–1839, 2002

Oster-Granite ML, Mcphie DL, Greenan J, Neve RL. Age-dependent neuronal and synaptic degeneration in mice transgenic for the C-terminus of the amyloid precursor protein. J Neurosci 16, 6732–6741, 1996

Osuntokun BO, Sahota A, Ogunniyi AO. Lack of an association between the ε4 allele of ApoE and Alzheimer disease in elderly Nigerians. Ann Neurol 38, 463–465, 1995

Ott A, Breteler MMB, Van Haskamp E et al. Smoking increases the risk of dementia: The Rotterdam study (abstract). Neurology 48 (Suppl A), 78, 1997

Ott BR, Lafleche G, Whelihan WM, Buongiorno GW, Albert MS, Fogel BS. Impaired awareness of deficits in Alzheimer disease. Alzheimer Dis Assoc Disord 10(2), 68–76, 1996

Ousset PJ, Viallard G, Puel M, Celsis P, Démonet JF, Cardebat D. Lexical therapy and episodic word learning in dementia of the Alzheimer type. Brain Lang 80, 14–20, 2002

Ozawa S, Kamiya H, Tsuzuki K. Glutamate receptors in the mammalian central nervous system. Prog Neurobiol 54, 581–618, 1998

Pachana NA, Boone KB, Miller BL, Cummings JL, Berman N. Comparison of neuropsychological functioning in Alzheimer's disease and frontotemporal dementia. J Int Neuropsychol Soc 2(6), 505–510, 1996

Paganini-Hill A, Henderson VW. Estrogen replacement therapy and risk of Alzheimer disease. Arch Intern Med 156, 2213–2217, 1996

Panisset M, Gauthier S, Moessler H, Windisch M and the Cerebrolysin Study Group. Cerebrolysin in Alzheimer's disease: a randomized, double-blind, placebo-controlled trial with a neurotrophic agent. J Neural Transm 109, 1089–1104, 2002

Parnetti L. Clinical pharmacokinetics of drugs for Alzheimer disease. Clin Pharmacokinet 29, 110–129, 1995

Papassotiropoulos A, Bagli M, Jessen F, Bayer TA, Maier W, Rao ML, Heun R. A genetic variation of the inflammatory cytokine interleukin-6 delays the initial onset and reduces the risk for sporadic Alzheimer's disease. Ann Neurol 45(5), 666–668, 1999

Papassotiropoulos A, Lütjohann D, Bagli M et al. Plasma 24S-hydroxycholesterol: a peripheral indicator of neuronal degeneration and potential state marker for Alzheimer's disease. Neuroreport 11, 1959–1962, 2000

Park SB, Coull JT, McShane RH, Young AH, Sahakian BJ, Robbins TW, Cowen PJ. Tryptophan depletion in normal volunteers produces selective impairment in learning and memory. Neuropharmacology 33, 575–588, 1994

Pasinetti GM. Cyclooxygenase and Alzheimer's disease: implications for preventive initiatives to slow the progression of clinical dementia. Arch Gerontol Geriatr 33, 13–28, 2001

Pasquier F. Neuropsychological features and cognitive assessment in frontotemporal dementia. In: Pasquier F, Lebert F, Scheltens P (eds), Frontotemporal dementia. ICG, Dordrecht, 49–69, 1996

Patrick DL, Erickson P. Health Status, Quality of Life and Health Related Quality of Life. Health status and health policy. New York: Oxford Press, 20–26, 1993

Patterson MB, Mack JL. Manual for the administration of the CERAD Behavioral Rating Scale for Dementia. Cleveland, OH: Case Western Reserve University, 1996

Patterson MB, Mack JL, Mackell JA et al. A longitudinal study of behavioral pathology across five levels of dementia severity in Alzheimer's Disease: the CERAD Behavior Rating Scale for Dementia. Alzheimer Dis Assoc Disord 11 (Suppl 2), S40–S44, 1997

Patterson MB, Bolger JP. Assessment of behavioural symptoms in Alzheimer disease. Alzheimer Dis Assoc Disord 8 (Suppl 3), 4–20, 1994

Paulsen JS, Salmon DP, Thal LJ, Romero R, Weisstein-Jenkins C, Galasko D, Hofstetter CR, Thomas R, Grant I, Jeste DV. Incidence of and risk factors for hallucinations and delusions in patients with probable AD. Neurology 54, 1965–1971, 2000

Payami H, Zareparsi S, Montee KR, Sexton GJ, Kaye JA, Bird TD, Yu CE, Wijsman EM, Heston LL, Litt M, Schellenberg GD. Gender difference in apolipoprotein E-associated risk for familial Alzheimer disease: a possible clue to the higher incidence of Alzheimer disease in women. Am J Human Genetics 58(4), 803–811, 1996

Pearson RC, Esiri MM, Hiorns RW, Wilcock GK, Powell TP. Anatomical correlates of the distribution of the pathological changes in the neocortex in Alzheimer disease. Proc Natl Acad Sci USA 82(13), 4531–4534, 1985

Pedersen WA, Kloczewiak MA, Blusztajn JK. Amyloid beta-protein reduces acetylcholine synthesis in a cell line derived from cholinergic neurons of the basal forebrain. Proc Natl Acad Sci USA 93, 8068–8071, 1996

Pepeu G. Preclinical pharmacology of cholinesterase inhibitors. In: E Giacobini (ed), Cholinesterases and Cholinesterase Inhibitors. Martin Dunitz, London, 145–155, 2000

Pepeu G, Marconcini-Pepeu I and Amaducci L. A review of phosphatidylserine pharmacological and clinical effects. Is phosphatidylserine a drug for the ageing brain? Pharmacol Res 33, 73–80, 1996

Pepin EP, Eslinger PJ. Verbal memory decline in Alzheimer's disease. Neurology 39, 1477–1482, 1989

Perego C, Vetrugno CC, De Simoni MG, Algeri S. Aging prolongs the stress-induced release of noradrenaline in rat hypothalamus. Neurosci Lett 157, 127–130, 1993

Perry EK, Tomlinson BE, Blessed G, Bergman K, Gibson PH, Perry RH. Correlation of cholinergic abnormalities with senile plaques and mental test scores in senile dementia. BMJ 2, 1457–1459, 1978

Perry RH, Irving D, Blessed G, Fairbairn A, Perry EK. Senile dementia of Lewy body type: a clinically and neuropathologically distinct form of Lewy body dementia in the elderly. J Neurol Sci 95, 119–139, 1990

Perry E, Martin-Ruiz C, Lee M, Griffiths M, Johnson M, Piggott M, Haroutunian V, Buxbaum JD, Nasland J, Davis K et al. Nicotinic receptor subtypes in human brain ageing, Alzheimer and Lewy body diseases. Eur J Pharmacol 393(1–3), 215–222, 2000

Perry RJ, Hodges JR. Differentiating frontal and temporal variant frontotemporal dementia from Alzhimer's disease. Neurology 54, 2277–2284, 2000

Perry RJ, Watson P, Hodges JR. The nature and staging of attention dysfunction in early (minimal and mild) Alzheimer's disease: relationship to episodic and semantic memory impairment. Neuropsychologia 38, 252–271, 2000

Persson G, Skoog I. Subclinical dementia: relevance of cognitive symptoms and signs. J Geriatr Psychiatry Neurol 5, 172–178, 1992

Petersen RC, Stevens JC, Ganguli M, Tangalos EG, Cummings JL, DeKosky ST. Practice parameter: early detection of dementia: mild cognitive impairment (an evidence-based review). Report of the Quality Standards Subcommittee of the American Academy of Neurology. Neurology 56(9), 1133–1142, 2001

Petersen RC, Smith GE, Waring SC, Ivnik RJ, Tangalos EG, Kokmen E. Mild cognitive impairment: clinical characterization and outcome. [erratum appears in Arch Neurol 56(6), 760, 1999]. Arch Neurol 56(3), 303–308, 1999

Petersen RC, Smith GE, Waring SC, Ivnik RJ, Kokmen E, Tangelos EG. Aging, memory, and mild cognitive impairment. Int Psychogeriatr 1, 65–69, 1997

Petersen RC, Smith GE, Ivnik RJ, Kokmen E, Tangalos EG. Memory function in very early Alzheimer's disease. Neurology 44(5), 867–872, 1994

Petersen RC. Mild cognitive impairment or questionable dementia. Arch Neurol 57, 643–644, 2000

Petersen LR, Petersen MJ. Short-term retention of individual items. J Exp Psychol 91, 341–343, 1959

Peterson LR, Peterson MJ. Short-term retention of individual verbal items. J Exp Psychol 58, 193–198, 1959

Peterson RC, Smith GE, Waring SC et al. Aging, memory and mild cognitive impairment. Int Psychogeriatr 9, 65–69, 1997

Peterson C. Changes in calcium's role as a messenger during aging in neuronal and nonneuronal cells. Ann NY Acad Sci 21, 279–293, 1992

Peterson RC. Mild cognitive impairment: Transition between aging and Alzheimer's disease. Neurologia 15, 93–101, 2000

Pfeifer R, Scheier C. Understanding Intelligence. Cambridge, Mass. MIT Press, 1999

Pelps ME, Mazziotta JC. Positron emission tomograpy: human brain function and biochemistry. Science 228, 799–809, 1985

Pierre U, Wood-Dauphinee S, Korner-Bitensky N, Gayton D, Hanley J. Proxy use of the Canadian SF-36 in rating health status of the disabled elderly. J Clin Epidemiol 51 (11), 983–990, 1998

Pike CJ, Overman MJ, Cotman CW. Amino-terminal deletions enhance aggregation of beta-amyloid peptides *in vitro*. J Biol Chem 270(41), 23895–23898, 1995

Pillon B, Deweer B, Agid Y, Dubois B. Explicit memory in Alzheimer's, Huntington's, and Parkinson's disease. Arch Neurol 50, 374–379, 1993

Pizzi M, Valeriio A, Arrighi V, Galli P, Belloni M, Ribola M, Alberici A, Spano P and Memo M. Inhibition of glutamate-induced neurotoxicity by a tau antisense oligonucleotide in primary culture of rat cerebellar granule cells. Eur J Neurosci 7, 1603–1613, 1995

Pociot F, Molvig J, Wogensen L, Worsaae H, Nerup J. A TaqI polymorphism in the human interleukin-1 beta (IL-1 beta) gene correlates with IL-1 beta secretion *in vitro*. Eur J Clin Invest 22(6), 396–402, 1992

Poduslo SE, Neal M, Schhwankhaus J. A closely linked gene to apolipoprotein E may serve as an additional risk factor for Alzheimer disease. Neurosci Lett 201, 81–83, 1995

Poduslo SE, Neal M, Herring K, Shelly J. The apolipoprotein C1 A allele as a risk factor for Alzheimer disease. Neurochem Res 23, 361–367, 1998

Poduslo SE, Shook B, Drigalenko E and Yin X. Lack of association of the two polymorphisms in alpha-2 macroglobulin with Alzheimer disease. Am J Med Genetics 100, 30–35, 2002

Poirier J. Apolipoprotein E in animal models of CNS injury and in Alzheimer disease. TINS 17, 525–530, 1994

Poirier J, Hess M, May PC, Finch CE. Astrocyclic apolipoprotein E mRNA and GFAP mRNA in hippocampus after entorhinal cortex lesioning. Brain Res Mol Brain Res 11, 97–106, 1991

Poirier J, Baccichet A, Dea D, Gauthier S. Cholesterol synthesis and lipoprotein reuptake during synaptic remodelling in hippocampus in adult rats. Neuroscience 55, 81–90, 1993

Poirier J, Delisle MC, Quition R, Aubert I, Farlow M, Lahiri D, Hui S, Bertrand P, Nalbantoglu J, Gilfix BM et al. Apolipoprotein E4 allele as a predictor of cholinergic deficits and treatment outcome in Alzheimer disease. Proc Natl Acad Sci USA 92, 12260–12264, 1995

Polinsky RJ. Clinical pharmacology of rivastigmine: a new generation acetylcholinesterase inhibitor for the treatment of Alzheimer's disease. Clin Ther 20, 634–647, 1998

Poller W, Faber JP, Klobeck G, Olek K. Cloning of the human alpha 2–macroglobulin gene and detection of mutations in two functional domains: the bait region and the thioester site. Hum Genet 88, 313–319, 1992

Poorkaj P, Sharma V, Anderson L, Nemens E, Alonso ME, Orr H, White J, Heston L, Bird TD, Schellenberg GD. Missense mutations in the chromosome 14 familial Alzheimer disease presenilin 1 gene. Hum Mutat 11, 216–221, 1998

Porer J. Apolipoprotein E in animal models of central nervous system injury and in Alzheimer disease. Trends Neurosci 17, 525–530, 1994

Porter RJ, Lunn BS, Walker LLM, Gray JM, Ballard CG, O'Brien JT. Cognitive deficit induced by acute tryptophan depletion in patients with Alzheimer disease. Am J Psychiatry 157, 638–640, 2000

Pratico D, Clark CM, Liun F, Lee VM-Y, Trojanowski JQ. Increased of brain oxidative stress in mild cognitive impairment. Arch Neurol 59, 972–976, 2002

Pratico D. F2-isopropanes: sensitive and specific non-invasive indices of lipid peroxidation *in vivo*. Atherosclerosis 147, 1–10, 1999

Pratico D, Clark CM, Lee VM-Y, Trojanowski JQ, Rokach J, FitzGerald GA. Increased 8,12-iso-iPF2a-VI in Alzheimer disease: correlation of a non-invasive index of lipid peroxidation with disease severity. Ann Neurol 48, 809–812, 2000

Press GA, Amaral DG, Squire LR. Hippocampal abnormalities in amnesic patients revealed by high-resolution magnetic resonance imaging. Nature 341, 54–57, 1989

Purandare N, Burns A, Craig S, Faragher B, Scott K. Depressive symptoms in patients with Alzheimer's disease. Int J Geriatr Psychiatry 16(10), 960–964, 2001

Rabins PV, Mace NL, Lucas MJ. The impact of dementia on the family. JAMA 248, 333–335, 1982

Rahman S, Sahakian BJ, Hodges JR, Rogers RD, Robbins TW. Specific cognitive deficits in mild frontal variant frontotemporal dementia. Brain 122, 1469–1493, 1999

Raji MA, Brady SR. Mirtazapine for treatment of depression and comorbidities in Alzheimer disease. Ann Pharmacother 35, 1024–1027, 2001

Ramirez-Duenas MG, Rogaeva EA, Leal CA, Lin C, Ramirez-Casillas GA, Hernandez-Romo JA, St George-Hyslop PH, Cantu JM. A novel mutation Leu171Pro mutation in presenilin-1 gene in a Mexican family with early onset Alzheimer disease. Annales de Génétique 41(3), 149–153, 1998

Randolph C. Implicit, explicit and semantic memory functions in Alzheimer disease and Huntington's disease. J Clin Exp Neuropsychol 13, (4), 479–494, 1991

Raskind MA, Sadowsky CH, Sigmund WR, Beitler PJ, Auster SB. Effect of tacrine on language, praxis and noncognitive behavioral problems in Alzheimer disease. Arch Neurol 54, 836–840, 1997

Raskind MA, Cyrus PA, Ruzicka BB, Gulanski BI for the Metrifonate Study Group. The effects of metrifonate on the Cognitive, behavioral and functional performance of Alzheimer's disease patients. J Clin Psychiatry 60, 318–325, 1999

Raz N, Gunning F, Head D, Dupuis J, McQuain J, Briggs S, Loken W, Thorton A, Acker J. Selective aging of the human cerebral cortex observed *in vivo*: differential vulnerability of the prefrontal gray matter. Cereb Cortex 7, 268–282, 1997

Ready RE, Ott BR, Grace J, Fernandez I. The Cornell-Brown Scale for Quality of Life in Dementia. Alzheimer Dis Assoc Disord 16, 109–115, 2002

Ready RE, Ott BR, Grace J, Cahn-Weiner DA. Apathy and executive dysfunction in mild cognitive impairment and Alzheimer disease. Am J Geriatr Psychiatry 11 (2), 222–228, 2003

Reiman EM, Caselli RJ, Yun LS, Chen K, Bandy D, Minoshima S, Thibodeau SN, Osborne D. Preclinical evidence of Alzheimer disease in persons homozygous for the epsilon 4 allele for apolipoprotein E. N Eng J Med 334, 752–758, 1996

Refolo LM, Pappolla MA, LaFrancois J, Malester B, Schmidt SD, Thomas-Bryant T, Tint GS, Wang R, Mercken M, Petanceska SS, Duff KE. A cholesterol-lowering drug reduces beta-amyloid pathology in a transgenic mouse model of Alzheimer's disease. Neurobiology of Disease 8(5), 890–899, 2001

Reid W, Broe G, Creasey H, Grayson D, McCuster E, Bennett H, Longley W, Rose M. Age at onset and pattern of neuropsychological impairment in mild early-stage Alzheimer Disease: A study of a community-based population. Arch Neurol 53 (10), 1056–1061, 1996

Reisberg B, Ferris SH, De Leon MJ, Sinaiko E, Franssen E, Kluger A, Mir P, Borenstein J, George AE, Shulman E et al. Stage-specific behavioral, cognitive, and *in vivo* changes in community residing subjects with age-associated memory impairment and primary degenerative dementia of the Alzheimer type. Drug Development Research 15(2–3), 101–114, 1988

Reisberg B, Ferris SH, de Leon MJ, Crook T. The Global Deterioration Scale for assessment of primary degenerative dementia. A J Psychiatry 139(9), 1136–1139, 1982

Reisberg B, Borenstein J, Salob SP, Ferris SH, Franssen E, Georgotas A. Behavioral symptoms in Alzheimer's disease: phenomenology and treatment. J Clin Psychiatry 48 (Suppl), 9–15, 1987

Reisberg B, Borenstein J, Franssen E, Shulman E, Steinberg G, Ferris SH. Remediable behavioral symptomatology in Alzheimer's disease. Hospital & Community Psychiatry 37(12), 1199–1201, 1986

Reisberg B, Schneider L, Doody R, Anand R, Feldman H, Haraguchi H, Kumar R, Lucca U, Mangone CA, Mohr E et al. Clinical global measures of dementia. Position paper from the International Working Group on Harmonization of Dementia Drug Guidelines. Alzheimer Dis Assoc Disord 11 (Suppl 3), 8–18, 1997

Reisberg B, Doody R, Stoffler A, Schmitt F, Ferris S, Mobius HJ. Memantine Study Group.

Memantine in moderate-to-severe Alzheimer's disease. N Engl J Med 348(14), 1333–1341, 2003

Reischies FM. Leichte kognitive Störung. In: Helmchen Gegenwart, vol. 4, Auflage. Berlin, Germany: Springer, 225–246, 1999

Relkin NR, Tanzi R, Breitner J, Farrer L, Gandy S, Haines J, Hyman B, Mullan M, Poirer J, Strittmatter W, Folstein M et al. Apolipoprotein E genotyping in Alzheimer disease: position statement of the National Institute on aging/Alzheimer's Association Working Group. Lancet 347(9008), 1091–1095, 1996

Reuter-Lorenz P, Jonides J, Smith E, Hartley A, Miller A, Marshuetz C, Koeppe R. Age differences in the frontal lateralization of verbal and spatial working memory revealed by PET. J Cog Neurosci 12, 174–186, 2000

Rey A. L'examen clinique en psychologie. Paris: Presses Universitaires de France, 1964

Richard F, Helbecque N, Neuman E, Guez D, Levy R, Amouyel P. APOE genotyping and response to drug treatment in Alzheimer disease. Lancet 349, 539, 1997

Richards M, Touchon J, Ledesert B, Ritchie K. Cognitive decline in aging: are AAMI and AACD distinct entities? Int J Geriatr Psychiatry 14, 534–540, 1999

Richardson JTE. Mental Imagery and Human Memory. Macmillan, London, 1980

Richardson JT. Imagery mnemonics and memory remediation. Neurology 42, 283–286, 1992

Riemenschneider M, Buch K, Schmolke M, Kurz A, Guder WG. Cerebrospinal protein tau is elevated in early Alzheimer disease. Neurosci Lett 212, 209–211, 1996

Rinne JO, Myllykylä T, Lönnberg P, Marjamäki P. A post-mortem study of brain nicotinic receptors in Parkinson's and Alzheimer disease. Brain Res 547, 167–170, 1991

Pipich DN, Petrill SA, Whitehouse PJ, Ziol EW. Gender differences in language of AD patients: a longitudinal study. Neurology 45, 299–302, 1995

Ritchie J, Touchon J. Mild cognitive mild: Conceptual basis and current nosological status. Lancet 355, 225–228, 2000

Roberts GW, Gentleman SM, Lynch A, Murray L, Landon M, Graham DI. Beta amyloid protein deposition in the brain after severe head injury: implications for the pathogenesis of Alzheimer's disease. J Neurol Neurosurg Psychiatry 57(4), 419–425, 1994

Rocca P, Cocuzza E, Marchiaro L, Bogetto F. Donepezil in the treatment of Alzheimer disease. Long-term efficacy and safety. Prog Neuropsychopharmacol Biol Psychiatry 26, 369–373, 2002

Rocca WA, Hofman A, Brayne C, Breteler MM, Clarke M, Copeland JR, Dartigues JF, Engedal K, Hagnell O, Heeren TJ et al. Frequency and distribution of Alzheimer's disease in Europe: a collaborative study of 1980–1990 prevalence findings. The EURODEM-Prevalence Research Group. Ann Neurol 30(3), 381–390, 1991

Rockwood K, Strang D, MacKnight C, Downer R, Morris JC. Interrater reliability of the Clinical Dementia Rating in a multicenter trial. JAGS 48, 558–559, 2000

Rogaev EI, Sherrington R, Rogaeva EA, Levesque G, Ikeda M, Liang Y, Chi H, Lin C, Holman K, Tsuda T et al. Familial Alzheimer disease in kindreds with missense mutations in a gene on chromosome 1 related to the Alzheimer disease type 3 gene. Nature 376, 775–778, 1995

Rogers SL, Perdomo C, Friedhoff LT. Clinical benefits are maintained during long-term treatment of Alzheimer disease with the acetylcholinesterase inhibitor E2020. Eur Neuropsychopharmacol 5, 386–387, 1995

Rogers J, Kirby LC, Hempelman SR, et al. Clinical trial of indomethacin in Alzheimer disease. Neurology 43, 1609–1611, 1993

Rogers SL, Doody RS, Pratt RD, Ieni JR. Long-term efficacy and safety of donepezil in the treatment of Alzheimer's disease: Final analysis of a US multicenter open-label study. Eur Neuropsychopharmacol 10, 195–203, 2000

Rogers SL, Friedhoff LT. The efficacy and safety of donepezil in patients with Alzheimer's disease results of a US multicenter, randomized, double-blind, placebo-controlled trial. The Donepezil Study Group. Dementia 7, 293–303, 1996

Rogers SL, Friedhoff LT. Long-term efficacy and safety of donepezil in the treatment of Alzheimer's disease: An interim analysis of the results of a US multicenter open label extension study. Eur Neuropsychopharmacol 8, 67–75, 1998

Rogers SL, Farlow MR, Doody RS, Mohs R, Friedhoff LT. A 24–week, double-blind, placebo-controlled trial of donepezil in patients with Alzheimer's disease. Donepezil Study Group. Neurology 50, 136–145, 1998

Roland PE. Brain Activation. New York, Wiley, 1993

Roman GC, Tatemichi TK, Erkinjuntti T, Cummings JL, Masdeu JC, Garcia JH, Amaducci L, Orgogozo JM, Brun A, Hofman A et al. Vascular dementia: diagnostic criteria for research studies. Report of the NINDS-AIREN International Workshop. Neurology 43(2), 250–260, 1993

Romas SN, Mayeux R, Tang M-X, Lantigua R, Medrano M, Tycko B, Knowles J. No association between a presenilin 1 polymorphism and Alzheimer disease. Arch Neurol 57, 699–702, 2000

Ron Brookmeyer et al. Projections of Alzheimer disease in the United States and the Public Health Impact of Delaying Disease Onset. Am J Pub Health 88, 1337, 1998

Rosen WG, Mohs RC, Davis KL. A new rating scale for Alzheimer disease. Am J Psychiatry 141, 1356–1364, 1984

Rosen WG, Terry RD, Fuld PA, Katzman R, Peck A. Pathological verification of ischemic score in differentiation of dementias. Ann Neurol 7(5), 486–488, 1980

Rosenberg RN, Baskin F, Fosmire JA, Risser R, Adams P, Sverlik D, Honig LS, Cullum M, Weiner MF. Altered amyloid protein processing in platelets of patients with Alzheimer disease. Arch Neurol 54, 139–144, 1997

Roses AD, Devlin B, Conneally PM et al. Measuring the genetic contribution of APOE in late-onset Alzheimer disease. Am J Hum Genet 57 (Suppl), A202, 1995

Roses AD. Apolipoprotein E genotyping in the differential diagnosis, not prediction of Alzheimer disease. Ann Neurol 38, 6–14, 1995

Rösler M, Anand R, Cicin-Sain A , Gauthier S, Agid Y, Dal-Bianco P, Stahelin HB, Hartman R, Gharabawi M. Efficacy and safety of rivastigmine in patients with Alzheimer disease: international randomised controlled trial. BMJ 318, 633–638, 1999

Rosser AE, Hodges JR. The Dementia Rating Scale in Alzheimer's disease, Huntington's disease and progressive supranuclear palsy. J Neurol 241, 531–536, 1994

Rossor M. Alzheimer's disease. BMJ 307, 779–782, 1993

Roth M, Tym E, Mountjoy CQ, Huppat SA, Hendrie H. CAMDEX: A standardized instrument for the diagnosis of mental disorders in the elderly with special reference to the early detection of dementia. Br J Psychiatr 149, 698–709, 1986

Roth M, Huppert FH, Tym E, Mountjoy CQ. CAMDEX, the Cambridge Examination for Mental Disorders of the Elderly. Cambridge University Press, Cambridge, 1988

Roth M, Mountjoy CL, Amrein R and the International Collaborative Study Group. Moclobamide in elderly patients with cognitive decline and depression: an international double-blind, placebo-controlled trial. Br J Psychiatry 168 (2), 149–157, 1996

Rubin EH, Morris JC, Grant EA, Vendegna T. Very mild senile dementia of the Alzheimer type. I. Clinical assessment. Arch Neurol 46, 379–382, 1989

Rubin DC. Autobiographical Memory. New York. Cambridge University Press, 1986

Rubin EH, Storandt M, Miller JP, Grant EA, Kinscherf DA, Morris JC , Berg L. Influence of age on clinical and psychometric assessment of subjects with very mild or mild dementia of the Alzheimer type. Arch Neurol 50, 380–383, 1993

Rubin DC, Schulkind MD. Properties of word cues for autobiographical memory. Psychology Reports 81, 47–50, 1997

Rubin DC, Baddeley AD. Telescoping is not time compression: A model of the dating of autobiographical events. Memory & Cognition 17, 653–661, 1989

Rubin DC. In: CP Thompson, DJ Herrman, D Bruce, JD Reed, DG Payne, MP Toglia (eds) Autobiographical Memory. Theoretical and Applied Perspectives. Erlbaum, Mahwah, NJ, 47–67, 1998

Rubin EH, Storandt M, Miller JP, Kinscherf DA, Grant EA, Morris JC, Berg L. A prospective study of cognitive function and onset of dementia in cognitively healthy elders. Arch Neurol 55(3), 395–401, 1998

Rudolph RL, Derivan AT. The safety and tolerability of venlafaxine hydrochloride: analysis of the clinical trial database. J Clin Psychopharmacol 16 (Suppl 2), S54–S59, 1996

Ruether E, Ritter R, Apecechea M, Freytag S, Windisch M. Efficacy of the peptidergic nootropic drug Cerebrolysin in patients with senile dementia of the Alzheimer type (SDAT). Pharmacopsychiatry 27 (1), 32–40, 1994

Ruether E, Ritter R, Apecechea M, Freytag S, Gmeinbauer R, Windischh M. Sustained improvements in patients with dementia of Alzheimer's type (DAT) – months after termination of Cerebrolysin therapy. J Neural Transm 107, 815–829, 2000

Ruether E, Husmann R, Kinzler E, Diabl E, Klinger D, Spatt J, Ritter R, Schmidt R, Taneri Z, Winterer W, Koper D, Kasper S, Rainer M, Moessler H. A 28-week, double-blind, placebo-controlled study with Cerebrolysin in patients with mild to moderate Alzheimer's disease. Int Clin Pharmacol 16, 253–263, 2001

Rugg MD, Fletcher PC, Chua PM, Dolan RJ. The role of the prefrontal cortex in recognition memory and memory for source. Neuroimage 10, 520–529, 1999

Russo C, Angelini G, Dapino D, Piccini A, Piombo G, Schettini G, Chen S, Teller JK, Zaccheo D, Gambetti P, Tabaton M. Opposite roles of apolipoprotein E in normal brains and in Alzheimer's disease. Proc Natl Acad Sci USA 95(26), 15598–15602, 1998

Rusted J, Sheppard L. Action-based memory in Alzheimer disease: a longitudinal look at tea making. Neurocase 8, 111–126, 2002

Sagar HJ, Cohen NJ, Sullivan EV, Corkin S, Growdon JH. Remote memory function in Alzheimer's disease and Parkinson's disease. Brain 111 (Pt 1), 185–206, 1988

Sahadevan S, Lim JPP, Tan NJL, Chan SW. Psychometric identification of early Alzheimer disease in an elderly Chinese population with differing educational levels. Alzheimer Dis Assoc Disord 16 (2), 65–72, 2002

Sahakian BJ, Morris RG, Evenden JL, Heald A, Levy R, Philpot M, Robbins TW. A comparative study of visuospatial memory and learning in Alzheimer-type dementia and Parkinson's disease. Brain 111 (Pt 3), 695–718, 1988

Sahakian B, Jones G, Levy R, Gray J, Warburton D. The effects of nicotine on attention, information processing and short-term memory in patients with dementia of Alzheimer type. Br J Psychiat 154, 797–800, 1989

Salmon DP, Shimamura AP, Butters N, Smith S. Lexical and semantic priming deficits in patients with Alzheimer disease. J Clin Exp Neuropsychology 10, 477–494, 1988

Salmon DP, Thal LJ, Butters N, Heindel WC. Longitudinal evaluation of dementia of the Alzheimer type: A comparison of 3 standardized mental status examinations. Neurology 40, 1225–1230, 1990

Salmon DP, Galasko D, Hansen LA, Masliah E, Butters N, Thal L et al. Neuropsychological deficits associated with diffuse Lewy body disease. Brain Cogn 31, 148–165, 1996

Salmon DP, Kwo-on-Yuen P, Heindel WC, Butters N, Thal LJ. Differentiation of Alzheimer's disease and Huntington's disease with the Dementia Rating Scale. Arch Neurol 46, 1204–1208, 1989

Samuel W, Terry RD, DeTeresa R, Butters N, Masliah E. Clinical correlates of cortical and nucleus basalis pathology in Alzheimer dementia. Arch Neurol 51(8), 772–778, 1994

Sano M, Morris J, Thal L and Members of the Alzheimer's Disease Cooperative Study. San Diego: Developing outcome measures for clinical trials by the Alzheimer's Disease Cooperative Study (ADCS): Standardized training for the Clinical Dementia Rating (CDR), initial results. Neurology 43, A291–A292, 1993

Salthouse TA. What do adult age differences in the digit symbol substitution test reflect? J Gerontol 47, P121–P128, 1994

Salthouse TA. The aging of working memory. Neuropsychology 8, 535–543, 1994

Sanders HI, Warrington EK. Retrograde amnesia in organic amnesic patients. Cortex 11, 397–400, 1975

Sano M, Ernesto C, Thomas RG, Klauber MR, Schafer K, Grundman M, Woodbury P, Growdon J, Cotman CW, Pfeiffer E et al. A controlled trial of selegiline, alpha-tocopherol, or both as treatment for Alzheimer's disease. The Alzheimer's Disease Cooperative Study. [comment]. N Engl J Med 336(17), 1216–1222, 1997

Satou T, Imano M, Akai F, Hashimoto S, Itoh T, Fujimoto M. Morphological observation of effects of Cere on cultured neural cells. Adv Biosci 87, 195–196, 1993

Saunders AM, Strittmatter WJ, Schmechel D, St George-Hyslop PH, Pericak-Vance MA, Joo SH, Rosi BL, Gusella JF, Crapper-MacLachlan DR, Alberts MJ et al. Association of apolipoprotein E allele epsilon 4 with late-onset Alzheimer disease. Neurology 43, 1467–1472, 1993

Schacter DL, Chiu CVP, Ochsner KN. Implicit memory: a selective review. Ann Rev Neuroscie 16, 159–182, 1993

Schacter DL, Buckner RL, Koutstaal W, Dale AM, Rosen BR. Late onset of anterior prefrontal activity during true and false recognition: an event-related fMRI study. Neuroimage 6, 259–269, 1997

Schacter DL. Memory. In: MI Posner (ed), Foundations of Cognitive Science. Cambridge MA: MIT Press, 1989

Schacter DL. Understanding implicit memory. Am Psychol 47, 559–569, 1992

Schacter DL, Rich SA, Stampp MS. Remediation of memory disorders: Experimental evaluation of the spaced-retrieval technique. J Clin Exp Neuropsychology 7, 79–96, 1985

Schacter DL, Glisky EL. Memory remediation: restoration, alleviation and the acquisition of domain-specific knowledge. In: Uzzell BP, Gross Y (eds), Clinical Neuropsychology of Intervention, Boston, Nijhoff, 257–282, 1986

Scheltens P, Leys D, Barkhof F, Huglo D, Weinstein HC, Vermersch P, Kuiper M, Steinling M, Wolters EC, Valk J. Atrophy of medial temporal lobes on MRI in *probable* Alzheimer's disease and normal ageing: diagnostic value and neuropsychological correlates. J Neurol Neurosurg Psychiatry 55(10), 967–972, 1992

Schenk D, Barbour R, Dunn W, Gordon G, Grajeda H, Guido T, Hu K, Huang J, Johnson-Wood K, Khan K et al. Immunization with amyloid-b attenuates Alzheimer-disease-like pathology in the PDAPP mouse. Nature 400, 173–177, 1999

Scheuner D, Eckman C, Jensen M, Song X, Citron M, Suzuki N, Bird TD, Hardy J, Hutton M, Kukuli W et al. Secreted amyloid beta-protein similar to that in the senile plaques of Alzheimer's disease is increased *in vivo* by the presenilin 1 and 2 and APP mutations linked to familial Alzheimer's disease. Nat Med 2(8), 864–870, 1996

Schmajuk NA, DiCarlo JJ. Stimulus configuration, classical conditioning, and hippocampal function. Psychological Rev 99(2), 268–305, 1992

Schmidt R, Freidl W, Fazekas F, Reinhart B, Grieshofer P, Koch M, Eber B, Schumacher M, Polmin K, Lechner H. The Mattis Dementia Rating Scale: normative data from 1,001 healthy volunteers. Neurology 44(5), 964–966, 1994

Schmolck H, Squire LR. Impaired perception of facial emotions following bilateral damage to the anterior temporal lobe. Neuropsychology 15 (1), 30–38, 2001

Schneider LS, Olin JT. Clinical global impressions in Alzheimer's clinical trials. Int Psychogeriat 8, 277–288, 1996

Schneider LS, Olin JT. Clinical global impressions in Alzheimer's clinical trials. Int Psychogeriatr 277–288, discussion 288–290, 1996

Schneider LS. Assessing outcomes in Alzheimer disease. Alzheimer disease and Associated Disorders 15 (1), S8–S18, 2001

Schneider L, Pollock VE, Lyness SA. A metaanalysis of controlled trials of neuroleptic treatment in dementia. J Am Geriatr Soc 38, 553–563, 1990

Schneider LS, Olin JT, Doody RS, Clark CM, Morris JC et al. Validity and reliability of Alzheimer's disease cooperative study-clinical global impression of change. Alzheimer Dis Assoc Disord 11 (Suppl 2), S22–S32, 1997

Schroder H, Giacobini E, Wevers A, Birtsch A, Schutz U. Nicotinic receptors in Alzheimer disease. In: EF Dominino (ed), Brain Imaging of Nicotine and Tobacco Smoking. Npp Books, Ann Harbor, MI 48106, 73–93, 1995

Schuff N, Amend D, Ezekiel F, Steinman SK, Tanabe J, Norman D, Jagust W, Kramer JH, Mastrianni JA, Fein G, Weiner MW. Changes of hippocampal N-acetyl aspartate and volume in Alzheimer disease. Neurology 49, 1513–1521, 1997

Schwartz J-C, Levesque D, Martres M-P, Sokoloff P. Dopamine D3 receptor: Basic and clinical aspects. Clin Neuropharmacol 16(4), 295–314, 1993

Scinto LF, Daffner DR. Early Diagnosis of Alzheimer Disease. Humana Press: NJ, 2000

Sclan SG, Saillon A, Franssen E, Hugonot-Diener L, Saillon A, Reisberg B. The Behavioral

Pathology in Alzheimer disease Rating Scale (BEHAVE-AD): Reliability and analysis of symptom category scores. Int J Geriat Psychiatry 11, 819–830, 1996
Scogin F, Richard HC, Keith S, McElreath I. Progressive and imaginal relaxation training for elders with subjective anxiety. Psychol Aging 7, 3–12, 1992
Scoville WB, Milner B. Loss of recent memory after bilateral hippocampal lesions. J Neurol Neuroscurg Psychiatry 20, 11–21, 1957
Sebastian MV, Menor J, Elosua R. Patterns of errors in short-term forgetting in AD and ageing. Memory 9, (4/5/6), 223–231, 2001
Segovia G, Porras A, Del Arco A, Mora F. Glutamatergic neurotransmission in aging: a critical perspective. Mech Aging Dev 122, 1–29, 2001
Seiger A, Nordberg A, von Holst H, Backman L, Ebendal T, Alafuzoff I, Amberla K, Hartvig P, Herlitz A, Lilja A et al. Intracranial infusion of purified nerve growth factor to an Alzheimer patient: the first attempt of a possible future treatment strategy. Behav Brain Res 57(2), 255–261, 1993
Selai CE, Trimble MR. Assessing quality of life in dementia. A review. Aging Ment Health,3, 101–111, 1999
Selai CE, Trimble MR, Rossor MN, Harvey RJ. The Quality of Life Assessment Schedule (QOLAS) – A new method for assessing quality of life in dementia. In: SM Albert, RG Logsdon (eds), Assessing Quality of Life in Alzheimer's Disease. New York: Springer Publishing Company, 31–48, 2000
Seletti B, Benkelfat C, Blier P, Annable L, Gilbert F, de Montigny C. Serotonin 1A receptor activation by flesonoxan in humans: body temperature and neuroendocrine responses. Neuropsychopharmacology 13, 93–104, 1995
Selkoe DJ. The cell biology of beta-amyloid precursor protein and presenilin in Alzheimer's disease. Trends Cell Biol 8, 447–453, 1998
Selkoe DJ. Clearing the brain's amyloid cobwebs. Neuron 32, 177–180, 2001
Selkoe DJ. Alzheimer disease: genes, proteins and therapy. Physiol Rev 81, 741–766, 2001
Selkoe DJ. Cell biology of the amyloid β-protein precursor and the mechanism of Alzheimer disease. Ann Rev Cell Biology 10, 373–403, 1994
Semon RS. The Mneme, Allen & Unwin, 1921
Seshadri S, Drachman DA, Lippa CF. Apolipoprotein E4 allele and the lifetime risk of Alzheimer disease. Arch Neurol 52, 1074–1079, 1995
Shallice T, Fletcher P, Frith CD, Grasby P, Frackowiak RS, Dolan RJ. Brain regions associated with acquisition and retrieval of verbal episodic memory. Nature 368(6472), 633–635, 1994
Shankar KK, Walker M, Frost D, Orrell MW. The development of a valid and reliable scale for rating anxiety in dementia (RAID). Aging Mental Health 3, 39–49, 1999
Shapira J, Cummings JL. Alzheimer disease: Changes in sexual behavior. Medical aspects of Human Sexuality, June, 32–36, 1989
Shen J, Bronson RT, Chen DF, Xia W, Selkoe DJ, Tonegawa S. Skeletal and CNS defects in presenilin-1-deficient mice. Cell 89, 629–639, 1997
Shenaut GK, Ober BA. Methodological control of semantic priming in Alzheimer's disease. Psychol Aging 11, 443–448, 1996
Sheng JG, Mrak RE, Griffin WST. Distribution of interleukin-1-immunoreactive microglia in

cerebral cortical layers: implications for neuritic plaque formation in Alzheimer disease. Neuropathol Appl Neurol 24, 278–283, 1998
Sheng JG, Mrak RE, Griffin WST. Glial-neuronal interactions in Alzheimer disease: progressive association of IL-1α microglia and S100β astrocytes with neurofibrillary tangle stages. J Neuropathol Exp Neurol 56, 285–290, 1997
Sheng JG, Zhou XQ, Mrak RE, Griffin WST. Progressive neuronal injury associated with neurofibrillary tangle formation in Alzheimer disease. J Neuropathol Exp Neurol 57, 323–328, 1998b
Sheng JG, Mrak RE, Griffin WST. Interleukin-1α expression in brain regions in Alzheimer disease: correlation with neuritic plaque distribution. Neuropathol App Neurol 21, 290–301, 1995
Sherrington R, Rogaev EI, Liang Y, Rogaeva EA, Levesque G, Ikeda M, Chi H, Lin C, Li G, Holman K et al. Cloning of a gene bearing missense mutations in early-onset familial Alzheimer's disease [comment]. Nature 375(6534), 754–760, 1995
Shibata N, Ohnuma T, Takahashi T, Baba H, Ishizuka T, Ohtsuka M, Ueki A, Nagao M, Arai H. Effect of IL-6 polymorphism on risk of Alzheimer disease: Genotype-phenotype association study in Japanese cases. Am J Med Genet (Neuropsychiatric Genetics) 114, 436–439, 2002
Sigurdsson EM, Lee JM, Dong XW, Hejna MJ, Lorens SA. Bilateral injections of amyloid-b 25–35 into the amygdala of young Fisher rats: behavioral, neurochemical, and time dependent histopathological effects. Neurobiol Aging 18, 591–608, 1997
Sigurdsson EM, Scholtzova H, Mehta PD, Frangione B, Wisniewski T. Immunization with a nontoxic/nonfibrillar amyloid-b homologous peptide reduces Alzheimer's disease-associated pathology in transgenic mice. Am J Pathol 159, 439–447, 2001
Silva AJ, Kogan JH, Frankland PW, Kida S. CREB and memory. Ann Rev Neurosci 21, 127–148, 1998
Silwinski M, Lipton RB, Buschke H and Stewart W. The effects of preclinical dementia on estimates of normal cognitive functioning. J Gerontol 51B, P217–P225, 1996
Simons JS, Graham KS, Galton CJ, Patterson K, Hodges JR. Semantic knowledge and episodic memory for faces in semantic dementia. Neuropsychology 15, 101–114, 2001
Simons M, Keller P, Dichgans J and Schulz JB. Cholesterol and Alzheimer's disease: is there a link? Neurology 57, 1089–1093, 2001
Simpson PM, Surmont DJ, Wesnes KA and Wilcock GK. The Cognitive Drug Research Computerized Assessment System for the demented patients: A validation study. Int J Geriatric Psychiatry 6, 95–102, 1991
Simpson PM, Wesnes KA, Christmas L. A computerized system for the assessment of drug-induced performance changes in young, elderly and demented populations. Brit J Clin Pharmac 27, 711–712P, 1989
Singh VK. Neuroautoimmunity: pathogenic implications for Alzheimer's disease. Gerontology 43, 79–94, 1997
Sinha S, Anderson JP, Barbour R, Basi GS, Caccavello R, Davis D, Doan M, Dovey HF, Frigon N, Hong J et al. Purification and cloning of amyloid precursor protein β-secretase from human brain. Nature 402, 537–540, 1999
Sirigu A, Duhamel JR, Cohen L, Pillon B, Dubois B, Agid Y. The mental representation of and movements after parietal cortical damage. Science 273, 1564–1568, 1996

Sjögren M, Hesse C, Basun H, Köl G, Thostrup H, Kilander L, Marcusson J, Edman A, Wallin A, Karlsson I et al. Tacrine and rate of progression in Alzheimer's disease – relation to ApoE allele genotype. J Neural Transm 108, 451–458, 2001

Skoog I, Kalaria RN, Breteler MMB. Vascular factors and Alzheimer disease. Alzheimer Dis Assoc Disord 13 (Suppl 3), 106–114, 1999

Slooter JC, van Duijn CM. Genetic epidemiology of Alzheimer disease. Epidemiol Rev 19, 107–119, 1997

Small GW, Rabins PV, Barry PP, Buckholtz NS, DeKosky ST, Ferris SH, Finkel SI, Gwyther LP, Khachaturian ZS, Lebowitz BD et al. Diagnosis and treatment of Alzheimer disease and related disorders. Consensus statement of the American Association for Geriatric Psychiatry the Alzheimer's Association, and the American Geriatrics Society. JAMA 278(16), 1363–1371, 1997

Small GW, Kuhl DE, Riege WH, Fujikawa DG, Ashford JW, Metter EJ, Mazziotta JC. Cerebral glucose metabolic patterns in Alzheimer's disease. Effect of gender and age at dementia onset. Arch Gen Psychiatry 46(6), 527–532, 1989

Small GW. The Memory Bible: An Innovative Strategy for Keeping the Brain Young. London: Penguin, 2002

Small GW. Neuroimaging and genetic assessment for early diagnosis of Alzheimer's disease. J Clin Psychiatry 57 (Suppl 14), 9–13, 1996

Small SA, Perera GM, DeLaPaz R, Mayeux R, Stern Y. Differential regional dysfunction of the hippocampal formation among elderly with memory decline and Alzheimer's disease. Ann Neurol 45(4), 466–472, 1999

Small BJ, Fratiglioni L, Viitanen M, Winblad B and Beckman L. The course of cognitive impairment in preclinical Alzheimer disease. Three- and 6-year follow-up of a population-based sample. Arch Neurol 57, 839–844, 2000

Small BJ, Herlitz A, Fratiglioni L, Almkvist O, Backman L. Cognitive predictors of incident Alzheimer's disease: a prospective longitudinal study. Neuropsychology 11(3), 413–420, 1997

Smith MA, Hirai K, Hsiao K, Pappolla MA, Harris PL, Siedlak SL, Tabaton M, Perry G. Amyloid-beta deposition in Alzheimer transgenic mice is associated with oxidative stress. J Neurochemistry 70(5), 2212–2215, 1998

Smith EE, Jonides J, Koeppe RA, Awh E, Schumacher EH and Minoshima S. Spatial *versus* object working memory: PET investigations. J Cogn Neurosci 7, 337–356, 1995

Smith EE, Jonides J, Koeppe RA. Dissociating verbal and spatial working memory using PET. Cereb Cortex 6, 11–20, 1996

Smith MA, Sayre LM, Monnier VM, Perry G. Oxidative posttranslational modifications in Alzheimer disease – a possible pathogenic role in the formation of senile plaques and neurofibrillary tangles. Mol Chem Neuropathol 28, 41–48, 1996

Smith CD, Malcein M, Meurer K, Schmitt FA, Markesbery WR, Pettigrew LC. MRI temporal lobe volume measures and neuropsychologic function in Alzheimer disease. J Neuroimaging 9 (1), 2–9, 1999

Snaedal I, Johannesson T, Jonsson JE, Gylfadottir G. The effects of nicotine in dermal plaster on cognitive functions in patients with Alzheimer's disease. Dementia 7, 47–52, 1996

Snowden JS, Goulding PJ, Neary D. Semantic dementia: a form of circumscribed cerebral atrophy. Behav Neurol 2, 167–182, 1989

Snowden JS, Griffiths HL, Neary D. Semantic-episodic memory interactions in semantic dementia: implications for retrograde memory function. Cognitive Neuropsychology 13, 1101–1137, 1996

Snowden JS, Neary D, Mann DMA. Fronto-temporal lobal degeneration: fronto-temporal dementia, progressive aphasia, semantic dementia. Clinical neurology and neurosurgery monographs. New York: Churchill Livingstone, 1996

Soininen H, Partanen VJ, Nelkala EL, Riekinen PJ. EEG findings in senile dementia and normal aging. Acta Neurol Scand 65, 59–70, 1982

Solomon PR, Herschoff A, Kelly B, Relin M, Brush M, De Veaux RD, Pendleburg WW. A 7 minute neurocognitive screening battery highly sensitive to Alzheimer disease. Arch Neurol 55, 349–355, 1998

Solomon PR, Pomerleau D, Bennett L, James J, Morse DL. Acquisition of the classically conditioned eyeblink response in humans over the lifespan. Psychol Aging 4, 34–41, 1989

Sparks DL, Kuo YM, Roher A, Martin T, Lukas RJ. Alterations of Alzheimer's disease in the cholesterol-fed rabbit, including vascular inflammation. Preliminary observations. Ann NY Acad Sci 903, 335–344, 2000

Spear NE. The Processing of Memories: Forgetting and Retention. Erlbaum, Hillsdale, New Jersey, 1978

Spear NE, Riccio DC. Memory: Phenomena and Principles. Allyn and Bacon, Needham Heights, Massachusetts, 1994

Spellacy F, Spreen O. A short form of the Token test. Cortex 5, 390–397, 1969

Spinnler H, Tognoni G. Standardizzazione e taratura italiana di test neuropsicologici. Ital J Neurol Sci (suppl. 8), 1987

Spinnler H, Della Sala S, Bandera R, Baddeley AD. Dementia, ageing, and the structure of human memory. Cognitive Neuropsychology 5, 193–211, 1988

Squire LR, Zola-Morgan S. The medial temporal lobe memory system. Science 253, 1380–1386, 1991

Squire LR, Alvarez P. Retrograde amnesia and memory consolidation: a neurobiological perspective. Curr Opin Neurobiol 5, 169–177, 1995

Squire LR. Memory and Brain. Oxford Univ Press, New York, 1987

Squire LR, Schmolck H, Stark SM. Impaired auditory recognition memory in amnesic patients with medial temporal lobe lesions. Learning & Memory 8, 252–256, 2001

Stahl SM. The new cholinesterase inhibitors for Alzheimer disease. Part 2. J Clin Psychiatry 61, 813–814, 2000

Stahl SM. Paying attention to your acetylcholine, pt. 1: structural organization of nicotinic receptors. J Clin Psychiatry 61, 547–548, 2000

Stahl SM. Paying attention to your acetylcholine, pt. 2: the function of nicotinic receptors. J Clin Psychiatry 61, 628–629, 2000

Stadelmann C, Bruck W, Bancher C, Jellinger K, Lassmann H. Alzheimer-disease-DNA fragmentation indicates increased neuronal vulnerability, but not apoptosis. Neuropathol Exp Neurol 57, 456–464, 1998

Stark CEL, Squire LR. Simple and associative recognition memory in the hippocampal region. Learning & Memory 8, 190–197, 2001

Stark M, Coslett HB, Saffran EM. Impairment of an egocentric map of locations: Implications for perception and action. Cognitive Neuropsychology 13, 481–523, 1996

Starkstein SE, Petracca G, Chemerinski E, Kremer J. Syndromic validity of apathy in Alzheimer's disease. Am J Psychiatry 158(6), 872–877, 2001

Steckler T, Sahgal A. The role of serotonergic – cholinergic interactions in the mediation of cognitive behaviour. Behav Brain Res 67, 165–199, 1995

Stern CE, Corkin S, Gonzalez RG, Guimaraes AR, Baker JR, Jennings PJ et al. The hippocampal formation participates in novel pisture encoding: evidence from functional magnetic resonance imaging. Proc Natl Acad Sci USA 93, 8660–8665, 1996

Stern Y, Sano M, Mayeux R. Effects of oral physostigmine in Alzheimer disease. Am J Psychiatry 142, 28–33, 1985

Stern Y, Liu X, Albert M. Brandt J. Jacobs DM. Del Castillo-Castaneda C. Marder K. Bell K. Sano M. Bylsma F. Modeling the influence of extrapyramidal signs on the progression of Alzheimer disease. Arch Neurol 53, 1121–1126, 1996

Stern Y, Mayeux R, Sano M, Hauser WA, Bush T. Predictors of disease course in patients with probable Alzheimer disease. Neurology 37, 1649–1653, 1987

Stern RG, Mohs RC, Davidson M, Schmeidler J, Silverman J, Kramer-Ginsberg E, Searcey T, Bierer L, Davis KL. A longitudinal study of Alzheimer disease: Measurement, rate and predictors of cognitive deterioration. Am J Psychiatry 151, 390–396, 1994

Stewart AL, Sherbourne CD, Brod M. Measuring Health-Related Quality of Life in older and Demented populations. In: B Spilker (ed), Quality of Life and Pharmacoeconomics in Clinical Trials. Philadelphia, PA. Lippincott-Raven Publishers, 1996

Stewart WF, Kawas C, Corrada M, Metter EJ. Risk of Alzheimer disease and duration of NSAIDs use. Neurology 48 (3), 626–632, 1997

Stigsby B, Johannesson G, Ingvar DH. Regional EEG analysis and regional cerebral blood flow in Alzheimer's and Pick's disease. Electroencephalogr Clin Neurophysiol 51, 537–547, 1981

St John P, Montgomery P. Are cognitively intact seniors with subjective memory loss more likely to develop dementia. Int J Geriatr Psychiatry 17, 814–820, 2002

Stoppe G, Brandt CA, Staedt JH. Behavioural problems associated with dementia: the role of newer antipsychotics. Drugs-Aging 14, 41–54, 1999

Storandt M, Hill R. Very mild senile dementia of the Alzheimer type. II. Psychometric test performance. Arch Neurol 46, 383–386, 1989

Storandt M. Age, ability level and method of administering and scoring the WAIS. J Gerontolo 32, 175–178, 1977

Strittmatter WJ, Saunders AM, Schmechel D, Pericak-Vance M, Enghild J, Salvesen GS, Roses AD. Apolipoprotein E: high-avidity binding to beta-amyloid and increased frequency of type 4 allele in late-onset familial Alzheimer disease, Proc Natl Acad Sci USA 90, 1977–1981, 1993

Strittmatter WJ, Weisgraber KH, Huang DY, Dong LM, Salvesen GS, Pericak-Vance M, Schmechel D, Saunders AM, Goldgaber D, Roses AD. Binding of human apolipoprotein E to synthetic amyloid beta peptide: isoform-specific effects and implications for late-onset Alzheimer disease. Proc Natl Acad Sci USA 90(17), 8098–8102, 1993

Stroop JR. Studies of interference in serial verbal reactions. J Exp Psychol 18, 643–662, 1935

Sturchler-Pierrat C, Abramowski D, Duke M, Wiederhold KH, Mistl C, Rothacher S, Ledermann B, Burki K, Frey P, Paganetti PA et al. Two amyloid precursor protein transgenic mouse models with Alzheimer disease-like pathology. Proc Natl Acad Sci USA 94(24), 13287–13292, 1997

Sullivan EV, Corkin S, Growden JH. Verbal and nonverbal short-term memory in patients with Alzheimer disease and in healthy elderly subjects. Dev Neuropsychol 2, 387–400, 1986

Sultzer DL. Neuroimaging and the origin of psychiatric symptoms in dementia. Int Psychogeriatr 8, 239–243, 1996

Sunderland T, Molchan S, Lawlor B, Martinez R, Mellow A, Martinson H, Putnam K, Lalonde F. A strategy of "combination chemotherapy" in Alzheimer's disease: rationale and preliminary results with physostigmine plus deprenyl. Int Psychogeriatr 4 (Suppl 2), 291–309, 1992

Sunderland T, Linker G, Mirza N, Putnam KT, Friedman DL, Kimmel LH, Bergeson J, Manetti GJ, Zimmermann M, Tang B, Bartko JJ, Cohen RM. Decreased β-amyloid1-42 and increased Tau levels in cerebrospinal fluid of patients with Alzheimer disease. JAMA 289, 2094–2103, 2003

Suzuki N, Iwatsubo T, Odaka A, Ishibashi Y, Kitada C, Ihara Y. High tissue content of soluble β1-40 is linked to cerebral amyloid angiopathy. Am J Pathol 145, 452–460, 1994

Suzuki N, Hardebo JE. The cerebrovascular parasympathetic innervation. Cerebrovasc Brain Metab Rev 5, 33–46, 1993

Sweatt JD. Memory mechanisms: the yin and yang of protein phosphorylation. Curr Biol 11, 391–394, 2001

Tabert MH, Albert SM, Borukhova-Milov L, Camacho Y, Pelton G, Liu X, Stern Y, Devanand DP. Functional deficits in patients with mild cognitive impairment. Prediction of AD. Neurology 58, 758–764, 2002

Talbot PR, Lloyd JJ, Snowden JS, Neary D, Testa HJ. A clinical role for 99mTc-HMPAO SPECT in the investigation of dementia? J Neurol Neurosurg Psychiatry 64(3), 306–313, 1998

Tariot PN, Cohen RM, Sunderland T, Newhouse PA, Yount D, Mellow AM, Weingartner H, Mueller EA, Murphy DL. L-deprenyl in Alzheimer's disease. Preliminary evidence for behavioral change with monoamine oxidase B inhibition. Arch Gen Psychiatry 44(5), 427–433, 1987

Tan ZS, Seshadri S, Beiser A, Wilson PWF, Kiel DP, Tocco M, D'Agostino RB, Wolf PA. Plasma total cholesterol level as a risk factor for Alzheimer disease. Arch Intern Med 163, 1053–1057, 2003

Tang M-X, Stern Y, Marder K, Bell K, Gurland B, Lantigua R, Andrews H, Feng L, Tycko B, Mayeux R. The APOE-epsilon4 allele and the risk of Alzheimer disease among African Americans, whites, and Hispanics. J Am Med Assoc 279(10), 751–755, 1998.

Tanzi RE, Gusella JF, Watkins PC, Bruns GA, St George-Hyslop P, Van Keuren ML, Patterson D, Pagan S, Kurnit DM, Neve RL. Science 235, 880–884, 1987

Tanzi RE. A genetic dichotomy model for the inheritance of Alzheimer disease and common age related disorders. J Clin Invest 104, 1175–1179, 1999

Tariot PN, Blazina L. The psychopathology of dementia. In: Handbook of Dementing Illnesses. New York, NY: Marcel Dekker Inc, 461–475, 1993

Tariot PN, Mack JL, Patterson MB, Edland SD, Weiner MF, Fillenbaum G, Blazina L, Teri L, Rubin E, Mortimer JA, Stern Y. Behavioral Pathology Committee of the Consortium to

Establish a Registry for Alzheimer's Disease. The Behavior Rating Scale of the Consortium to Establish a Registry for Alzheimer's Disease. Am J Psychiatry 152, 1349–1357, 1995

Tariot PN, Cummings JL, Katz IR, Mintzer J, Perdomo CA, Schwam EM, Whalen E. A randomized, double-blind, placebo-controlled study of the efficacy and safety of donepezil in patients with Alzheimer's disease in the Nursing Home Setting. JAGS 49, 1590–1599, 2001

Tariot PN, Solomon PR, Morris JC, Kershaw P, Lilienfeld S, Ding C and the Galantamine USA-10 Study Group. A 5–month, randomized, placebo-controlled trial of galantamine in AD. Neurology 54, 2269–2276, 2000

Taylor EM. The Appraisal of Children with Cerebral Deficits. Cambridge (MA): Harvard University Press, 1959

Tejani-Butt SM, Yang J, Pawlyk AC. Altered serotonin transporter sites in Alzheimer disease raphe and hippocampus. NeuroReport 6, 1207–1210, 1995

Teng EL, Hasegawa K, Homma A et al. The Cognitive Abilities Screening Instrument (CASI): a practical test for cross-cultural epidemiological studies of dementia. Int Psychogeriatr, 6, 45–58, 1994

Teresi J, Evans D. Cognitive assessment measures for chronic care populations. J Mental Health and Aging 2, 151–174, 1996

Teri L, Logsdon RG. Identifying pleasant activities for Alzheimer's disease patients: the Pleasant Events Schedule-AD. Gerontologist 31, 124–127, 1991

Teri L, Truax P, Logsdon RG et al. Assessment of behavioral problems in dementia: The Revised Memory and Behavior Problems Checklist. Psychol Aging 7, 622–631, 1992

Teri L, Logsdon RG, Uomoto J, McCurry SM. Behavioural treatment of depression in dementia patients: a controlled clinical trial. J Gerontol 52B (4), 159–166, 1997

Terry RD, Katzman R. Senile dementia of the Alzheimer's type. Ann Neurol 14, 497–506, 1983

Teunisse S, Derix MM, van Crevel H. Assessing the severity of dementia. Patient and caregiver. Arch Neurol 48, 274–277, 1991

Thal LJ, Masur DM, Blau AD, Fuld PA, Klauber MR. Chronic oral physostigmine without lecithin improved memory in Alzheimer disease. J Am Geriatr Soc 37, 42–48, 1989

Thal LJ, Lasker B, Sharpless NS, Bobotas G, Schor JM, Nigalye A. Plasma physostigmine concentrations after controlled-release oral administration (letter). Arch Neurol 46, 13, 1989

The Neuroinflammation Working Group. Inflammation and Alzheimer's disease. Neurobiol Aging 21, 383–421, 2000

Thibault O, Porter NM, Chen KC, Blalock EM, Kaminker PG, Clodfelter GV, Brewer LD, Landfield PW. Calcium dysregulation in neuronal aging and Alzheimer's disease: history and new directions. Cell Calcium 24, 417–433, 1998

Thompson CP, Skowronski JJ, Lee DJ. Telescoping in dating naturally occurring events. Memory & Cognition 16, 461–468, 1988

Thomsen T, Bickel U, Fischer JP, Kewitz H. Galantamine hydrobromide in a long-term treatment of Alzheimer disease. Dementia 1, 46–51, 1990

Tiberghien G. Psychologie de la mémoire humaine. In: R Bruyer, M Van der Linden M (eds),

Neuropsychologie de la mémoire humaine. Grenoble: Presses Universitaires de Grenoble, 9–37, 1991

Tierney MC, Snow WG, Reid DW et al. Psychometric differentiation of dementia: replication and extension of the findings of Storandt and coworkers. Arch Neurol 44, 720–722, 1987

Tierney MC, Szalai JP, Snow WG et al. Progression of probable Alzheimer's disease in memory-impaired patients: A prospective longitudinal study. Neurology 46, 661–665, 1996

Tiseo PJ, Perdomo CA, Friedhoff LT. Metabolism and elimination of 14C-donepezil in healthy volunteers: A single-dose study. Br J Clin Pharamcol 46 (Suppl 1), 19–24, 1998

Tobin SL, Chun N, Powell T and McConnell LM. The genetics of Alzheimer disease and the application of molecular tests. Genet Testing 3, 37–45, 1999

Tombaugh TN, McIntyre NJ. The Mini-Mental State Examination: a comprehensive review. J Am Gerontol Soc 40, 922–935, 1992

Tolbert SR, Fuller MA. Selegiline in treatment of behavioral and cognitive symptoms of Alzheimer disease. Ann Pharmacotherapy 30, 1122–1129, 1996

Tomlimson BE, Blessed G and Roth M. Observations on the brains of non-demented old people. J Neurol Sci 7, 331–356, 1968

Topper R, Gehrmann J, Banati R, Schwartz M, Block F, Noth J and Kreutzberg GW. Rapid appearance of beta-amyloid precursor protein immunoeactivity in glial cells following brain injury. Acta Neuropathol 89, 23–28, 1995

Touchon J, Ritchie K. Prodromal cognitive disorder in Alzheimer's disease. Int J Geriatr Psychiatry 14, 556–563, 1999

Tractenberg RE, Schafer K, Morris JC. Interobserver disagreements on Clinical Dementia Rating Assessment: Interpretation and implications for training. Alzheimer Dis Assoc Disord 15 (3), 155–161, 2001

Tranel D, Damasio AR. In: AD Baddeley, BA Wilson, FN Watts (eds), Handbook of Memory Disorders. Wiley, New York, 27–50, 1995

Trapp-Moen B, Tyrey M, Cook G, Heyman A, Fillenbaum GG. In-home assessment of dementia by nurses: Experience using the CERAD evaluations. The Gerontologist 41 (3), 406–409, 2001

Trojanowski JQ, Clark CM, Arai H, Lee V. Elevated levels of tau in cerebrospinal fluid, implications for the antemortem diagnosis of Alzheimer disease. Alzheimer's Dis Rev 1, 77–83, 1996

Tsuang D, Larson EB, Bowen J, McCormic W, Teri L, Nochlin D, Leverenz JB, Peskind ER, Lim A, Raskind MA et al. The utility of apolipoprotein E genotyping in the diagnosis of Alzheimer disease in a community-based case series. Arch Neurol 56, 1489–1495, 1999

Tully MW, Lambros Matrakas K, Musallam K. The eating behavior scale: A simple method of assessing functional ability in patients with Alzheimer disease. J Nutrition, Health & Aging 2 (2), 119–121, 1998

Tulving E. Elements of Episodic Memory. London: Oxford Univ. Press, 1983

Tulving E, Kapur S, Craik FIM, Moscovitch M, Houle S. Hemispheric encoding/retrieval assymetry in episodic memory: positron emission tomography findings. Proc Natl Acad Sci USA 91, 2016–2020, 1994

Tulving E. Memory and consciousness. Can Psychologist 26, 1–12, 1985

Tulving E, Markowitsch HJ, Craik FIM, Habib R, Houle S. Functional neuroanatomy of encoding and retrieval of pictorial information in memory. Cereb Cortex 6, 71–79, 1996

Tulving E, Markowitsch HJ, Craik FIM, Habib R, Houle S. Novelty and familiarity activations in PET studies of memory encoding and retrieval. Cereb Cortex 6, 71–79, 1996

Tulving E. How many memory systems are there? Am Psychologist 60, 385–398, 1985

Tune L, Ross C. Delirium. In: CE Coffey, JL Cummings (eds), Textbook of Geriatric Neuropsychiatry. Washington, DC. American Psychiatric Press, 351–368, 1994

Tuokko H, Vernon-Wilkinson R, Weir J et al. Cued recall and early identification of dementia. J Clin Exper Neuropsychol 13, 871–879, 1991

Turvey MT. On peripheral and central processes in vision: inferences from an information-processing analysis of masking with patterned stimuli. Psychol Rev 80 (1), 1–52, 1973

Tyler LK, Moss HE. Towards a distributed account of conceptual knowledge. Trends in Cognitive Sciences 5(6), 244–252, 2001

Uddman R, Edvinsson L. Neuropeptides in the cerebral circulation. Cerebrovasc Brain Metab Rev 1, 230–252, 1989

Ulrich J, Johannson-Locher G, Seiler WO, Stähelin HB. Does smoking protect from Alzheimer disease? Alzheimer-type changes in 301 unselected brains from patients with known smoking history. Acta Neuropathol 94, 450–454, 1997

US Food and Drug Administration. Guidance for industry: providing clinical evidence of effectiveness for human drug and biological products, 1998

Vaidya C, Gabrieli JDE, Monti LA, Tinklenberg JR, Yesavage JA. Dissociation between two forms of conceptual priming in Alzheimer's disease. Neuropsychology 13, 516–524, 1999

Vallar G, Shallice T. Neuropsychological impairments of short-term memory. Cambridge: Cambridge University Press, 1990

Vallar G, Papagno C. Neuropsychological impairments of short-term memory. In: AD Baddeley, BA Wilson, FN Watts (eds), Handbook of Memory Disorders, Chichester, UK, Wiley, 1995

Van Broeckhoven CL. Molecular genetics of Alzheimer disease: identification of genes and gene mutations. Eur Neurol 35, 8–19, 1995

Van der Linden M, Cornil V, Meulemans T, Ivanoiu A, Salmon E, Coyette F. Acquisition of a novel vocabulary in an amnesic patient. Neurocase 7(4), 283–293, 2001

Van Duijn CM, Havekes LM, van Broeckhoven C, de Kniff O, Hofman A. Apolipoprotein E genotype and association between smoking and early onset Alzheimer disease. BMJ 310, 627–631, 1995

Van Duijn CM, de Knijff P, Wehnert A et al. The apolipoprotein e2 allele is associated with an increased risk of early-onset Alzheimer disease and a reduced survival. Ann Neurol 37, 605–610, 1995

Van Dyck C, Newhouse P, Falk WE, Mattes JA. Extended-release physostigmine in Alzheimer disease. Arch Gen Psychiatry 57, 157–164, 2000

Vane JR, Botting RM. Mechanism of action of anti-inflammatory drugs. Int J Tissue React 20, 3–15, 1998

Vanhalle C, Van der Linden M, Belleville S, Gilbert B. Putting names to faces: Use of spaced retrieval strategy in a patient with dementia of the Alzheimer type. ASHA Special Interest Division 2: Neurophysiology and neurogenic speech and language disorders, 8, 17–21, 1998

Van Hoesen GW, Augustinack JC, Redman SJ. Ventromedial temporal lobe pathology in dementia, brain trauma and schizophrenia. Ann NY Acad Sci 877, 575–594, 1999

Van Leeven F. Single and multiple transgenic mice as models for Alzheimer disease. Prog Neurobiol 61, 305–312, 2000

Vanley CT, Aguilar MJ, Kleinhenz RJ, Lagios MD. Cerebral amyloid angiopathy. Human Pathology 12(7), 609–16, 1981

Vassar R. The β-secretase, BASE. J Mol Neurosci 17, 157–170, 2001

Vasterling JJ, Seltzer B, Watrous WE. Longitudinal assessment of deficit unawareness in Alzheimer's disease. Neuropsychiatry Neuropsychol Behav Neurol 10, 197–202, 1997

Vega GL, Weiner MF, Lipton AM, von Bergmann K, Lütjohann D, Moore C, Svetlik D. Reduction in levels of 24S-hydroxycholesterol by statin treatment in patients with Alzheimer disease. Arch Neurol 60, 510–515, 2003

Verkratsky A, Toescu EC. Calcium and neuronal ageing. Trends Neurosci 2, 2–7, 1998

Vickers JC, Dickson TC, Adlard PA et al. The cause of neuronal degeneration in Alzheimer's disease. Prog Neurobiol 60, 139–165, 2000

Visser PJ, Verhey FRJ, Ponds RWH, Jolles J. Diagnosis of preclinical Alzheimer's disease in a clinical setting. Int Psychogeriatr 13 (4), 411–423, 2001

Vogt C, Vogt O. Importance of neuroanatomy in the field of neuropathology. Neurology 1, 205–218, 1951

Volicer L, Berman SA, Cipolloni PB, Mandell A. Persistent vegetative state in Alzheimer disease. Does it exist? Arch Neurol 54(11), 1382–1384, 1997

Wahrle S, Das P, Nyborg AC, McLendon C, Shoji M, Kawarabayashi T, Younkin LH, Younkin SG, Golde TE. Cholesterol-dependent gamma-secretase activity in buoyant cholesterol-rich membrane microdomains. Neurobiology of Disease 9(1), 11–23, 2002

Waldemar G, Winblad B, Engedal K et al. Donepezil benefits patients with either mild or moderate Alzheimer's disease over one year. Neurology 54 (Suppl 3), A470. Abstract, 2000

Waldemar G. Functional brain imaging with SPECT in normal aging and dementia: methodological, pathophysiological, and diagnostic aspects. Cerebrovasc Brain Metabol Rev 7, 89–130, 1995

Wang Q-S, Zhou J-N. Retrieval and encoding of episodic memory in normal aging and patients with mild cognitive impairment. Brain Res 924, 113–115, 2002

Warrington EK. The selective impairment of semantic memory. Q J Exp Psychol 27, 635–657, 1975

Watson JM, Balota DA, Sergent-Marshall SD. Semantic, phonological and hybrid veridical and false memories in healthy adults and in dementia of the Alzheimer's type. Neuropsychology 15, 254–267, 2001

Weaver CE Jr, Marek P, Park-Chung M, Tam SW, Farb DH. Neuroprotective activity of a new class of steroidal inhibitors of the N-methyl-D-aspartate receptor. Proc Natl Acad Sci USA 94(19), 10450–4, 1997

Wechsler D. WAIS-R Manual. New York: Psychological Corporation, 1981

Weiner MF, Tractenberg RE, Jin S, Gamst A, Thomas RG, Koss E, Thal LJ. Assessing

Alzheimer's disease patients with the Cohen-Mansfield Agitation Inventory: scoring and clinical implications. J Psychiatric Res 36, 19–25, 2002

Weiner MF. Martin-Cook K. Foster BM. Saine K. Fontaine CS. Svetlik DA. Effects of donepezil on emotional/behavioral symptoms in Alzheimer's disease patients. J Clin Psychiatry 61(7), 487–492, 2000

Weiner MF, Koss E, Patterson M, Jin S, Teri L, Thomas R, Thal LJ, Whitehouse P. A comparison of the Cohen-Mansfield agitation inventory with the CERAD behavioral rating scale for dementia in community-dwelling persons with Alzheimer's disease. J Psychiatr Res 32, 347–351, 1998

Weingartner HJ, Grafman J, Boutelle W, Kaye W, Martin P. Forms of cognitive failure. Science 221, 380–382, 1983

Weinstein EA, Friedland RP, Wagner EE. Denial/unawareness of impairment and symbolic behavior in Alzheimer disease. Neuropsychiatry Neuropsychol Behav Neurol 7, 176–184, 1994

Weksler ME, Relkin N, Turkenich R, LaRusse S, Zhou L, Szabo P. Patients with Alzheimer disease have lower levels of serum anti-amyloid peptide antibodies than healthy elderly individuals. Experimental Gerontology 37, 943–948, 2002

Welsh K, Butters N, Hughes J, Mohs R, Heyman A. Detection and staging of dementia in Alzheimer disease. Use of the neuropsychological measures developed for the Consortium to Establish a Registry for Alzheimer disease. Arch Neurol 49, 448–452, 1992

Welsh K, Butters N, Hughes J, Mohs R, Heyman A. Detection of abnormal memory decline in mild cases of Alzheimer's disease using CERAD neuropsychological measures. Arch Neurol 48, 278–281, 1991

Welsh KA, Butters N, Mohs RC, Beekly D, Edland S, Fillenbaum G, Heyman A. The Consortium to Establish a Registry for Alzheimer's Disease (CERAD). Part V. A normative study of the neuropsychological battery. Neurology 44(4), 609–614, 1994

Welsh-Bohmer KA, Mohs RC. Neuropsychological assessment of Alzheimer disease. Neurology 49, S11–S13, 1997

Wenk GL, Stoher JD, Quintana G, Mobley G, Wiley RG. Behavioral, biochemical, histological and electrophysiological effects of 192 Ig-saporin injection into basal forebrain of rat. J Neurosci 14, 5986–5995, 1994

Wesnes K, Simpson PM and Christmas L. The assessment of human information processing abilities in psychopharmacology. In: I Hindmarch, PD Stonier (eds), Human Psychopharmacology: Measures and Methods, Volume 1. Chichester: Wiley, 79–92, 1987

West RL. An application of prefrontal cortex function theory to cognitive aging. Psychol Bull 120, 272–292, 1996

Weyer G, Erzigkeit H, Kanowski S, Ihl R, Hadler D. Alzheimer disease assessment scale: Reliability and validity in a multicenter clinical trial. Int Psychogeriatr 9 (No.2), 123–138, 1997

Whatmough C, Chertkow H, Murtha S, Hanratty K. Dissociable brain regions process object meaning and object structure during picture naming. Neuropsychologia 40, 174–186, 2002

Wheeler MA, Stuss DT, Tulving E. Toward a theory of episodic memory: the frontal lobes and autonoetic consciousness. Psychol Bull 121, 331–354, 1997

White H, Levin ED. Chronic four week nicotine skin patch treatment effects on cognitive performance in Alzheimer disease. Psychopharmacology 143, 158–165, 1999

Whitehouse PJ, Price DL, Stubble RG, Clark AW, Coyle JT, DeLong MR. Alzheimer disease and senile dementia: loss of neurons in the basal forebrain. Science 215, 1237–1239, 1982

Whitehouse PJ, Price DL, Clark AW, Coyle JT, DeLong MR. Alzheimer disease: Evidence for selective loss of cholinergic neurons in the nucleus basalis. Ann Neurol 10, 122–126, 1981

Whitehouse PJ, Orgogozo JM, Becker RE, Gauthier S, Pontecorvo M, Erzigkeit H, Rogers S, Mohs RC, Bodick N, Bruno G, Dal-Bianco P. Quality-of-life assessment in dementia drug development. Position paper from the International Working Group on Harmonization of Dementia Drug Guidelines. Alzheimer Dis Assoc Disord 11 (Suppl 3), 56–60, 1997

Whitford GM. Alzheimer disease and serotonin: a review. Neuropsychobiology 15, 133–142, 1986

Whitley RJ, Lakeman F. Herpes simplex virus infections of the central nervous system: therapeutic and diagnostic considerations. Clin Infect Dis 20, 414–420, 1995

Williams JMG, Scott J. Autobiographical memory in depression. Psychol Med 18, 689–695, 1988

Wilson AL, Langley LK, Monley J, Bauer T, Rottunda S, Mcfalls E, Kovera C, Mccarten JR. Nicotine patches in Alzheimer disease: pilot study on learning, memory, and safety. Pharmacol Biochem Behav 51, 509–514, 1995

Wilson BA. Rehabilitation of Memory. Guildford Press, New York, 1987

Wilson RS, Bacon L, Fox J, Kazniak A. Primary and secondary memory in dementia of the Alzheimer type. J Clin Neuropsychol 5, 337–344, 1983

Wilson BA, Patterson KE. Rehabilitation and cognitive neuropsychology: does cognitive psychology apply? J Appl Cogn Psychol 4, 247–260, 1990

Wilson RS, Kaszniak AW, Fox JH. Remote memory in senile dementia. Cortex 17, 41–48, 1981

Wilson PW, Anderson KM, Harris WB, Castelli WP. Determinants of change in total cholesterol and HDL-C with age: the Framingham Study. J Gerontol 49, M252–M257, 1994

Wimo A, Wetterholm AL, Mastey V, Winblad B. Evaluation of the health care resource utilization and caregiver time in anti-dementia drug trials – a quantitative battery. In: A Wimo, G Karlsson, B Winblad (eds), Health Economics of Dementia. Chichester: John Wiley & Sons, 465–493, 1998

Winblad B, Poritis N. Memantine in severe dementia: Results of the M-9 BEST study (benefit and efficacy in severely demented patients during treatment with memantine). Intern J Geriatric Psychiatry 14, 135–146, 1999

Winblad B. Maintaining functional and behavioral abilities in Alzheimer disease. Alzheimer Dis Assoc Disord 5 (1), S34–S40, 2001

Winblad B, Engedal K, Soininen H et al. Donepezil enhances a global function, cognition and activities of daily living compared with placebo in a one-year, double blind trial in patients with mild to moderate Alzheimer disease (poster). Presented at the Int Psychogeriatr Association, Vancouver, Canada, August 15–20, 1999

Winblad B, Engedal K, Soininen H, Verhey F, Waldemar G, Wetterholm A-L, Haglund A, Subbiah P and the Donepezil Nordic Study Group. A 1-year, randomized, placebo-controlled study of donepezil in patients with mild to moderate AD. Neurology 57, 489–495, 2001

Winblad B, Brodaty H, Gauthier S et al. Pharmacotherapy of Alzheimer's disease: is there a need to redefine treatment success? Int J Geriat Psychiatry 16, 653–666, 2001

Wolfe MS, Xia W, Ostaszewski BL, Diehl TS, Kimberly WT, Selkoe DJ. Two transmembrane aspartates in presenilin-1 required for presenilin endoproteolysis and gamma-secretase activity. Nature 398(6727), 513–517, 1999

Wolff SD, Balaban RS. Magnetization transfer contrast (MTC) and tissue water proton relaxation *in vivo*. Magn Reson Med 10, 135–144, 1989

Wolkowitz OM, Kramer JH, Reus VI, Costa MM, Yaffe K, Walton P, Raskind M, Peskind E, Newhouse P, Sack D et al. DHEA-Alzheimer's Disease Collaborative Research. DHEA treatment of Alzheimer's disease: a randomized, double-blind, placebo-controlled study. Neurology 60(7), 1071–1076, 2003

Wolozin B, Kellman W, Ruosseau P, Celesia GG, Siegel G. Decreased prevalence of Alzheimer disease associated with 3-hydroxy-3-methyglutaryl coenzyme A reductase inhibitors. Arch Neurol 57(10), 1439–1443, 2000

Wonnacott S. The paradox of nicotinic acetylcholine receptor upregulation by nicotine. Trends Pharmacol Sci 216–219, 1990

Woodruff-Pak DS, Finkbiner RG. Larger nondeclarative than declarative deficits and memory in human aging. Psychol Aging 10, 416–426, 1995

Woodruff-Pak DS, Thompson RF. Classical conditioning of the eyeblink response in the delay paradigm in adults aged 18–83 years. Psychol Aging 3, 219–229, 1988

Woodruff-Pak. Eyeblink classical conditioning differentiates normal aging from Alzheimer disease. Integrative Physiological and Behavioral Science 36 (2), 87–108, 2001

World Health Organization. International Classification of mental disorders. ICD-10, Chapt V (F). Clinical descriptions and diagnostic guidelines: Geneva, Switzerland, 1992

Wragg M, Hutton M, Talbot C. Genetic association between intronic polymorphism in presenilin-1 gene and late-onset Alzheimer disease: Alzheimer disease Collaborative Group. Lancet 347, 509–512, 1996

Wragg RE, Jeste DV. Overview of depression and psychosis in Alzheimer disease. Am. J Psychiatry 146, 577–587, 1989

Wright LK. The impact of Alzheimer disease on the marital relationship. The Gerontologist 31, No. 2, 1991

Xie W, Chipman JG, Robertson DL, Erikson RL and Simmons DL. Expression of a mitogen-responsive gene encoding prostaglandin synthase is regulated by mRNA splicing. Proc Natl Acad Sci USA 88, 2692–2696, 1991

Xu HX, Gouras GK, Greenfield JP, Vincent B, Naslund J, Mazzarelli L, Fried G, Jovanovic JN, Seeger M, Relkin NR et al. Estrogen reduces neuronal generation of Alzheimer β-amyloid peptides. Nat Med 4(4), 447–451, 1998

Yamada K, Tanaka T, Han D, Senzaki K, Kameyama T, Nabeshima T. Protective effects of idebenone and alpha-tocopherol on β-amyloid-(1-42)-induced learning and memory deficits in rats: implication of oxidative stress in β-amyloid-induced neurotoxicity *in vivo*. Eur J Neurosci 11, 83–90, 1999

Yan SD, Fu J, Soto C, Chen X, Zhu H, Al-Mohanna F, Collison K, Zhu A, Stern E, Saido T et al. An intracellular protein that binds amyloid-beta peptide and mediates neurotoxicity in Alzheimer's disease. Nature 389(6652), 689–695, 1997

Yasuda M, Maeda K, Hashimoto M, Yamashita H, Ikejiri Y, Bird TD, Tanaka C, Schellenberg GD. A pedigree with a novel presenilin-1 mutation at a residue that is not conserved in presenilin-2. Arch Neurol 56, 65–69, 1999

Yasuno F, Suhara T, Nakayama T, Ichimiya T, Okubo Y, Takano A, Ando T, Inoue M, Maeda J, Suzuki K. Inhibitory effect of hippocampal 5-HT_{1A} receptors on human explicit memory. Am J Psychiatry 160, 334–340, 2003

Yates FA. The Art of Memory. Routledge, Kegan Paul, London, 1966

Yermakova AV, Rollins J, Callahan LM, Rogers J, O'Banion MK. J Neuropathol Exp Neurol 58, 1135, 1999

Yesavage JA. Relaxation and memory training in 39 elderly patients. Am J Psychiatry 141, 778–781, 1984

Yesavage JA, Brink TL, Rose TL, Lum O, Huang V, Adey M, Leirer VO. Development and validation of a geriatric depression screening scale: a preliminary report. J Psychiatric Res 17(1), 37–49, 1982–83

Yin JCP, Del Vecchio M, Zhou H, Tully T. CREB as a memory modulator: induced expression of a dCREB2 activator isoform enhances long-term memory in drosophila. Cell 81, 107–115, 1995

Yoshitake T, Kiyohara Y, Kato I et al. Incidence and risk factors of vascular dementia and Alzheimer's disease in a defined elderly Japanese population: the Hisayama Study. Neurology 45, 1161–1168, 1995

Zanetti O, Binetti G, Magni E, Rozzini L, Bianchetti A, Trabucchi M. Procedural memory stimulation in Alzheimer disease: impact of a training programme. Acta Neuropsychologica Scandinavica 95, 152–157, 1997

Zappoli R, Versari A, Paganini M, Arnetoli G, Muscas GC, Gangemi PF, Arneodo MG, Poggiolini D, Zappoli F, Battaglia A. Brain electrical activity (quantitative EEG and bit-mapping neurocognitive CNV components), psychometrics and clinical findings in presenile subjects with initial mild cognitive decline or probable Alzheimer-type dementia. ItalJ Neurol Sci 16(6), 341–376, 1995

Zarit SH. Methodological consideration in caregiver intervention and outcome research. In: E Light, G Niederehe, BD Lebowitz (eds), Stress Effects on Family Caregivers of Alzheimer's Patients. Research and Interventions. New York, Springer, 351–369, 1994

Zaudig M. A new systematic method of measurement and diagnosis of "mild cognitive impairment" and dementia according to ICD-10 and DSM-III-R criteria. Int Psychogeriatr 2, 203–219, 1992

Zeiss AM, Davies HD, Wood M and Tinklenberg JR. The incidence and correlates of erectile problems in patients with Alzheimer disease. Arch Sex Behav 19, No. 4, 1990

Zheng H, Jiang M, Trumbauer ME, Sirinathsinghji DJ, Hopkins R, Smith DW, Heavens RP, Dawson GR, Boyce S, Conner MW et al. beta-Amyloid precursor protein-deficient mice show reactive gliosis and decreased locomotor activity. Cell 81(4), 525–531, 1995

Zipfel GJ, Babcock DJ, Lee JM, Choi DW. Neuronal apoptosis after CNS injury: the roles of glutamate and calcium. J Neurotrauma 17, 857–869, 2000

Zubenko GS, Moossy J, Martinez AJ, Rao G, Claassen D, Rosen J, Kopp U. Neuropathologic and neurochemical correlates of psychosis in primary dementia. [erratum appears in Arch Neurol 49(10), 1064. 1992]. Arch Neurol 48(6), 619–624, 1991

Subject index